POPULATION, AGRICULTURE, AND BIODIVERSITY

Problems and Prospects

POPULATION, AGRICULTURE, AND BIODIVERSITY

Problems and Prospects

EDITORS

J. PERRY GUSTAFSON
University of Missouri, Columbia

PETER H. RAVEN
Missouri Botanical Garden

PAUL R. EHRLICH
Stanford University

UNIVERSITY OF MISSOURI PRESS
Columbia

University of Missouri Press, Columbia, Missouri 65211
Printed and bound in Malaysia

Library of Congress Cataloging-in-Publication Data

Names: Gustafson, J. Perry, editor. | Raven, Peter H., editor. | Ehrlich, Paul R., editor.
Title: Population, agriculture, and biodiversity: problems and prospects / edited by J. Perry Gustafson, Peter H. Raven, and Paul R. Ehrlich.
Description: Columbia : University of Missouri, 2020. | Includes bibliographical references and index.
Identifiers: LCCN 2019034806 (print) | LCCN 2019034807 (ebook) | ISBN 9780826222022 (hardcover) | ISBN 9780826274403 (ebook)
Subjects: LCSH: Food security. | Food supply. | Agrobiodiversity conservation. | Agriculture--Environmental aspects. | Population--Environmental aspects.
Classification: LCC HD9000.5 .P627 2020 (print) | LCC HD9000.5 (ebook)
LC record available at https://lccn.loc.gov/2019034806
LC ebook record available at https://lccn.loc.gov/2019034807

ISBN: 978-0-8262-2202-2

This paper meets the requirements of the American National Standard for Permanence of Paper for Printed Library Materials, Z39.48, 1984.

Interior design by Aubrey Khan, Neuwirth & Associates

Typefaces: Acumin, Frutiger, and Minion

We dedicate this book to the memory of Norman E. Borlaug, and to the thousands of trainees, students, farmers, and scientists whom he mentored during his long and productive life, and to all present and future scientists, students, trainees, and farmers helping to feed the world.

CONTENTS

CONTRIBUTOR BIOGRAPHIES

MATHEW ABRAHAM is the Assistant Director of the Tata-Cornell Institute for Agriculture and Nutrition at the Charles H. Dyson School of Applied Economics and Management at Cornell University. He has a PhD from the Department of International Economics and Management, Copenhagen Business School, Denmark. Prior to that, he worked as a researcher at the Centre for Public Policy at the Indian Institute of Management, Bangalore. He has over a decade of experience working in agriculture and food security. His research interests are in agricultural markets, collective action in agriculture, food security in developing countries, institutional innovation in rural development, and social entrepreneurship.

KYM ANDERSON is George Gollin Professor Emeritus in the School of Economics at the University of Adelaide (where he has been affiliated since 1984) and an Honorary Professor at the Australian National University's Crawford School of Public Policy (where he was a Research Fellow 1977–1983 and a part-time Professor of Economics 2012–2018). In two periods of extended leave he served as deputy head of economic research at the GATT (now WTO) Secretariat in Geneva (1990–1992) and as Lead Economist (Trade Policy) at the World Bank in Washington, D.C. (2004–2007). Since post-graduate studies at the University of Chicago and Stanford University, he has published more than four hundred articles and forty books. During 2010–2017 he served on the Board of Trustees of the International Food Policy Research Institute (Washington, D.C.), chairing it from 2015–2017. He is a recipient of an honorary doctor of economics degree from the University of Adelaide. In 2015 he became a Companion of the Order of Australia (AC).

JAMES N. BLIGNAUT is a freelance consultant and Professor Extraordinaire at the Stellenbosch University (School of Public Leadership) and Honorary

Research associate of the South African Environmental Observation Network, Pretoria. He holds a PhD in Economics from the University of Pretoria. He is an environmental and natural resource economist that focuses on the economics of restoration, and the restoration of economics, across a broad spectrum of subjects, including regenerative and conservation agriculture. From 1996 to 2017 he was Professor in Economics at the University of Pretoria and has led a team with respect to the development of a payments for ecosystem services scheme in Rwanda (2018–2019) involving conservation agriculture and rural development.

KATE A. BRAUMAN is the Lead Scientist for the Global Water Initiative at the University of Minnesota's Institute on the Environment. She also leads the Institute's impact goal on achieving safe water. Brauman's research integrates hydrology and land use with economics and policy to better understand how water use by people affects the environment and our ability to live well in it. Through projects as diverse as payments for watershed services, global variation in "crop per drop," and worldwide trends in water consumption and availability, Brauman works with the Global Water Initiative to find sustainable solutions to pressing water issues. Brauman received her doctorate from Stanford University and her undergraduate degree from Columbia University. She is currently a Coordinating Lead Author for the Global Assessment of the Intergovernmental Platform on Biodiversity and Ecosystem Services.

DEREK BYERLEE is an adjunct professor in the Global Human Development Program in the School of Foreign Services at Georgetown University and formerly, Editor in Chief of the journal *Global Food Security*. He is an internationally recognized agricultural economist who specializes in agricultural development and food security. He has held senior positions at Michigan State University; the International Maize and Wheat Improvement Center (CIMMYT) in Mexico and South Asia; and the World Bank, where he worked as Lead Economist, Rural Strategy and Policy. He finished his career in the Bank by co-directing preparation of the their flagship 2008 World Development Report on food and agriculture, the first on agriculture since 1982. Since leaving the World Bank, he has continued working with a number of international organizations with an emphasis on investment in agribusiness, large-scale farming, and intensifications and land use. He served as chair of the Standing Panel on Impact Assessment of the CGIAR's Science Council since 2009; member and then chair of the Technical Advisory Committee of the Global Agricultural and Food Security Program, 2012–2017; and member of the Board of Trustees of the International Food Policy

Research Institute. He was elected a Fellow of the American Association of Agricultural Economists in 2004. He holds a PhD in Agricultural Economics from Oregon State University and D.Econ. (hon.) from the University of Adelaide.

MICHAEL J. CHRISTOFFERS is an Associate Professor of plant sciences at North Dakota State University (NDSU). He was raised on a farm in Minnesota where his family produced and sold certified seed, sparking an interest in agriculture and genetics. He pursued those interests at NDSU, where he obtained a BS in crop and weed sciences. He then attended the University of Missouri where he obtained a PhD in genetics, working in the area of wheat molecular biology. His current research at NDSU involves weed genetics, where he primarily investigates herbicide resistance and emerging genetic technologies to combat weeds. Michael teaches face-to-face and online courses in introductory genetics for undergraduates in both science and non-science disciplines. He also teaches a graduate course in the action and fate of herbicides.

ANNE H. EHRLICH is a Senior Research Scientist Emerita at Stanford University. Over more than fifty years, she carried out field and lab research and co-authored technical articles in population biology as well as teaching courses on environmental policy. She also has written and co-authored books and articles on issues of public concern such as population growth, environmental protection, the environmental consequences of nuclear war, and climate disruption. She is a member of the American Academy of Arts and Sciences, and an honorary Fellow of the California Academy of Sciences. She is the recipient of honorary doctorates from Bethany College and Oregon State University, the United Nations Sasakawa Environmental Prize, the Heinz Award for Environmental Achievement, and the Tyler Prize for Environmental Achievement.

PAUL R. EHRLICH is Bing Professor of Population Studies Emeritus and President, Center for Conservation Biology, Stanford University. He has carried out field, laboratory, and theoretical research on the dynamics and genetics of insect populations, the evolutionary interactions of plants and herbivores, the behavioral ecology of birds and reef fishes, the effects of crowding on human beings, human cultural evolution, and health problems related to industrialization. He is author or co-author of more than 1,100 scientific papers and articles and over forty books. Ehrlich is a member of the U.S. National Academy of Sciences, the American Philosophical Society, and a

Foreign Member of the Royal Society. Among his many other honors is the Royal Swedish Academy of Sciences, Crafoord Prize (an explicit replacement for the Nobel Prize). He has appeared on more than a thousand TV and radio programs and was a correspondent for NBC News.

NINA V. FEDOROFF is Evan Pugh Professor Emerita, Penn State University. She received her PhD in Molecular Biology from the Rockefeller University and has also served on the faculties of the Carnegie Institution for Science, the Johns Hopkins University, and the King Abdullah University of Science and Technology in Saudi Arabia. Fedoroff served as the Science and Technology Adviser to the Secretary of State and to the Administrator of the U.S. Agency for International Development (USAID) from 2007 to 2010. She pioneered in the field of plant molecular biology, cloning, and characterizing maize transposable elements. Fedoroff has published three books and more than 170 scientific papers. She is a member of the U.S. National Academy of Sciences and a 2006 National Medal of Science laureate.

TONY FISCHER is a crop agronomist/physiologist, currently an Honorary Fellow at CSIRO Agriculture and Food, Canberra, Australia. He graduated as B.Ag.Sci. (Univ. of Melbourne) and PhD (Univ. of California, Davis). His career included research scientist and manager in Australia (NSW Agriculture, CSIRO, Australian Centre for International Agricultural Research) and at CIMMYT in Mexico, first as an agronomist/physiologist (1970–1975), then as Director of the Wheat Program (1988–1995). Also there was board and advisory service with the CGIAR (ICARDA, IRRI, CIMMYT) and the Grains Research and Development Corporation (Canberra). He continues to add to a long publication record (over four hundred citations in 2018), including a comprehensive 2014 book on increasing crop yields and global food security (http://aciar.gov.au/publication/mn158). He is a fellow of the Australian Institute of Agriculture, the Australian Academy of Technological Sciences and Engineering, and the American Societies of Agronomy and Crop Science.

DAVID I. GUSTAFSON is an independent scientist who uses modeling to help food systems meet human nutrition needs in more sustainable ways. His academic training was in chemical engineering (Stanford, BS, 1980; University of Washington, PhD, 1983). He worked thirty years in private industry (Shell, Rhône-Poulenc, Monsanto), and now leads research projects at two non-profits: Conservation Technology Information Center and ILSI Research Foundation. Gustafson's early career focused on predicting agricultural impacts on water quality. He subsequently developed new modeling approaches to pollen-mediated gene flow and the population genetics of insect and weed resistance. Beginning in 2007, Gustafson began leading efforts to understand climate adaptation and mitigation imperatives in the global agri-food system. He has served on various national and international teams looking at this issue, including the Executive Secretariat of the U.S. Government's National Climate Assessment Development and Advisory Committee, which issued the third such congressionally mandated assessment in 2014.

J. PERRY GUSTAFSON holds a BS in Crop Science and a MS in Agronomy from Colorado State University, Ft. Collins, and a PhD in Genetics from the University of California, Davis. From 1972 to 1981, Gustafson served on the faculty of the University of Manitoba, Winnipeg, where he worked on a wheat-rye hybrid. In 1981 he was recruited by the U.S. Department of Agriculture's Agricultural Research Service (USDA-ARS) to work on wheat genetics at the University of Missouri, Columbia. Over three decades with the USDA-ARS and MU, he developed a research program with a particular focus on understanding the genetic mechanisms that control gene expression in wheat. He has acted as an advisor and consultant for the USDA's Foreign Agricultural Service (FAS), the United Nations' International Atomic Energy Agency (IAEA), and various international agriculture research centers. He is the recipient of many awards, honors, and medals in recognition of his research, teaching, and broader service to agriculture.

WENDY HARWOOD is a Senior Scientist at the John Innes Centre, Norwich. Her group works on the genetic modification of crop plants, in particular wheat and barley, and uses genetic modification as a tool to better understand genes important in the development of improved agricultural crops. Wendy's group also works to improve the technology for genetic modification, to increase understanding of the genetic modification event, and to provide data for use in the safety assessment of GM crops. She has a first-class degree in biology and a PhD in plant transformation.

RUSSELL L. JONES is a Fellow in Environmental Safety at Bayer, specializing in the behavior of crop protection products in soil and water. Research interests include the development of management practices for minimizing movement of crop protection products to ground and surface water, design and conduct of field research studies, and the application of environmental modeling to risk assessment. He has also been an active participant in several SETAC and FOCUS working groups of regulatory, industry, and academic scientists providing guidance for conduct of field studies and modeling to support registration of crop protection products. He received a BS and MS from the Massachusetts Institute of Technology and a PhD from the University of California, Berkeley, all in chemical engineering. For the past forty-four years, he has been employed by Bayer and its predecessor companies.

STEPHANIE R. LARSON is the Director, University of California Cooperative Extension (UCCE), Sonoma County and promotes research and extension programs in Sonoma County, along with being the Livestock and Range Management Advisor. She addresses climate change, food production and ecological and economical management of working landscapes in Sonoma and Marin Counties. She has a BS and MS in animal science and a PhD in rangeland management. Dr. Larson assists local livestock producers to improve production and marketing of livestock, conducting research in the wise stewardship of lands, humane methods of predator control, and sustainable agriculture for the livestock industry. Her program documents and integrates the ecosystem services provided by rangelands, highlighting the benefits of using grazing animals as tools to address vegetation management reduced fire fuels, improved habitat and increased forage production. She is a licensed certified rangeland manager; she brings public and private land owners and managers together to make science-based decisions and policies to manage working landscapes for the benefit of all users. Her efforts are bringing more economic opportunities to local agriculture producers, making their operations more sustainable and economically viable.

LUKE T. MACAULAY is an Assistant Cooperative Extension Specialist in Rangeland Planning and Policy based in the Department of Environmental Science, Policy, and Management at the University of California, Berkeley. He earned his PhD and MA in range management from the University of California, Berkeley, and completed a BA in a Great Books program and Spanish from University of Notre Dame. Before his graduate studies, he worked for several years as the spokesman for the federal prosecutor's office in San Francisco. There he learned about the importance of communications in achieving goals.

GEMMA MOLERO is a wheat scientist working at the International Maize and Wheat Improvement Centre since 2011, after completing her PhD in Plant Ecophysiology at University of Barcelona. Her main research activities consist of studying adaptive traits that enhance yield under different environments, with special emphasis on photosynthesis and phenotyping of genetic resources. She is coordinating the photosynthesis research experiments at the IWYP-HUB (International Wheat Yield Partnership) while also serving as the leader of MasAgro-Trigo, a Mexican initiative aiming to increase yield potential and climate resilience of wheat. She has developed panels of germplasm to capture diversity in expression of radiation use efficiency (RUE) enabling the identification of genetic basis of RUE and related traits. Her research in photosynthesis and assimilate partitioning is used routinely to design strategic crosses in physiological pre-breeding at CIMMYT. She has published papers on plant physiology, plant response to abiotic stress, and the development of specific tools in high-ranking international journals. Molero is also in charge of coordinating activities from visitors and doctorate students to CIMMYT's experimental station in Ciudad Obregon, Mexico, and is involved in project management and funding activities.

EVIATAR NEVO is a leader in the field of evolutionary biology and is among the founders of modern evolutionary theory. He is currently Professor of Evolutionary Biology at the University of Haifa, Israel. He established and directed the Institute of Evolution (IOE, 1973–2008) and the International Graduate Center of Evolution (IGSE) (2004–2008), both of which are world-unique. His scientific career has involved interdisciplinary studies of biodiversity evolution, adaptation, and speciation across life based on the science he advanced of Evolutionary Functional Ecological Genetics and Genomics across life from viruses and bacteria through fungi, plants, animals, and humans. Nevo has published approximately 1,500 scientific papers, more than a thousand of them peer reviewed, and thirty-five books. In 2000, he was elected to the National Academy of Sciences USA and he won the Israeli Prize of Life Sciences in 2016.

AVI PEREVOLOTSKY is a retired researcher at the Agricultural Research Organization, The Volcani Center, Bet Dagan, Israel. He holds both a BS in biology and a MS in zoology from Tel Aviv University, and a PhD in Ecology from the University of California, Davis. In his research activity, he investigated management aspects of grazing and afforested systems under Mediterranean climate. He also carried out research on nature conservation. Since 1985 he has served as a scientific consultant for the Rothschild Foundation, including leading the program in agro-ecology. During 1998 to 2003 he served as the chief scientist of the Israeli Nature and Parks Authority. He is currently in the process of publishing a book in agro-ecology.

STUART L. PIMM is the Doris Duke Chair of Conservation at the Nicholas School of the Environment, Duke University. He is a leading expert on the causes of modern extinctions and what can be done to prevent them. His practical activities include being the President and Founder of SavingSpecies, a non-profit that helps conservation groups in species-rich countries restore land and habitat connections. With over three hundred publications, many in science and nature, he has been honored with the Tyler Prize for Environmental Achievement (2010) and the Heineken Prize from the Royal Netherlands Academy of Arts and Sciences (2006). His books include *The Balance of Nature? Ecological Issues in the Conservation of Species and Communities*, and *The World According to Pimm: A Scientist Audits the Earth.*

PRABHU PINGALI is a professor in the Charles H. Dyson School of Applied Economics and Management at Cornell University, with a joint appointment in the Division of Nutritional Sciences, and the Founding Director of the Tata-Cornell Institute for Agriculture and Nutrition (TCI). Prior to joining Cornell, he was the Deputy Director, Agricultural Development Division of the Bill & Melinda Gates Foundation, from 2008 to 2013. He served as the director of the UN Food and Agriculture Organization's Agriculture and Development Economics Division from 2002 to 2007. In addition, he worked with the CGIAR for fifteen years (from 1987 to 2002), first with IRRI in the Philippines and then with CIMMYT in Mexico. Pingali is a member of the U.S. National Academy of Sciences and an AAEA Fellow. He has more than three decades' worth of experience working with some of the leading international agricultural development organizations as a research economist, development practitioner, and senior manager. Pingali has written thirteen books and more than 120 refereed journal articles and book chapters on food policy.

FRANCISCO PINTO is an associate scientist specialist in remote sensing at the International Maize and Wheat Improvement Center (CIMMYT) in Mexico. He holds a BS and a MS in Agricultural Sciences from the University of Chile, and a PhD in Agricultural Sciences from the University of Bonn, Germany. He conducted his PhD project at the Institute of Bio- and Geosciences, Plant Sciences (IBG-2) at the Research Center Jülich, where he gained experience on the use of remote sensing for plant phenotyping and the study of plant photosynthesis using sun-induced chlorophyll fluorescence. He was appointed by CIMMYT in 2016, when he joined the wheat physiology group at the Global Wheat Program. Since then, he has been working in the development of a High-throughput phenotyping platform for wheat research and breeding in the CIMMYT experimental station at Ciudad Obregon, Sonora, Mexico.

PETER H. RAVEN, a leading botanist and conservationist, is president of the Missouri Botanical Garden and George Engelmann Professor of Botany Emeritus at Washington University in St. Louis. In 2001, he received the National Medal of Science, the highest award for scientific accomplishment in the United States. Raven served for twelve years as Home Secretary of the National Academy of Sciences, to which he was elected in 1977, and is also a member of the academies of science in a number of other countries, including China, Russia, the UK, Brazil, and Australia. He was co-editor of the forty-nine-volume *Flora of China* and has written or co-authored numerous books and publications, including *Biology of Plants*, the internationally best-selling textbook in botany. He has held Guggenheim and John D. and Catherine T. MacArthur Foundation fellowships.

MATTHEW REYNOLDS is a Distinguished Scientist at the International Maize and Wheat Improvement Centre, where his professional goals are to develop and transfer technologies to increase productivity of wheat cropping systems worldwide with a special focus on developing countries. Impacts include a new generation of advanced lines based on physiological breeding approaches to widen the wheat gene pool, increased understanding of yield potential and adaptation of wheat to drought and heat stress, development of high throughput phenotyping methodologies, and capacity building. To further these goals he has been active in developing global collaborations to tap into the expertise of plant scientists worldwide—such as the International Wheat Yield Partnership, and the Heat and Drought Wheat Improvement Consortium. He currently leads the community of practice on crop modelling for the CGIAR Big Data in Agriculture platform, and is a board member of the Global Plant Council. He has published widely in the area of crop physiology and adapting crops to climate change and has mentored graduate students through his affiliations with universities worldwide. He has also served as a consultant for a number of public and private institutions, and in 2018 was appointed to the Mexican Academy of Science.

ROBERT P. SABBA is a research specialist in the Department of Plant Sciences at North Dakota State University in Fargo. He holds a BS in biochemistry from the State University of New York at Stony Brook, and a PhD in biology from Syracuse University in New York. Dr. Sabba has worked on the physiology and genetics of picloram resistance in yellow starthistle, the mechanism of cellulose-inhibiting herbicides, and on improving the breeding and storability of the potato. His current research centers on understanding the genetic basis of herbicide resistance in weed species, such as foxtail, kochia, waterhemp, and wild oat.

SIVAKUMAR SUKUMARAN is a lead scientist working on spring wheat pre-breeding for yield potential, drought and heat stress at the International Maize and Wheat Improvement Center (CIMMYT), Mexico. He holds a BS in Agriculture from Kerala Agricultural University, Thrissur, Kerala, India and a MS in Plant Breeding and Genetics from Tamil Nadu Agricultural University, Coimbatore, Tamil Nadu, India. After his PhD in Plant Breeding and Genetics from Kansas State University, Sukumaran was recruited in 2013 to CIMMYT as a post-doctoral fellow in the wheat physiology group to work on Genetic analysis of wheat adaptation. Currently, Sukumaran leads the spring wheat pre-breeding in wheat physiology and seeds of discovery projects in the Global Wheat Program at CIMMYT.

ALISON L. VAN EENENNAAM is a Cooperative Extension Specialist in the field of Animal Genomics and Biotechnology in the Department of Animal Science at University of California, Davis. She received a Bachelor of Agricultural Science from the University of Melbourne in Australia, and both an MS in animal science, and a PhD in genetics from UC Davis. Her publicly funded research and outreach program focuses on the use of animal genomics and biotechnology in livestock production systems. She has published extensively on livestock genetic improvement, and given over six hundred invited presentations to audiences globally. A passionate advocate of science, Dr. Van Eenennaam was the recipient of the 2014 Council for Agricultural Science and Technology (CAST) Borlaug Communication Award, the 2016 BIO Future Maker Award, and in 2017 was elected as a Fellow of the American Association for the Advancement of Science (AAAS).

VARSHA VIJAY is a conservation ecologist and a postdoctoral research fellow at the National Socio-Environmental Synthesis Center (SESYNC). At SESYNC she focuses on evaluating the effectiveness of protected areas in meeting ecological and socio-economic goals of conservation. In particular she is interested in the balance of human health and economic needs with those of threatened and rare species or habitats. She holds a BS and PhD in Environmental Sciences from Duke University. Her research program has focused on understanding both direct and indirect effects of agricultural activity on habitat loss and biodiversity. Examples include the expansion of oil palm in the tropics and human-cheetah conflict in Southern Africa. She is a recipient of the Brower Youth Award for Environmental Leadership.

AMY E. YOUNG is a staff research associate at the University of California, Davis in the laboratory of Dr. Alison Van Eenennaam in the Department of Animal Science. She earned a BS in biological science from Colorado State University and a MS in genetics from the University of California, Davis. Her research interests and background are in animal genetics, genomics, and health of companion animals, horses, and livestock. She has conducted research in companion animal veterinary genetics and equine pharmacogenetics at the UC Davis School of Veterinary Medicine. Most recently, she has been involved with research and contributed to outreach efforts on beef cattle genetics and animal biotechnology.

ACKNOWLEDGMENT

The editors of this volume, the staff of the University of Missouri Press, and all of the contributing authors gratefully acknowledge the volunteer efforts of Melody Kroll, of the University of Missouri, Columbia, who contributed extensively to the development and improvement of this book. It is not exaggeration to say that were it not for her tireless efforts liaising between the editors of the volume, editors at the Press, and contributing authors across the globe, this book never would have seen print. For all of her assistance, and unflagging good humor, we thank her sincerely.

POPULATION, AGRICULTURE, AND BIODIVERSITY

Problems and Prospects

INTRODUCTION

Peter H. Raven, J. Perry Gustafson, and Paul R. Ehrlich

ABOUT 11,000 YEARS AGO, SOME of our ancestors began to gather and sow seeds of their food plants around their camps for later harvest. This agricultural revolution gave them access to a much more secure food supply than they had enjoyed previously. At that time, the global population of humans amounted to perhaps a million people, with only about 100,000 of them in Europe. What had happened earlier on our 4.57-billion-year-old planet? Life originated some 4 billion years ago, in the form of bacteria, with some of its more complex representatives moving onto land starting about 400 million years ago. Flowering plants and placental mammals both originated in the early Cretaceous Era, about 130 million years ago. Their evolutionary pathways subsequently affected each of them in complex ways. Primates, the order of mammals to which we belong, first appeared near the end of the Cretaceous Era, perhaps 75 million years ago, with humans diverging in Africa from a common ancestor with the other African apes some 6 to 8 million years ago. The earliest known fossils of our own species, Homo sapiens, occur in 300,000-year-old rocks from Morocco. Our ancestors migrated out of Africa at least 60,000 years ago, spreading rapidly throughout Eurasia and to Australia and reaching the Americas at least 15,000 years ago.

People therefore inhabited all six habitable continents by the time agriculture was developed in Eurasia (more than once); a little later and independently in the Americas; but not known in Australia until Europeans arrived. It was the deployment of agriculture that facilitated an enormous increase in our population size over fewer than five hundred subsequent generations. This increase led in turn to our dominance on Earth, and thus to all of the positive and negative consequences that accompanied this growth.

We now number about 8.0 billion people and are projected to increase to 9.9 billion by the year 2050 (www.prb.org). We are currently using natural productivity 1.75 times faster than nature can regenerate its sources, up from

0.7 times fifty years ago. This amounts to our using the productivity of 1.75 copies of planet Earth, which of course is not renewable. Of high significance is that some 60 percent of our global footprint results from carbon sequestration, a clear indication that we need to make the transition away from coal, oil, and gas as fuel sources as rapidly as possible or risk climate disruption destroying all our other sources of productivity.

Our current dominance on the planet is marked by very high levels of inequity between nations and classes of people; we must address these levels of inequity if we are ever to attain global stability. The British charity Oxfam has estimated that the eight richest people in the world currently control as much wealth as the 3.6 million poorest among us. An examination of the papers in this volume indicate substantially different opinions about the number of poor people on earth and the number who are hungry or even starving, as well as the trends in both groups. Ehrlich and Ehrlich (this volume) conclude that there currently are about 800 million people seriously underfed and perhaps 2 billion who are deficient in at least one micronutrient. About half of us live on less than $2.50 per day and some 80 percent on less than $10 per day, with a similar proportion of our total population living in countries where the income differentials are widening.

Some people have argued and continue to argue that human consumption rather than the size of our population is the key factor in destroying the ecological balance of the earth. Clearly, the more people, the more consumption, no matter what else is true. Certainly relatively wealthy people are playing the dominant role in driving problems like global climate disruption, but to claim that numbers don't matter in a world where we and our domestic animals constitute about 95 percent of the biomass of all vertebrate animals on Earth clearly amounts to a strangely unrealistic view of the actual situation. Indeed, there is no clear evidence that the Earth can even support the existing human population for long, much less a larger one. If our goal is environmental stability we really need to limit population growth.

The Problem

As the title of this collection of essays suggests, the main threads of our discussion are (1) the relation between human population growth and consumption; (2) the need to increase food supplies to feed the world population, and (3) the chances for avoiding the extinction of a major proportion of the biodiversity that collectively makes possible our survival on earth. Those relationships are highly intertwined, and changes in each of them are increasingly blocking our chances to achieve environmental stability on our fragile planet.

In the agricultural chapters offered in this volume, we consider primarily the improvement of agricultural practices and the crops themselves in the context of their influence on global sustainability. Improved and more efficient crop production would lead to increased production of plant-derived foods while reducing the negative consequences of our agriculture on global sustainability. We comment on animal agriculture in this context because it occupies an area about three times larger than that on which crops are cultivated, increases greatly the global production of greenhouse gases, and consumes far more resources than crops in the course of producing human food. It would be desirable to review many aspects of animal agriculture in the context of human population and consumption levels, as well as the effects of grazing and animal production on biodiversity (as in the contribution of Young et al., this volume).

Let us consider agriculture and food production first. Estimates of the amount of additional food necessary to feed 9.9 billion people in thirty years range between 70 percent and 100 percent more than is being produced in 2018. Reaching those goals would necessitate an annual increase in food production of 2.5 to 3.0 percent. Our actual progress has recently been in the 0.4 to 1.0 percent increase per year (Fischer, this volume), with a number of serious challenges lying ahead. The best strains of individual crops grown in the most favorable places have often achieved a yield approximating the necessary general increase, but how can we achieve such yields more widely, especially in a rapidly changing world? Despite optimistic projections of feeding the world population offered at least as early as the 1950s, and even taking into account the successes of the Green Revolution from the 1960s onward, we are still not coming close to adequately feeding the current world population. At the same time that the human population is increasing rapidly, people want to consume more and more (and especially more and more meat), and the problem of hunger and starvation remains intractable or continues to grow.

Among the most significant factors that stand in the way of improved agricultural sustainability and productivity are climate disruption, water supply, the need to use the available arable land properly, plant diseases, and our inefficient use of fertilizers and pesticides.

In general, climate disruption has emerged as one of the most important of these challenges. The nations of the world have thus far have not been able to agree to sufficiently effective measures to halt or even limit it effectively on a global scale. It is ironic, in this connection, that agriculture itself contributes about 21 percent of the greenhouse gases emitted because of human activities! Indeed, emissions related to agriculture contribute about 72 percent of

N_2O and 39 percent of CH_4, methane, the latter a particularly potent greenhouse gas (Abraham and Pingali, this volume). In this connection, the deployment of both mitigation and adaptation strategies will be of key importance generally—and certainly for the sustainability and productivity of agriculture. A trend away from animal agriculture would be extremely helpful in limiting the effect of agricultural systems in causing climate disruption and an important way to limit its other consequences as well.

Closely related to the problem of growing climate disruption is that of pollution affecting crops directly (D. I. Gustafson and R. L Jones, this volume). Agriculture as practiced today has a direct effect on the quality of surface water through the dissemination of the chemicals applied, contributes substantially to greenhouse gas emission, and drives an enormous loss of nutrients and top soil. Excess nitrogen fertilizer applied in agriculture and in private households has a highly negative effect on life as a whole, and must be limited far beyond what current practice yields. As we intensify our efforts to improve the way we practice agriculture, it will be necessary to find ways to limit these negative effects. As we approach this outcome, we will need to improve both research productivity and legal restrictions to maintain the quality of the global environment, including such matters as marine pollution in the Gulf of Mexico, in Australia's Great Barrier Reef, and in many other areas.

Brauman (this volume) reviews water, clearly one of the very most critical limiting factors in facilitating agricultural productivity, in this context. Brauman carefully points out the importance for every locality of the dynamic relationship between water supply and water use, and the major overall need for making the agricultural and domestic use of water as sustainable and focused as it can possibly become. This is especially true in the face of all the other uses for water that are important for human life, since agriculture is the major consumer of water at present.

As Fedoroff (this volume) discusses, it takes between 500 and 2,000 liters of water to produce a kilogram of grain, with about half of that product fed to animals. Since about 90 percent of what the animals consume supports their own metabolism, only about one-tenth of the food value in the grain is present in the animals eventually harvested for food. The animals are also a highly significant source of greenhouse gases. Particularly those people emerging from poverty are often eager to include much more animal protein in their diets than had been earlier possible. Fish and other aquatic organisms are more efficient in converting plant to animal protein than mammals, and can therefore form a part of diet with less damage done to the earth's stability. In addition, it is difficult to envision the widespread adoption of vegetarianism, as desirable as that might be from a planetary point of view. Perhaps the synthetic production of animal protein, as Fedoroff suggests, may turn out to provide at least a part of the solution to this highly intractable problem.

The most damaging groups of agricultural pests are insects, fungi, and weeds, so whatever we can do to control them will continue to be very important. Christoffers and Sabba (this volume) have reviewed the area both historically and with a special emphasis on the increase in insecticide, fungicide, and herbicide resistance as factors in substantially improving crop productivity and sustainability on a global basis. Although we have used a few kinds of chemicals to benefit crops for several thousand years, it was the commercialized organic compounds for these purposes from the late nineteenth century onward that has both revolutionized agriculture and led to major problems in the field. As we have treated insects, fungi, and weeds with these compounds, resistance has become widespread and losses amounting to billions of dollars per year have resulted. Christoffers and Sabba review methods developed to combat these resistant strains, with future benefits enhanced by crop rotation, reducing overdependence on particular chemicals, limiting the size of the areas of application, gene stacking, the development of RNA interference, and the breeding of stronger, more competitive crops. Unfortunately, global climate disruption presents serious challenges in this area too: Deutsch et al. (2018) project that global yield losses to rice, maize, and wheat to insects will increase by 10 to 25 percent per degree of global mean surface warming.

A number of the papers presented here discuss the opportunities and difficulties that we face in increasing our crop production. In relation to possible yield increases, Fischer analyzes the possible increase in grain crops (and some others) that may result from the application of new breeding techniques and other advanced agronomic practices. He urges the expansion of existing collaborative efforts between breeders, agronomists, and farmers, but points out that the increased privatization of agricultural research and development runs counter to the implementation of such partnerships.

Concerning some of the major outputs of these generally industry-based efforts, Fedoroff (this volume) discusses the future role of genetics, including GMOs and CRISPR technologies in making possible the efficient production of plants and animals with desired agronomic characteristics. In addition, Harwood (this volume) reviews the state of the art for such methods of crop improvement, and decries the irrational opposition that retards progress in these areas. It is incredible—indeed we consider it immoral—to argue that the products thus produced pose dangers for humans and nature generally.

Denying them to a large proportion of the human population is to increase starvation and malnutrition for reasons, which not a single Academy of Sciences or Medicine, or similar group, would agree. Partly because of their hatred of industry, a number of political groups are causing enormous damage to huge numbers of their fellow human beings. As Fedoroff brings out so well, it is high time to cease these activities and be as rational as we can if we are to have any hope whatever. Indeed, it is ironic that the imposition of unreasonable legal restrictions on using these contemporary methods, clearly progressive improvements on methods humans have employed for thousands of years, has made the production of improved crops so expensive that often only industries can afford to work with them!

Reynolds et al. (this volume) review the opportunities and limitations intrinsic in physiological breeding of plants for climate disruption, a topic that will clearly become increasingly important as the years pass. The combination of genomics with high-throughput phenotyping has made possible the selection of complex physiological features that are of critical importance in facilitating this adjustment, accelerating the rate of selection considerably. We can enhance the role of exotic material in breeding, thus making it possible to make improvements in such features as crop photosynthesis, drought resistance, and other important adaptive traits in ways that seemed inaccessible only a few years ago.

Nevo (this volume) offers an illuminating discussion of the enormous and underexploited genomic resources available for the improvement of our major cereal crops. Exciting resources are now available for precise gene transfer, substitution, and rearrangement. Combined with the exciting approaches to physiological breeding of crops reviewed by Reynolds and his colleagues, their promise becomes even greater. As we take up these techniques, the vast genetic diversity that Nevo has described for the relatives of cereals clearly presents an important model to take into approach while working with other groups. Considering these approaches together, it becomes apparent that the future is not hopeless, but that we must greatly strengthen our efforts to improve the situation our descendants and we are facing while there is still time to do so. Indeed, all organisms, including those that are being lost to us so rapidly, may have genes that will help us to improve our crops and enhance global stability. Therefore, reserving them to the extent we can emerges as one of the very most important contributions that we will be able to make to future generations.

Other techniques that certainly have the potential for increasing agricultural yield while maintaining stability also deserve wider attention. For example, we are starting to develop precision farming based on satellite imagery using GPS methods; this strategy usually leads to yield increases in particular situations. On a smaller scale, some crops are being grown in closed environments, variations on the traditional theme of greenhouse cultivation, but often with the plants much more closely crowded than has been customary. In addition, even in urban buildings, aquaponics can prove highly productive depending on the circumstances and the design of the growing areas.

An enormous but often overlooked aspect limiting agricultural productivity is post-harvest loss to insects, fungi, and unsuitable storage conditions. In India, the post-harvest loss of semi-perishable foods amounts to a shocking 40 percent, in a country where so many are so hungry. In the U.S., 10 percent or more of stored grain may be lost. Clearly, vast stores of food would become available to feed people if we could find effective ways to control these losses; in general, however, they do not receive the attention that they deserve.

The problem of misdistribution of food and other forms of wealth within and especially among countries presents an enormous obstacle to "feeding the world." Among the twenty-eight poorest countries, twenty-one are in Sub-Saharan Africa, with the others including Afghanistan, Haiti, Nepal, Tajikistan, and Yemen. GNP per capita is decreasing in all of these countries, and at the same time, they are reaching a point where there is nothing left for them to export. We need to ask ourselves what is the likelihood of the richer countries helping them substantially with financial support or food aid. They are often preoccupied with enriching themselves and, as Ehrlich and Ehrlich (this volume) have emphasized, many of those richer countries are acquiring agricultural land to produce additional food for themselves in many of the world's poorest countries. The overall situation looks especially dire in Sub-Saharan Africa, with a population projected to more than double from its present 1 billion people to a staggering 2.2 billion people over the next thirty years. How much do we who live in the world's wealthy countries really care about what is happening enough to do something about it?

Compounding this problem is the fact that people who are very poor or even starting to emerge from poverty often do not have the funds to take advantage of the advances in agricultural practices. For example, they probably will not be able to purchase improved seeds, even if they well understand that those seeds would provide a higher yield and better life. They also cannot afford advanced agricultural equipment, such as the planting equipment that poor Indian farmers in the Punjab could use to drill in their wheat seeds without burning the stubble from the preceding crop of rice, thus helping to avoid the lethal smog that seasonally envelops northern India. This relationship tends to become a vicious cycle, since poor farmers generally live in poor countries where the wealthy benefit most from whatever resources might be available. In the poorest countries, of course, there is increasingly no chance at all to break out of poverty without significant international assistance in amounts that have not been forthcoming in the past.

Anderson (this volume) has contributed a useful discussion of trade as a factor in global security. Since we will always be unable to distribute food production among nations of varying size and geographical position on a highly diverse planet, trade is extremely important and, at the same time, overregulated and otherwise limited at present. Current trends are running counter to the ideal situation she describes, and it will always be difficult to overcome national greed as a factor in the maintenance of trade barriers. Indeed, Blignaut and Perevolotsky (this volume) have pointed out plausibly that enough food is probably being produced already to feed our current population, and that inadequate distribution is in fact the cause for most present hunger and starvation. Both a growing shortage and an unwillingness to make part of what we produce available for others seems poised to get worse unless we are willing to change our attitudes and practices into something that is both efficient and just. Indeed, the more we implicitly insist that every country produce its own food, the more environmental damage will result overall.

Drought and other negative factors related to climate disruption are clearly increasing instability in poorer parts of the world, such as the Middle East. Instability also will pose important obstacles in many regions, thus increasingly blocking our efforts to improve agricultural yields substantially. These is simply no room in the world for the massive shifts in agriculture that will be required if we continue on our present course. In other words, the days when a modern Moses could lead his people to a fertile land and a subsequent King David could simply take over agriculture by killing and driving away the people living there are gone. Today, there is no extra space, and the world is divided into competitive nations which often forget that we are all exploiting the same planet, like it or not.

Biological Extinction

As mentioned above, our ancestors evolved in a biologically rich world, one that had become somewhat cooler and drier than it had been previously. Undisturbed ecosystems are relatively stable in their particular regions. They make it possible for water to circulate through them and return to the atmosphere; for the maintenance of soils; for a balanced assembly of species to coexist, albeit with continual additions and subtractions over the long run;

and for individual species, including humans, to find their own niche within them. Because it is irreversible, however, the loss of biodiversity is the most important problem that we are facing today—but few realize that fact, much less take action adequate to prevent it.

Pimm and Vijay (this volume) estimate that taxonomists have named and described 1.7 million species to date; others have calculated that as many as 12 million species are likely to exist (not counting a possible trillion species of microbes). The vast majority of species remain unknown to us. We estimate that over the most recent tens of millions of years, the rate of extinction per year has been approximately 0.1 species per million per year. The overall number of species appears to have increased slowly since the major extinction event at the end of the Cretaceous Era, with the loss of species constituting a kind of brake on the rate at which newly evolved species increase the overall total. We estimate the current rate of extinction of the non-microbes at about 100 species/million/year, a 1,000 times rate increase over the past. The loss of individual populations is growing much more rapidly than the loss of species, with major decreases in bird and insect populations being chronicled widely. Therefore, it has become likely that global extinction rates will soon reach an unimaginable crescendo, with disastrous consequences for the remaining life on Earth, including ours. Our seeming inability to attack the problem of global climate disruption effectively is already affecting biodiversity, and the problem can only grow as long as we continue to exercise our greed instead of our common sense.

Human influence on the extinction of species has grown along with major increases in our numbers and consumption rates over the past 11,000 years. The most important single factor has been habitat destruction, with the spread of agriculture as its most important driving force. Considering that more than 11 percent of the Earth's surface is devoted to crop production and an additional 20 to 30 percent to grazing, most of which is unsustainably practiced on natural landscapes, it is a simple matter to see why this is so. Young et al. (this volume) present a detailed analysis of the impact of grazing, which is very substantial, on the chances of biodiversity survival and thereby present an implicit argument for curtailing our meat consumption. As pointed out by Fedoroff (this volume), the inherently resource-intensive production of livestock is perhaps the greatest challenge to achieving food security or environmental balance.

Consider what the occupation of Europe for ever-increasing numbers of people must have meant for this process. The continent was essentially completely deforested by 1830, when coal attained lasting, widespread use as a fuel and the forests grew back. Especially considering that the Mediterranean parts of Europe are an important global "hot spot" for biodiversity, one can only imagine how many species have disappeared since the first agriculturists invaded Europe from the east, conquered the relatively few people that were living there at that time, and converted large portions of the continent for human purposes. Building a population of some 500 million people with some of the highest consumption rates in the world has been as costly to the natural world as it has been beneficial in developing the many benefits of civilization and spreading them, along with destructive harvesting and agricultural practices, worldwide.

For the tropics, the challenges are especially great (Byerlee, this volume). At least two-thirds of the biodiversity in the world lives there, its species the most poorly known in any region of the world. There are a number of individual conservation efforts aimed at different regions of the tropics, but for the most part, they do not address human needs of the populations living in the same areas effectively. Without long-term environmental stability in a given region, many conservation efforts will be doomed to failure. Most industrialized nations assign top priority to obtaining tropical resources for their own use as rapidly as possible, with much less attention to the needs of the poor citizens of the same countries.

As an example of what is taking place in the tropics, a quarter of all lowland forests there have been cleared over the past quarter century for a variety of reasons, both internal and external; the deforestation rate in many countries is accelerating. Byerlee (this volume) is certainly correct in suggesting the intensification of land use in particular areas as an important strategy for avoiding the destruction of all ecosystems in the tropics. At the same time, however, the continuing rapid increase of the numbers of people who live there, many of them desperately poor, poses major challenges for implementing rational land use practices. This is particularly true when investors are able to make money from the extensification, rather than the intensification, of agriculture and other land uses. Byerlee presents a number of factors of key importance for attaining sustainability in the tropics, all of which will involve a great deal of mutual understanding and an overall global commitment to achieving them.

The worst imaginable outcome from the loss of biodiversity and habitats lies in the possible collapse of ecosystems in various parts of the world. The progressive loss will have a devastating effect on the atmosphere, on weather, on water circulation, on soil preservation, on agricultural productivity, and in many additional ways that will make human lives much more difficult than they are today. The many products that we derive from other living organisms will be lost with them, and it will take millions of years, with or without us, for the planet to regain its ecological stability.

We can and must do whatever we can to limit the loss of biodiversity while finding ways to adequately feed the world's population. Broadly speaking, the accomplishment of these goals will require both attaining and sustaining a stable human population; limiting consumption worldwide to a socially just, equitable level; empowering women and children worldwide; and arresting global disruption. The improvement of agriculture has the potential to help or hurt in the achievement of these necessary steps: Limiting the damage that it does is one of the most important ways we can adjust to reality. If we can achieve a stable population at a sustainable level, we would be opening up the possibility of saving our civilization, which is teetering on the brink of destruction about two hundred generations onward from the time humans first invented a written language. In this effort, we hope that considering the many and diverse views presented in this volume will prove a solid contribution to that end.

References

Deutsch, C. A., Tewksbury, J. J., Tigchelaar, M., Battisti, D. S., Merrill, S. C., Huey, R. B. and Naylor, R. I. (2018). Increase in crop losses to insect pests in a warming climate. *Science, 361*, 916–923.

POPULATION, RESOURCES, AND THE ENVIRONMENT

The Situation in 2019

Paul R. Ehrlich and Anne H. Ehrlich
Stanford University

In this chapter, today's situation is summarized in comparison with that in 1968, when *The Population Bomb* (Ehrlich, 1968) was published. The human predicament is now much more serious, since the human population has more than doubled in size, key resources are much more depleted, and environmental deterioration is substantially more advanced. The evidence today suggests that levels of hunger and malnutrition, though reduced since 1968, will likely increase as the human population grows, climate change advances, and biodiversity, a critical part of civilization's life-support systems, is rapidly degraded. A "perfect storm" of environmental problems now threatens to lead to a collapse of civilization.

Those threats include climate disruption, loss of biodiversity (and thus ecosystem services), land-use change and resulting degradation, global toxification, ocean acidification, decay of the epidemiological environment, increasing depletion of groundwater and other critical resources, and resource wars (which could go nuclear). This is not just a list of problems, it is an interconnected complex of problems resulting from interactions within and between what can be thought of as two gigantic complex adaptive systems: the biosphere system and the human socio-economic system. The manifestations of this interaction are often referred to as "the human predicament." That predicament is getting continually and rapidly worse, driven by overpopulation, massive inequities, overconsumption among the middle class and rich, and the use of environmentally malign technologies and social-economic-political arrangements to service the consumption. And organized steps that might ameliorate the situation are, unhappily, nowhere in sight.

This chapter is a review of some environmental aspects of the predicament, with a focus on the least-discussed major driver of the problems, overpopulation and continued population growth.

Population Growth and Human Nutrition

A central issue of the human predicament a half-century ago (as now) was the prospect of feeding a growing population. In 2019 Earth is carrying some 7.7 billion people instead of the ~3.5 billion it had in 1968 and is projected to add more than 2 billion more individuals by 2050. Some 200–300 million people have died of starvation or hunger-related disease in the last fifty years, but not the greater numbers anticipated in the 1970s based on the work of agricultural specialists such as Lester Brown (1963), William Paddock (1967), and Arthur Hopcraft (1968) in the 1960s.[1]

Farmers in developed and developing countries, especially richer farmers, adopted "green-revolution" technologies, based on improved grain strains, abundant water, and heavy use of petrochemicals and fossil fuels, much more rapidly than most analysts had expected in the 1960s. As a result, grain production surged ahead of population growth. In addition, international actions created a global system of food storage and transfer that could avert famine when crops failed in any region of the world. And with the spread of family planning programs, birth rates in many nations began declining significantly in the 1970s, thus easing the growth in demand for food.

But the long-term consequences of the green revolution remain in doubt, and chronic food shortage is still a problem in many developing countries, while overnutrition, micronutrient malnutrition, and over-processed diets increasingly plague both poor and rich nations (Kahn and Ehrlich, 2018; Tanumihardjo et al., 2007). Today, primarily because of the lack of food distribution, about 800 million people are seriously underfed and, according to the FAO, as many as 2 billion people are micronutrient malnourished (Ramakrishnan, 2002). Thus it remains to be seen whether the much-maligned 1968 statement in *The Population Bomb* that "the battle to feed all of humanity is over" will be disproven. We doubt whether humanity will satisfactorily be able to feed 11 billion people around the end of this century. Since, after sixty years of assurances that the food problem would be solved, today we're still not feeding 7.6 billion. Indeed, the number of undernourished people in the world has been rising since 2014, reaching an estimated 815 million in

1 Estimates are difficult to make, especially since there appear to be no data on deaths associated with micronutrient malnutrition.

2016 (FAO, 2017), and several billion suffer significant micronutrient malnourishment (Scott, 2017).

Most analysts think doubling food production would be required to feed a 35 percent bigger and still growing human population adequately by 2050 (Ray et al., 2013; Tilman et al., 2011). For any chance of success, humanity will need to stop expanding land area for agriculture (to preserve ecosystem services); raise yields where possible; increase efficiency in use of fertilizers, water, and energy; become more vegetarian; reduce food wastage; stop wrecking the oceans; significantly increase investment in sustainable agricultural research; and move feeding everyone to the very top of the policy agenda. All of these tasks will require revolutionary changes in human behavior (Godfray and Garnett, 2014; Ehrlich and Harte, 2015b) long recommended but thus far elusive. Perhaps more critical, there may be insurmountable biophysical barriers to increasing yields—indeed, to avoiding *reductions* in yields—in the face of climate disruption.

Most people fail to realize the urgency of the food situation because they don't understand the agricultural system and its complex, non-linear connections to the drivers of environmental deterioration. The system itself, for example, is a major source of greenhouse gas emissions and thus is an important driver of climate disruption, which in turn is a serious threat to food production. More than a millennium of change in temperature and precipitation patterns is now entrained, with the prospect of more crop-threatening severe storms, droughts, heat waves, and floods—all of which are already evident. Thus maintaining—let alone expanding—food production will be ever more difficult in decades ahead.

Furthermore, agriculture and human enterprises are the leading causes of losses of biodiversity and the critical ecosystem services supplied to agriculture itself and other human enterprises, as well as a major source of global toxification, both of which pose additional risks to food production. The threat to food production of climate disruption alone means that humanity's entire system for mobilizing energy needs to be rapidly transformed in an effort to hold atmospheric warming well below a lethal 3°C rise in global average temperature (e.g., Anderson, 2015). It also means we must alter much of our water-handling infrastructure to provide the necessary flexibility to bring water to crops in an environment of constantly changing precipitation patterns.

Some, of course, suggest the battle can still end in victory (e.g., Foley, 2011), and we hope they are correct. But we have severe doubts in the face of the potentially disastrous (Mann, 2009), already evident, and intensifying effects of climate disruption and resultant water shortages (e.g., Aarhus

University, 2014; Liu and Yang, 2012; Mehra et al., 2001). Food production is also threatened by increasing losses of biodiversity (Ceballos et al., 2015a; Ceballos et al., 2017)—especially of enemies of crop pests (Boyles et al., 2011) and pollinators (Insu et al., 2015), which can have serious nutritional impacts (Smith et al., 2015). The projected addition of some 4 billion people by the end of the century will further crowd out other organisms as indicated by the increasing biomass of both *Homo sapiens* and livestock (Bar-On et al., 2018). Increased food production is also threatened by the degradation of soils (Milman, 2015; Montgomery, 2012), overdrafting of fossil groundwater (e.g., Barringer, 2011; Brauman et al., 2016; Mankin et al., 2015), and ocean acidification, pollution, and overfishing (e.g., Britten et al., 2015), all partially the result of overpopulation.

The controversial original Green Revolution (Singh, 2000) certainly was a medium-term rescue for many hungry people after 1970, but the revolution's lead scientist Norman Borlaug was deeply concerned about population growth. He featured the population issue in his 1970 Nobel Peace Prize address, saying:

> There can be no permanent progress in the battle against hunger until the agencies that fight for increased food production and those that fight for population control unite in a common effort. Fighting alone, they may win temporary skirmishes, but united they can win a decisive and lasting victory to provide food and other amenities of a progressive civilization for the benefit of all mankind.[2]

The needed unity unfortunately has not emerged, and the continued serious maldistribution of food supplies explains overall levels of hunger today (Ehrlich, P. R. and Harte, 2015a).

With no prospect that the maldistribution is likely to be corrected soon, it is surely no accident that "ag-investment" is spreading. Nations like China, India, South Korea, and Saudi Arabia are buying land in Africa, Brazil, and other developing nations (Cotula, 2009, 2012, Klare, 2012) at least partially in an attempt to ensure that their own populations will be fed as the food situation tightens (Bräutigam and Zhang, 2013). Like many other investors, they are anticipating a time, sometimes described as "peak soil" (Hunt and

2 We thank Ms. Margaret Breinholt, Norman Borlaug's niece, for locating the manuscript of the speech for us in the Borlaug archives: http://borlaug.tamu.edu/archive/arc_nobel.php. It is instructive that Borlaug used that now much-maligned term "population control." It was clear to him that keeping population size from having a negative impact on society, just like seeing to it that people were adequately fed, was in the end an important governmental function.

McFarlane, 2014), when population size and growth make fertile and well-watered land a premier resource for investment, including environmental consequences that likely will increase the value of that investment.[3] An analysis of the global land trade network found, ominously, "that the structure of the network is potentially prone to propagating crises (e.g., if importing countries become dependent on crops exported from their land trading partners). This network analysis approach can be used to quantitatively analyze and understand telecoupled systems (those with human and natural systems connected) as well as to anticipate and diagnose the potential effects of telecoupling" (Seaquist et al., 2014).

An array of factors makes the human predicament today much more precarious than it was in 1968. Of course, the most important is that more than twice as many people are now living on Earth as were then. Whatever population size, together with tightening resource constraints and increasing environmental degradation, could cause widespread societal collapse, humanity is a lot closer to it now. *Homo sapiens* is rapidly approaching the end of being able to exploit the fossil-fuel bonanza that powered the vast expansion of civilization over the past century and a half (Holdren, 2008). That alone makes collapse much more likely, and initiating steps toward humane population shrinkage right away much more imperative.

Empirically, the most effective way to move toward global population shrinkage involves providing women everywhere with rights, opportunities, and education equal to those provided to men (today these rights are not truly equal in any nation, are severely lacking in many, and are still being restricted in the United States) and providing all sexually active people with access to modern contraception and safe back-up abortion (Dasgupta and Ehrlich, 2013; Ehrlich and Ehrlich, 2012). Dealing with the flip side of overpopulation, overconsumption, is more complex in theory (Ehrlich, 2014), but transforming consumptive behavior has been shown to be achievable, as demonstrated by the mobilizations and demobilizations at the beginning and end of World War II (Ehrlich and Ehrlich, 2010, 2012). Of course, the war posed an obvious and immediate existential threat, which induced affected populations to change their consumptive habits, if only temporarily. We might hope that the existential threats posed today by climate disruption, faltering agriculture, rising seas, increasing floods, droughts, and wildfires, widespread toxification, and resource depletion would also change behavior, but the danger is not yet perceived widely or clearly enough to produce the needed effect.

3 See farmlandgrab.org.

Population and Environment

One would think that most educated people now understand that the larger the size of a human population, *ceteris paribus*, the more destructive its impact is on the environment. The degree of overpopulation is best indicated (conservatively) by ecological footprint analysis (Wackernagel and Rees, 1998) which shows that to support today's population sustainably at current patterns of consumption would require roughly another half an Earth-size planet, and to do so at the U.S. level would take four to five more Earths. Population size and growth (connected to human consumptive behavior) is a major driver of virtually every environmental dilemma facing humanity.

For this discussion, climate disruption can stand in for almost all of them. Each person added to the world's population increases the flow of greenhouse gases (GHGs) into the atmosphere through that person's daily consumption of goods and services, and especially through increasing the need for food. The food system itself contributes roughly 30 percent of GHGs emitted (Bellarby et al., 2008), yet the biggest threat to food production is escalating climate disruption. Indeed, that disruption is thought to have contributed, through an extreme drought, in causing the conflict in Syria (Kelley et al., 2015) and thus in Europe's migration crisis and international terrorism. Climate disruption also apparently was involved through drought and rising commodity prices in generating the Arab Spring and general instability in the heavily food-importing nations of the Middle East and North Africa (Perez, 2013).

We find it difficult today to be optimistic about the nutritional future of humanity, especially since effective action to prevent serious climate change was delayed for decades after the likely consequences were well understood (Intergovernmental Panel on Climate Change, 1996). In 2015 the International Energy Agency reported that promised reductions in greenhouse gas emissions, as indicated in national pledges made for that year's United Nations Climate Conference, were collectively too modest to keep global temperature increases below 2°C over preindustrial levels (Reed, 2015), given that we're already halfway there. The prospects of effective global actions darkened when the Trump kakistocracy handed global leadership on the issue to China by pulling out of the Paris accord (Sanger and Perlez, 2017), pushed by the growing Koch-funded plutocracy (Mayer, 2016, p. 259).

Action has been delayed far too long, and the race to eliminate greenhouse gas emissions before climate disruption, along with other factors in the human predicament, severely compromises civilization's population-food-resource-environment relationships will surely be a very close one

(Anderson, 2015), and the probable outcome remains very uncertain (Peters, 2018).

Population and Economics

A central cause for concern now with population size and growth is the persistence and near ubiquity of "faith-based macroeconomics"—faith in continuous economic expansion through technology, markets, and the so-called knowledge economy. For example, Nobel Laureate for Economics, Paul Krugman, with whose short-term views we frequently concur, almost always writes about growth as a cure for economic problems, especially unemployment. His disregard for the constraints imposed by the biophysical world is typical of his discipline, which is seemingly unaware of the degree to which humanity is coopting the flows of energy that underpin the life-support systems of humankind (Vitousek et al., 1997; Vitousek et al., 1986). Krugman has explicitly attacked the idea of limits to growth (Czech, 2014), although recent analyses indicate that the study of that name (Meadows et al., 1972) has been remarkably accurate in characterization of the behavior of the global system since its publication (Bardi, 2011; Day et al., 2009; Hall and Day Jr., 2009; Turner, 2008). These analyses highlight the consequences of such crazy ideas such as growing an economy at 3 to 5 percent per year over decades (or forever) as most innumerate economists, corporate leaders, and politicians believe possible. Too many "educated" people do not realize that in the real world a short history of exponential growth does not imply a long future of such growth.

A second fundamental problem with macroeconomics that accompanies its faith that growth can continue forever in a confined system is that people make rational choices. Economic growth can, in some circumstances, temporarily ameliorate unemployment, but pro-growth proponents stubbornly refuse to recognize increasing biophysical constraints. They also ignore the potential for reorganizing the economy on a more equitable basis and the possibility of making important and, in our view, necessary changes to manage unemployment problems such as shortened work weeks.

Nevertheless, the fallacy of unending growth has been well recognized for a long time. Distinguished economist Kenneth Boulding wrote in 1966: "Anyone who believes that exponential growth can go on forever in a finite world is either a madman or an economist" (Boulding, 1966). Well, the madmen and most economists are still at it. Of course, some economists believe that there are limits to how many people Earth can support, but that those limits are far in the future. In part that may be because most economists are under the delusion that climate disruption is the only existential threat to civilization.

In contrast, since Boulding's time, a small group of ecological economists have been struggling to persuade the discipline to address these real-world constraints (see, e.g. Arrow et al., 2004; Daly, 1973; Dasgupta, 2001; Goulder and Kennedy, 1997; Heinberg, 2011; Jackson, 2009). Yet most faith-based economists ignore even the expanding global debt Ponzi scheme—a faith-based edifice that depends on perpetual growth to assure that the claims on real things (resources, labor skills) the debt represents can be repaid. Ironically, widespread rational realization that this is no longer true could bring on a global financial collapse (Heinberg, 2011; May et al., 2008).

The second fallacious belief, that human choices are rational, is simply wrong; emotions and bounded rationality are critical to decision making (Bechara and Damasio, 2005; Livet, 2010) and play important roles in both reproductive and consumptive choices. Much of the failure of macroeconomics to become evidence-based (Ehrlich, 2008) can be blamed on the training provided to graduate students in economics departments, where the basics of environmental science, in our view increasingly central to most economic outcomes, are largely neglected. That failure, among other things, allows many economists to consider climate change as a mostly distant threat and to disregard the nature of other major environmental challenges and their interactions with both climate disruption and especially demographic trends (Harte, 2007), the consequences of which they totally misapprehend.

For example, a recent mantra of faith-based macroeconomics is that birth rates must be increased to prevent a population from aging—having a growing proportion of elderly people—raising the worry that there will be too few workers to support pensioners. This idea is especially common among European politicians, who do not realize that only the impossible, keeping the population growing forever, could prevent the population from aging unless highly immoral (and socially unacceptable) policies were employed (e.g., euthanasia).

Although the social and environmental benefits of ending population growth far outweigh the costs of an aging population, most of the political commentary on the subject has been, in technical terms, "spherically senseless": ridiculous from all viewpoints (Ehrlich and Ehrlich, 2006; Spijker and MacInnes, 2013). Indeed, population shrinkage is clearly needed if nations with aging populations, immersed in a world full of young, hungry, and mobile people, are to have any chance of achieving some kind of stability.

Equally perverse and pervasive is the so-called "demographic transition theory," which says economic development will automatically cause birth rates to decline to match lower death rates and eventually result in a constant population size. Historically, a widespread decline in death rates occurred in conjunction with industrialization, followed by a decline in birth rates, but the causes of these patterns are complex. In any case, there is no theory that dictates the evolution of a stationary population, and little empirical evidence that economic growth alone will automatically bring and keep birth rates low (Campbell and Potts, 2012). Indeed, most evidence shows the opposite: Falling birthrates lead to economic expansion. Meanwhile, thanks in part to economic growthmanship, much too little has been done in many parts of the world to remove barriers to family planning (Campbell et al., 2006), which clearly facilitates low fertility rates and provides many other benefits (Campbell and Potts, 2012).

Population and Resources

The faith-based economic system further ignores that many, if not most, resources are scarcer or less accessible in 2018 than they were in 1968 (and earlier) (Bardi, 2014; Heinberg, 2007; Klare, 2012). Human beings, being smart, have tended to pick the low-hanging fruit first. This is most easily seen by looking at EROI—the energy return on investment—for how many units of energy must be put into discovering, mining, drilling, and transporting each unit of energy (or other resource) positioned for use (Bardi et al., 2011; Hall et al., 2009; Hall et al., 2008).[4]

A classic example of declining EROI is obtaining oil. The first oil well went down some 69.5 feet from the surface to hit oil in Pennsylvania in 1859, while the Macondo well that blew up in the Gulf of Mexico in 2010 was drilled beneath a mile of water and struck oil nearly two miles below the sea floor. A more dramatic example is that when people first started using metal, essentially pure copper was lying around on the surface in places like Cyprus, while today copper is mined in some cases at a depth of more than a mile, and for ores that contain less than 1 percent copper. Beyond EROI is the complete exhaustion of a resource, exemplified, sadly, by human-caused extinctions. European immigrants to North America enjoyed billions of passenger pigeons as a nearly inexhaustible food resource. But "nearly" wasn't good enough.

A similar example can be seen in a resource only recognized as such recently—the capacity of the atmosphere to absorb human effluents. Since

4 A better indicator, if it were available, would be the real resource cost of extracting (and using) various natural resources (inclusive of external costs).

1968 atmospheric levels of climate-changing CO2 have risen from 325 to over 400 parts per million (ppm). The atmospheric sink into which humanity could safely pour its carbon and nitrogen emissions is a resource that civilization as a whole has used with no thought to any limits until quite recently. Now exploitation of that resource is exceeding its safe limit.

Efficiency of resource use in many cases has increased in response to market forces and regulation, but it is often counterbalanced, partly or entirely, by growth in use. Between 1975 and 2015 the fuel economy of American cars more than doubled, but the number of cars increased by more than 50 percent. In about the same period, the global population of vehicles roughly quadrupled.[5] Increased efficiency in resource use is often accompanied by growth of consumption; when that growth is a result, it is famously named "Jevon's Paradox" by economists (Polimeni et al., 2008).

The paradox is that as energy production has become more efficient (and each unit of energy cheaper), more energy-utilizing products have inevitably come on the market. Moreover, as alternative energy resources become available, depending on policy choices it could produce a drop in the price of fossil fuels, which in turn could lead to more purchases of the products that use (a lot of) those fuels. Thus the drop in the price of oil on the world market in 2015–2016 encouraged Americans to switch back to buying SUVs instead of fuel-efficient small cars and hybrids. So technological advances can have counterintuitive results—as when computers did not lead to a "paperless office," but to more use of paper (York, 2006). Without commensurate changes in norms and incentives, advances in energy technologies could hinder the effort to mitigate climate disruption.

Furthermore, there are clearly recognized thermodynamic limits to efficiency (although in many cases there is a long way to go). With regard to vehicles, although fuel use efficiency draws the most attention, issues like the pollution from tires abrading on roads, the toxic effects of tire-based artificial turf (Katz, 2007), the impacts of paving over valuable farmland and biodiversity reserves for cities and roads, the need for exotic materials (such as rare earths) and energy inputs in the manufacture of high-tech cars, and severance of other resources for automobile construction are generally ignored.

The resource situation today is made much more complex by the nation state system. It is sad that seemingly few people understand the consequences of a simple fact: Natural resources are more or less randomly distributed among nations ("how did our oil get under their sand?"). If that were understood, a crucial issue in "international" relations, highlighted by a sure-

5 From ~250 million to ~1000 million between 1970 and 2010.

to-explode migration crisis, would be one now rarely considered question in public discourse: "Are borders ethical?"

Population and Consumption: Two Sides of the Same Coin

The role of population size and growth in the human predicament is still, sadly, plagued by erroneous ideas. One, the "Pearce Fallacy," is named after environmental journalist Fred Pearce, who repeatedly claims that overconsumption is a much larger contributor to environmental deterioration than overpopulation (Pearce, 2009). This is roughly like claiming that the length of a rectangle is a much larger contributor to its area than its width. One must note, in this case, however, that the problems of changing the width (population size) of the scale of the human enterprise rectangle are very different from those of changing the length (per capita consumption).

Unabated population growth in most circumstances prevents the successful "development" of societies and retards increasing per capita consumption, keeping many people from becoming more prosperous. What typically happens is that, after reduction of high death rates of infants and children, a nation's population grows rapidly for a while. Then, when family planning has been introduced, there follows a period of slackening population growth and a concomitant rise in per capita consumption. The rapid growth of population and consumption do not occur simultaneously, but the result is a gigantic amount of consumption and, sadly, the destruction of human life-support systems. Of course, if family planning is inadequate, there will still be a gigantic increase in consumption from unabated growth, but not per capita. And population or consumption growth in super-consuming nations is super-dangerous (Bradshaw and Ehrlich, 2015).

Because humanely ending population growth and starting the, in our view necessary, population reduction will take a long time (Bradshaw, Corey, and Brook, 2014), it is a huge error to postpone or divert attention away from dealing with human population growth. For example, George Monbiot correctly points to the increase in the population of domesticated animals (but ignores the significant impacts of dogs and cats), while suggesting that those concerned about the size of the human population are racist (Monbiot, 2015). Some are, but many are simply aware that the rich folk of the world, most of them white, represent a huge environmental threat because of their high per capita consumption (e.g., Ehrlich and Ehrlich, 1989).

In any case the time to start a popular movement to correct that failure and direct cultural evolution toward providing the "foresight intelligence" and the agricultural, environmental, and demographic planning that markets cannot supply is badly needed. Then analysts (and society) might stop treating population growth as an unchangeable "given" and consider the nutritional, health, and environmental benefits of humanely ending growth of the global population well below 10 billion and starting a slow decline. That would be a win-win for society, and at the very least would reduce the number of people suffering if a collapse comes.

The critical importance of increasing the inadequate current action on the demographic driver of the predicament can be seen in the decades that will be required to change the ultimate peak size of the population humanely and sensibly (Bradshaw and Brook, 2014). In contrast, remember, we know from such things as the World War II mobilizations that consumption patterns can be altered dramatically in as little as a year, given appropriate incentives.

Governance, Institutions, and Collapse

Since publication of *The Population Bomb*, the discourse on the human predicament has changed. Besides new attention to the limits to growth and the development of evidence-based ecological economics, there is also rising discussion of ways to avoid the possible collapse of civilization. On the other hand, unabated ignorance still gives testimony to the general failure of the U.S. educational system to deal with demographic issues. A prime example from major U.S. media was a ludicrous *New York Times* screed by David Brooks, calling on Americans to cheer up because "Over the next forty years, the U.S. population will surge by an additional 100 million people, to 400 million" (Brooks, 2010). Equal ignorance of the population-resource-environment situation was shown in 2012 by an article also in the *New York Times* by Ross Douthat, "More Babies, Please" (Douthat, 2012), and one by Rick Newman in *U.S. News and World Report*, "Why a falling birth rate is a big problem," both additional signs of the utter failure of the U.S. educational system. Sadly, the evidence of educational failure persists in 2018 (McGurn, 2018).

Another crucial point is that *Homo sapiens*, both genetically and culturally, has always been a small-group animal. Hunter-gatherer populations generally had about 90 to 220 members (Aiello and Dunbar, 1993), and even today the number of an individual's associates tends to be about that size (Ehrlich, 2000). Yet people today are struggling to live in populations of millions to billions, and not doing too well at it. As has been asserted, only a change of society as profound and far-reaching as the agricultural revolution seems to hold much hope of avoiding a collapse (Ehrlich and Harte, 2015a). It might provide a slim hope of a change so profound as to cause the disappearance of most of the features of the industrial age and the myths that sustain it.

A Way Forward?

Perhaps a good approach would be for the United Nations to convene a global "constitutional convention" that would address the problem of how diverse states could organize themselves politically to address global existential problems without losing valued aspects of their individual identities. It could simultaneously actually convene parallel sets of negotiating conferences on ecosystems/biodiversity, agriculture, population, and so on. The long-ago ozone negotiations could provide a model for work in the other areas. Unfortunately, no sense of urgency has emerged in those areas. Looking at both interconnections and realistic solutions, negotiations might accomplish much. In such a process, long-standing resentments and political problems between nations could be ironed out better than by a general convention with the goal of modernizing global governance alone. Now that almost all nations are finally convinced that climate change is a shared threat, possibly other threats could also be accepted and addressed without twenty years of altercation, building on the climate success. As one major example, the Millennium Assessment of Ecosystems study could be repeated regularly (as does the IPCC) as an adjunct to the solutions-oriented deliberations.

Of course, this is an ambitious undertaking and would require a considerable boost in the UN's finances as well as general support by people everywhere—support that encouraged the organization to tackle really tough questions. Even though parts of the framework for all this already exist in the United Nations, the organization is scorned in many places (especially the U.S.), some of its operations are trivial or wasteful (as is typical of all governments), and most Americans are unaware of most of its activities. And the issues that nations must learn to face in their interactions in the UN include overpopulation and overconsumption by the rich. These should be vigorously pursued in striving to achieve the UN Sustainable Development Goals, and even greater vigor than already shown should be applied to pursuing equal rights for women.

Many of the issues in the governance convention would be quite similar to those that were addressed by the Federalists and Anti-Federalists in the famous debates over the draft U.S. Constitution in 1787–1788 (Bradshaw and Ehrlich, 2015, p. 192). With some success in resolving sovereignty issues, nations might consider an international "Manhattan Project" to tackle selected global problems with the utmost speed. The central international framework of the UN needs to be strengthened, given more authority in some areas, and needs to require more cooperation on many issues now neglected. The 2015 consensus agreement on climate in Paris, however inadequate at that moment, at least gave some hope that the world society was moving in the right direction. Sadly Donald Trump's efforts to destroy humanity's life-support systems reduced the value of that advance.

Utterly utopian and impractical? Perhaps. But nothing is more impractical now than continuing with business as usual. Humanity has a long history of revolutionary cultural change. We hope it will be continued at a greatly accelerated pace, moving in a progressive direction.

Acknowledgments

We thank Andy Beattie, Gretchen Daily, Lisa Daniel, Tim Daniel, Partha Dasgupta, Joan Diamond, Larry Goulder, Charlie Hall, John Harte, Mel Harte, John Holdren, Tad Homer-Dixon, Sandra Kahn, Chandran Nair, Simon Levin, Mark Mancall, Graham Pyke, Lee Ross, Kirk Smith, Chris Turnbull, and Wren Wirth for helpful comments.

References

Aarhus University. (2014, July 29). World-wide water shortage by 2040. *ScienceDaily*. Retrieved from www.sciencedaily.com/releases/2014/200 s7/140729093112.htm.

Aiello, L. C., and Dunbar R. I. M. (1993). Neocortex size, group size, and the evolution of language. *Current Anthropology 34*:184–193.

Anderson, K. (2015). Duality in climate science. *Nature Geoscience, 8*, 898.

Arrow, K., et al. (2004). Are we consuming too much? *Journal of Economic Perspectives, 18*, 147–172.

Bar-On, Y. M., Phillips, R., and Milo, R. (2018). The biomass distribution on Earth. *Proceedings of the National Academy of Sciences, 115*(25), 6506–6511

Bardi, U. (2011). *The limits to growth revisited*. Springer.

Bardi, U. (2014). *Extracted: how the quest for mineral wealth is plundering the planet*. Chelsea Green.

Bardi, U., Lavacchi, A., and Yaxley, L. (2011). Modelling EROEI and net energy in the exploitation of nonrenewable resources. *Ecological Modelling, 223*, 54–58.

Barringer, F. (2011, May 30). Ground water depletion detected from space. *The New York Times*.

Bartlett, A. (1993). The arithmetic of growth: methods of calculation. *Population and Environment: A Journal of Interdisciplinary Studies, 14*, 359–387.

Bechara, A., and Damasio, A. R. (2005). The somatic marker hypothesis: A neural theory of economic decision. *Games and Economic Behavior, 52*, 336–372.

Bellarby, J., Foereid, B., and Hastings, A. (2008). *Cool farming: climate impacts of agriculture and mitigation potential*. Greenpeace.

Boulding, K. E. (1966). The economics of the coming spaceship Earth. In Jarrett, H. (ed.) *Environmental quality in a growing economy*, pp. 3–14. Baltimore: Johns Hopkins University Press.

Boyles, J. G., Cryan, P. M., McCracken, G. F., and Kunz, T. H. (2011). Economic importance of bats in agriculture. *Science, 332*, 41–42.

Bradshaw, C., and Ehrlich, P. R. (2015). *Killing the koala and poisoning the prairie: Australia, America, and the environment*. University of Chicago Press.

Bradshaw, C. J. A., and Brook, B. W. (2014). Human population reduction is not a quick fix for environmental problems. *Proc. Natl. Acad. Sci. USA, 111*, 16610–16615.

Brauman, K. A., Richter, B. D., Postel, S., Malsy, M., and Flörke, M. (2016). Water depletion: An improved metric for incorporating seasonal and dry-year water scarcity into water risk assessments. *Elem. Sci. Anth., 4*, p. 000083. DOI.

Bräutigam, D., and Zhang, H. (2013). Green dreams: myth and reality in China's agricultural investment in Africa. *Third World Quarterly, 34*, 1676–1696.

Britten G. L., Dowd, M., and Worm, B. (2015). Changing recruitment capacity in global fish stocks. *Proc. Natl. Acad. Sci. USA. Retrieved from* http://www.pnas.org/content/early/2015/12/09/1504709112.full.pdf.

Brooks, D. (2010, April 6). Relax, we'll be fine. *The New York Times.*

Brown, L. (1963). Man, land & food: looking ahead at world food needs. U.S. Department of Agriculture.

Campbell, M., and Potts, M. (2012). *Do economists have frequent sex?* The Population Press.

Campbell, M., Sahin-Hodoglugil, N. N., and Potts, M. (2006). Barriers to fertility regulation: A review of the literature. *Studies in Family Planning, 37*, 87–98.

Carson, R. (1962). *Silent spring.* Houghton Mifflin.

Ceballos, G., Ehrlich A. H., and Ehrlich, P. R. (2015a). *The annihilation of nature: human extinction of birds and mammals.* Johns Hopkins University Press.

Ceballos, G., and Ehrlich, P. R. (2018). Misunderstanding the sixth mass extinction. *Science*, 360, 1080-1081

Ceballos, G., and Ehrlich, P. R., Barnosky, A. D., García, A., Pringle, R. M., and Palmer, T. M. (2015b). Accelerated modern human–induced species losses: Entering the sixth mass extinction. *Science Advances, 1*, e1400253.

Ceballos, G., Ehrlich, P. R., and Dirzo, R. (2017). Biological annihilation via the ongoing sixth mass extinction signaled by vertebrate population losses and declines. *Proceedings of the National Academy of Sciences, 114*, E6089–E6096.

Cotula, L. (2009). *Land grab or development opportunity? agricultural investment and international land deals in Africa.* Iied.

Cotula, L. (2012). The international political economy of the global land rush: A critical appraisal of trends, scale, geography and drivers. *The Journal of Peasant Studies, 39*, 649–680.

Cribb, J. (2014). *Poisoned planet: how constant exposure to man-made chemicals is putting your life at risk.* Allen and Unwin.

Czech, B. (2014, October 15). Paul Krugman on limits to growth: beware the bathwater. *Huffpost.* Retrieved from https://www.huffpost.com/entry/paul-krugman-on-limits-to_b_5988532.

Daily, G. C., and Ehrlich, P. R. (1996). Impacts of development and global change on the epidemiological environment. *Environment and Development Economics, 1*, 309–344.

Daly, H. E., et al. (1973). *Toward a steady-state economy.* W.H. Freeman and Co.

Dasgupta, P. (2001). *Human well-being and the natural environment.* Oxford University Press.

Dasgupta, P, and Ehrlich, P. R. (2013). Pervasive externalities at the population, consumption, and environment nexus. *Science, 340*, 324–328.

Day, J., John, W., Hall, C. A. S., Yáñez-Arancibia, A., Pimentel, D., Ibáñez-Marti, C., and Mitsch, W. J. (2009). Ecology in times of scarcity. *BioScience, 59*, 321–331.

Dennig, F., Budolfson, M. B., Fleurbaey, M., Siebert, A., and Socolow, R. H. (2015). Inequality, climate impacts on the future poor, and carbon prices. *Proc. Natl. Acad. Sci. USA.* doi: 10.1073/pnas.1513967112 7 December.

Douthat, R. (2012). More babies please. *The New York Times.*

Ehrlich P. R. (2014). Human impact: the ethics of I=PAT. *Ethics in Science and Environmental Politics, 14*, 11–18.

Ehrlich, P. R. 1968. *The population bomb.* Ballantine Books.

Ehrlich, P. R. (2000). *Human natures: genes, cultures, and the human prospect.* Island Press.

Ehrlich, P. R. (2008). Key issues for attention from ecological economists. *Environment and Development Economics, 13*, 1–20.

Ehrlich, P. R., and Ehrlich, A. H. (1989). Too many rich folks. *Populi, 16*, 20–29.
Ehrlich, P. R., and Ehrlich, A. H. (2006). Enough already. *New Scientist, 191*, 46–50.
Ehrlich, P. R. (2010). The culture gap and its needed closures. *International Journal of Environmental Studies, 67*, 481–492.
Ehrlich, P. R. (2012). Solving the human predicament. *International Journal of Environmental Studies, 69*, 557–565.
Ehrlich, P. R. (2013). Can a collapse of civilization be avoided? *Proceedings of the Royal Society B*. Retrieved from http://rspb.royalsocietypublishing.org/content/280/1754/20122845.
Ehrlich, P. R., and Harte, J. (2015a). Food security requires a new revolution. *International Journal of Environmental Studies* doi: 10.1080/00207233.2015.1067468:1-13.
Ehrlich, P. R. (2015b). Opinion: to feed the world in 2050 will require a global revolution. *Proc. Natl. Acad. Sci. USA, 112*, 14743–14744.
Ehrlich, P. R., Harte, J., Harwell, M. A., Raven, P. H., Sagan, C., Woodwell, G. M., et al. (1983). Long-term biological consequences of nuclear war. *Science, 222*, 1293–1300.
FAO: Food and Agriculture Organization of the United Nations (2017). How close are we to #Zero-Hunger?
Farrell, J. (2015). Corporate funding and ideological polarization about climate change. *Proc. Natl. Acad. Sci. USA* doi: 10.1073/pnas.1509433112 1-6.
Foley, J. A. (2011). Can we feed the world and sustain the planet? A five-step global plan could double food production by 2050 while greatly reducing environmental damage. *Scientific American*, 60, 60-65.
Godfray, H. C. J., and Garnett, T. (2014). Food security and sustainable intensification. *Phil. Trans. R. Soc. B, 369*(60), 60–65. 20120273.
Goulder, L. H., and Kennedy, D. (1997). Valuing ecosystem services: philosophical bases and empirical methods. In Daily, G. C. (ed.), *Nature's services*, pp. 23–47. Washington, D.C.: Island Press.
Hall, C. A. S., Balogh, S., and Murphy, D. J. R. (2009). What is the minimum EROI that a sustainable society must have? *Energies, 2*, 25–47.
Hall, C. A. S., and Day Jr., J. W. (2009). Revisiting the limits to growth after peak oil. *American Scientist*, 97: 230–237.
Hall, C. A. S., Powers, R., and Schoenberg, W. (2008). Peak oil, EROI, investments and the economy in an uncertain future. In Pimentel, D. (ed.), *Biofuels, solar and wind as renewable energy systems*, pp. 109–132. Berlin: Singer.
Harte, J. (2007). Human population as a dynamic factor in environmental degradation. *Population and Environment, 28*, 223–236.
Heinberg, R. (2007). *Peak everything*. New Society Publishers.
Heinberg, R. (2011). *The end of growth: adapting to our new economic reality*. New Society Publishers.
Holdren, J. P. (2008). Science and technology for sustainable well-being. *Science, 319*, 424–434.
Hopcraft, A. (1968). *Born to Hunger*. Macmillan.
Hunt, N., and McFarlane, S. (2014, July 17). "Peak soil" threatens future global food security. *Reuters*. Retrieved from https://www.reuters.com/article/us-peaksoil-agriculture/peak-soil-threatens-future-global-food-security-idUSKBN0FM1HC20140717.
Insu, K., Lonsdorf, E., Williams, N., Brittain, C., Isaacs, R., Gibbs, J., and Ricketts, T. (2015). Modeling the status, trends, and impacts of wild bee abundance in the United States. *Proc. Natl. Acad. Sci. USA*. doi: 10.1073/pnas.1517685113.
Intergovernmental Panel on Climate Change (IPCC) (1996). *Climate change 1995—the science of climate change: contribution of working group I to IPCC Second Assessment Report*. Cambridge University Press.
Jackson, T. (2009). Prosperity without growth: economics for a finite planet. *Earthscan*.
Kahn, S., and Ehrlich, P. R. (2018). *Jaws: The story of a hidden epidemic*. Stanford University Press.
Katz, A. (2007). Artificial turf full of toxins that can cause cancer. *UConn Health Today*. Retrieved from http://today.uchc.edu/headlines/2007/nov07/toxins.html.
Kelley, C. P., Mohtadi, S., Cane, M. A., Seager, R., and Kushnir, Y. (2015). Climate change in the Fertile Crescent and implications of the recent Syrian drought. *Proc. Natl. Acad. Sci. USA, 112*, 3241–3246.

Klare, M. T. (2012). *The race for what's left: the global scramble for the world's last resources.* Metropolitan Books.

Kristensen, H. M., McKinzie, M., and Postol, T. A. (2017). How U.S. nuclear force modernization is undermining strategic stability: The burst-height compensating super-fuze. *Bulletin of the Atomic Scientists.* Retrieved from https://thebulletin.org/2017/03/how-us-nuclear-force-modernization-is-undermining-strategic-stability-the-burst-height-compensating-super-fuze/.

Levin, S. (1999). *Fragile dominion.* Perseus Books.

Liu, J, and Yang, W. (2012). Water sustainability for China and beyond. *Science, 337,* 649–650.

Livet, P. (2010). Rational choice, neuroeconomy, and mixed emotions. *Proc. Royal Soc. B, 365,* 259–269.

Mankin, J. S., Viviroli, D., Singh, D., Hoekstra, A. Y., and Diffenbaugh, N. S. (2015). The potential for snow to supply human water demand in the present and future. *Environmental Research Letters, 10,* 114016.

Mann, M. E. (2009). Defining dangerous anthropogenic interference. *Proc. Natl. Acad. Sci. USA, 106,* 4065–4066.

Mann, M. E. (2012). *The hockey stick and the climate wars: dispatches from the front lines.* Columbia University Press.

May R. M., Levin, S., Sugihara, G. (2008). Ecology for bankers. *Nature, 451,* 893–895.

Mayer, J. (2016). *Dark money: The hidden history of the billionaires behind the rise of the radical right.* Anchor.

McGurn, W. (2018, April 30). The population bomb was a dud. *The Wall Street Journal.* Retrieved from https://www.wsj.com/articles/the-population-bomb-was-a-dud-1525125341.

Meadows, D. H., Meadows, D. L., Randers, J., Behrens III, W. W. (1972). *The limits to growth.* Universe Books.

Mehra, P., Downie, M., Pita, M. C., LM. W. (2001). Pharyngeal airway space after counterclockwise rotation of the maxillomandibular complex. *Am. J. Dentofacial Orthop., 120,* 154–159.

Milman, O. (2015, December 2). Earth has lost a third of arable land in past 40 years, scientists say. *The Guardian.* Retrieved from https://www.theguardian.com/environment/2015/dec/02/arable-land-soil-food-security-shortage.

Monbiot, G. (2015, November 15). There's a population crisis all right. But probably not the one you think. *The Guardian.* Retrieved from https://www.theguardian.com/commentisfree/2015/nov/19/population-crisis-farm-animals-laying-waste-to-planet.

Montgomery, D. R. (2012). *Dirt: the erosion of civilizations.* University of California Press.

Newman, R. (2012). Why the falling birth rate is a big problem. *U.S. News & World Report, 3.*

Oreskes, N., and Conway, E. M. (2010). *Merchants of doubt: how a handful of scientists obscured the truth on issues from tobacco smoke to global warming.* Bloomsbury Press.

Paddock, W., and Paddock, P. (1967). *Famine, 1975!: America's decision: who will survive?* Little, Brown.

Pearce, F. (2009). Consumption dwarfs population as main environmental threat. *Yale Environment, 360,* 20545–20567.

Perez, I. (2013, March 4). Climate change and rising food prices heightened Arab Spring. *Scientific American.* Retrieved from https://www.scientificamerican.com/article/climate-change-and-rising-food-prices-heightened-arab-spring/.

Peters, G. P. (2018). Beyond carbon budgets. *Nature Geoscience,* 1.

Polimeni, J. M., Mayumi, K., Giampietro, M., and Alcott, B. (2008). The jevons paradox and the myth of resource efficiency improvements. *Earthscan.*

Ramakrishnan, U. (2002). Prevalence of micronutrient malnutrition worldwide. *Nutrition Reviews, 60,* S42–S56.

Raspail, J. (1975). *The camp of the saints.* Scrivners.

Ray, D. K., Mueller, N. D., West, P. C., and Foley, J. A. (2013). Yield trends are insufficient to double global crop production by 2050. *PLoS ONE, 8,* e66428.

Reed, S. (2015, November 9). Shift to lower-carbon energy is too slow, I.E.A. report warns. *The New York Times.* Retrieved from https://www.nytimes.com/2015/11/10/business/energy-environment/shift-to-lower-carbon-energy-is-too-slow-report-warns.html?smid=tw-share&_r=0.

Rudd, R. L. (1964). *Pesticides and the living landscape.* Madison: University of Wisconsin Press.

Sanger, D. E., and Perlez, J. (2017). Trump hands the Chinese a gift: the chance for global leadership. *The New York Times.*

Scott, P. (2017). Global panel on agriculture and food systems for nutrition: food systems and diets: facing the challenges of the 21st century. *Food Security: The Science, Sociology and Economics of Food Production and Access to Food, 9,* 653–654.

Seaquist, J., Johansson, E. L., and Nicholas, K. A. (2014). Architecture of the global land acquisition system: applying the tools of network science to identify key vulnerabilities. *Environmental Research Letters, 9*(11), 114006.

Singh R. B. (2000). Environmental consequences of agricultural development: a case study from the Green Revolution state of Haryana, India. *Agriculture, Ecosystems and Environment, 82,* 97–103.

Smith, M. R., Singh, G. M., Mozaffarian, D., and Myers, S. S. (2015). Effects of decreases of animal pollinators on human nutrition and global health: a modelling analysis. *The Lancet.* doi: 10.1016/S0140-6736(15)61085-6.

Spijker, J., and MacInnes, J. (2013). Population ageing: the timebomb that isn't? *British Medical Journal.* doi: 10.1136/bmj.f6598.

Tainter, J. A. (1988). *The collapse of complex societies.* Cambridge University Press.

Tanumihardjo, S. A., Anderson, C., Kaufer-Horwitz, M., Bode, L., Emenaker, N. J., Haqq, A. M., Satia, J. A., Silver, H. J., and Stadler, D. D. (2007). Poverty, obesity, and malnutrition: an international perspective recognizing the paradox. *Journal of the Academy of Nutrition and Dietetics, 107,* 1966–1972.

Tilman, D., Balzer, C., Hill, J., and Befort, B. L. (2011). Global food demand and the sustainable intensification of agriculture. *Proc. Natl. Acad. Sci. USA, 108,* 20260–20264.

Toon, O., Robock, A., Turco, R. P., Bardeen, C., Oman, L., and Stenchikov, G. (2007). Consequences of regional-scale nuclear conflicts. *Science, 315,* 1224–1225.

Turner, G. M. (2008). A comparison of *The Limits to Growth* with 30 years of reality. *Global Environmental Change, 18,* 397–411.

Vitousek, P. M., Ehrlich P. R., Ehrlich A. H., and Matson P. A. (1986). Human appropriation of the products of photosynthesis. *BioScience, 36,* 368–373.

Vitousek, P. M., Mooney, H. A., Lubchenco, J., and Melillo, J. M. (1997). Human domination of earth's ecosystem. *Science, 277,* 494–499.

Wackernagel, M., and Rees, W. (1998). Our ecological footprint: reducing human impact on the earth. New Society Publishers.

Wilkinson, R., Pickett K. (2009). *The spirit level: why more equal societies almost always do better.* Penguin Books.

York, R. (2006). Ecological paradoxes: William Stanley Jevons and the paperless office. *Human Ecology Review, 13,* 143–147.

FOOD FOR 10 BILLION

Possibilities and Probabilities

Nina V. Fedoroff
Evan Pugh Professor Emerita, Penn State University

Writing about the prospects for sustainably expanding agriculture to feed the roughly 10 billion humans expected in coming decades puts me in mind of the portentous opening of Dickens's classic *A Tale of Two Cities*:

> It was the best of times, it was the worst of times, it was the age of wisdom, it was the age of foolishness, it was the epoch of belief, it was the epoch of incredulity, it was the season of Light, it was the season of Darkness, it was the spring of hope, it was the winter of despair, we had everything before us, we had nothing before us.

The twentieth century saw a dramatic expansion of agricultural production driven by both increases in the area under cultivation and by improvements in agricultural technology, principally mechanization, synthetic nitrogen fertilizer, and genetically improved seeds. Despite the predictions of catastrophists from Malthus to Ehrlich, the food supply grew more rapidly than population, reducing the fraction of people experiencing chronic hunger from half to a sixth even as our numbers grew from 3 to 6 billion by mid-century (Ehrlich, 1968; Malthus, 1798).

While most of the best agricultural land on earth was already under cultivation by 1960, agronomic and genetic improvements continued to power yield growth through the rest of the century. Most recently, the genetic and genomic revolutions of the late twentieth century made it possible to use molecular techniques to genetically modify crops to resist pests and pathogens and to tolerate herbicides. Together with GPS-guided farm machinery, these developments promoted more sustainable agriculture by decreasing its environmental impact, even as they continued to push yields up.

Today, we have enough food to meet the world's needs. Indeed, we have an extraordinary global food system that brings food from all over the planet to consumers who can afford to buy it. The food price spike of 2008 and the resurgence of high food prices in recent years have had little impact on the affluent citizens of the developed world who spend a small fraction of their income on food. This has given rise to the popular view, particularly in rich countries, that there's plenty of food, were it to be equitably distributed, even for a population of 9 or 10 billion (Bittman, 2014).

Yet even today, fluctuating food prices are having a profound impact on the world's poorest people, many of whom spend half or more of their income on food. During the food price crisis of 2008, there were food riots in more than thirty countries. Spiraling food prices drive the world's poorest into chronic hunger even in a world of relative plenty. Indeed, the demise of governments during the Arab Spring had its roots in high food prices. Increasingly, however, food price spikes find their underlying explanations in the severe weather events occurring in parts of the world distant from the poor countries most impacted by them (Werrell and Femia, 2013). Such severe weather events are harbingers of the coming impact of climate warming on the global food supply.

The robust annual increases in agricultural productivity seen during the second half of the twentieth century are shrinking, even turning negative in parts of the world. Indeed, current yield growth trends are simply insufficient to keep up with growing demand (Ray et al., 2013). As well, the rapid expansion of agriculture over the past century to feed today's population has had a devastating impact on biodiversity (Ayensu et al., 1999). Hence there is an acute need to curb the deleterious impact of agriculture on biodiversity and the services provided by complex ecosystems (Cardinale et al., 2012). Taken together with growing pressures on land and fresh water from other human activities, it is becoming clear that, using current agricultural technology, we will exceed the biophysical limits of the Earth to produce enough food sustainably for a population of the magnitude and affluence anticipated by century's end. Indeed, we are arguably already exceeding those limits.

There is another, largely cultural, development that is impacting humanity's ability to feed itself: urbanization. By the end of the twentieth century, a mere 1 to 2 percent of the developed world's population was supplying the entire population's food needs through highly productive, large-scale agriculture systems. Industrialization and the growth of large-scale, highly mechanized agriculture have driven an unprecedented demographic shift of people to cities over the past two centuries. Today, more than 80 percent of the U.S. population lives in cities, as does more than half of the world's population. This fraction is expected to rise to 70 percent by the middle of the twenty-first century.

The result is that most people are far removed from agriculture and the vast processing and distribution system that supplies the world's food. A nostalgic view of pre-industrial agriculture as more "natural" has taken hold among urban elites in parallel with concerns about the impact of human activity on biodiversity and the environment. This has led to the codification of largely nineteenth-century inefficient small-scale farming procedures under the rubric of "organic farming." The still-growing organic food industry is based on the false claims that food produced by the archaic organic farming rules is more "healthful" than food grown by conventional methods and that organic farming has a lower environmental impact than conventional farming.

These myths, deliberately fostered by both environmental activists and the organic food marketers for economic reasons, affect the ability of agriculture to adopt and use both new technology, primarily biotechnology, and efficiencies of scale. And yet, paradoxically, there is little doubt that farming methods and materials must continue to become more efficient and sophisticated if we are to realize humanity's hope of sustainable global food security for the anticipated population of 9 to 10 billion.

In the ensuing sections, I will briefly overview the major scientific and technological advances that underlie the rapid expansion of agriculture and, as a result, of population in the twentieth century. I will then outline the kinds of further technological innovations that are likely to make the greatest contribution to increasing the food supply sustainably in the face of climate warming. I will finish with a discussion of the cultural barriers that must be surmounted to reach sustainable food security.

Historical Perspective

For most of our evolutionary history, our numbers were small and we were mobile hunter-gatherers. We spent our time finding and capturing enough food to feed ourselves and our closest kin. Then sometime between 10,000 and 20,000 years ago—maybe even more—that started to change. We began to shape plants and animals to our own advantage and settled down to grow and herd them (Diamond, 1987). The process by which we have modified plants and animals to suit our needs, traditionally called "domestication," is a process of genetic modification (Doebley et al., 2006). Early peoples selected variant organisms—plants, animals, and microbes—with useful traits, such as seeds that adhere to plants until they are harvested and animals tame enough to herd. Domestication is a process of modification that is possible because of the genetic variation constantly arising in all living organisms.

While hunter-gatherers were quite sophisticated in their resource management, it was systematic planting and harvesting of crops that marks the origin of what we now call "agriculture" (Harris, 1996). Agriculture allowed people to produce more food than they consumed; cities and civilization followed. Simply put, human civilization emerged because we figured out how to produce surplus food in the mild climate of the Holocene. For the next ten millennia, people built cities and civilizations, wore out the land, invaded their neighbors or abandoned the cities and civilizations, eventually rebuilding on fresh land (Rimas and Fraser, 2010). It was often the fertility of the land that determined how long a civilization lasted. Plants extract nutrients from the soil and crop yields decline, making it harder and harder to produce enough food as the human population expands (Diamond, 1987).

Concern about access to sufficient food, today called "food security," is as old as mankind. Thomas Malthus's famous *Essay on Population*, published in 1798, crystallized the problem of balancing food and human population for the modern era (Malthus, 1798). Malthus's view was that humanity was doomed to food insecurity because our numbers increased exponentially, while our ability to produce food could only increase linearly.

Curiously, Malthus penned his essay at about the time that science began to play a major role in boosting agricultural productivity. Late-eighteenth-century milestones were Joseph Priestley's discovery that plants emit oxygen and Nicholas-Théodore de Saussure's definition of the chemical composition of plants (de Saussure, 1804; Priestly, 2013). Malthus could not have envisioned the extraordinary increases in productivity that the integration of science and technology into agricultural practice would stimulate over the ensuing two centuries.

Both organic and mineral fertilization of plants has been practiced since ancient times. Farmers knew that certain chemicals and biological materials, ranging from fish and oyster shells to manure and bones, stimulated plant growth (Bear, 1938; Russel and Williams, 1977). Justus von Liebig made important contributions to the study of plant nutrient requirements, understanding that biological sources of nitrogen could be replaced with purely chemical sources (Smil, 2004). But supplying nitrogen in the forms that plants use remained a major limitation until the development, early in the twentieth century, of the Haber-Bosch process for fixing atmospheric nitrogen (Smil, 2004). Today agriculture in the developed world relies primarily on chemical fertilizers. Indeed, the global human population could not have grown from roughly 1 billion at the turn of the nineteenth century to today's 7.6 billion without synthetic nitrogen fertilizer.

Crop Domestication

Humans practiced genetic modification long before chemistry entered agriculture, transforming inedible wild plants into crop plants, wild animals into domestic animals, and harnessing microbes to produce everything from cheese to wine, beer, and bread. Oddly, it is only our contemporary methods of bending organisms' genetic constitution to suit our needs that are today recognized as genetic modification, known in common parlance by the abbreviations "GM" (genetically modified), "GMO" (genetically modified organism), or "GE" (genetically engineered). Yet all of the useful, heritable traits nurtured by people in organisms constitute "domestication" and all are the result of genetic modifications. Each microbe, crop, and animal has its own interesting history. To take just one example, a fundamental trait that distinguishes wild from domesticated plants is the retention of mature seeds on the plant. Plants have many mechanisms for seed dispersal, but it is much easier for people to harvest seeds that remain attached to the plant at maturity. Hence one of the earliest steps in grain crop domestication was the identification of mutations—genetic changes—that prevent seed dispersal (Fuller and Allaby, 2009).

Corn (*Zea mays* subsp. mays), also known as maize, remains one of our most spectacular feats of genetic modification. Its huge ears, packed with starch and oil, provide one of humanity's most important sources of food and feed. Corn bears little resemblance to its wild progenitor, teosinte (*Zea mays* subsp. parviglumis). Indeed, when teosinte was first discovered in 1896, it was assigned to a different genus. By the 1920s, it was known that teosinte and corn readily produce fertile hybrids and are best classified as different subspecies of a single species, but controversies about their relationship and the origin of corn continued throughout most of the twentieth century. The key genetic changes that transformed teosinte into corn appear to have happened in the Balsas River Valley in Mexico some nine thousand years ago (Jaenicke-Despres et al., 2003). The mutations that converted teosinte, a plant with hard, inedible seeds, into modern corn altered just a handful of genes that control plant architecture and the identity of reproductive organs. Remarkably, once these mutations had been brought together in an early corn plant, they stayed together and spread very rapidly, moving from Mexico into the American southwest by three thousand years ago (Jaenicke-Despres et al., 2003).

Among the many other traits altered during domestication of plants are the size and shape of leaves, tubers, berries, fruits, and grains, as well as their abundance, toxicity, and nutritional value. The changes are often in genes coding for proteins that regulate the expression of many other genes (Doebley et al., 2006). Differences in nutrient composition among varieties of the same crop are caused by mutations in genes coding for proteins in a number of different biosynthetic pathways. Thus, for example, sweet corn harbors mutations that prevent the conversion of sugar to starch in the kernel (Boyer and Shannon, 1983).

Modern Crop Improvement

The genetic revolutions of the twentieth century boosted crop productivity immeasurably. Austrian monk Gregor Mendel's pioneering observations on inheritance were published in 1865, but did not get wide attention until a half-century later (Carlson, 1966). A simple demonstration project to illustrate Mendelian inheritance led to the rediscovery of hybrid vigor, a long-known phenomenon whose incorporation into crop breeding resulted in a dramatic expansion of the corn ear and, thereby, crop yield (Crow, 1998).

However, when corn hybrids were first introduced in the U.S. during the 1930s, they faced resistance and criticism similar to that leveled at contemporary GM crops. The hybrids were complex to produce and agriculture experiment stations were not interested. Eventually a company was formed to produce hybrid seed. But farmers accustomed to planting seed from last year's crop saw no reason to buy it. It was only when farmers realized the yield benefits and the drought-resistance of hybrid corn during the 1934–1936 dust-bowl years that they began to adopt hybrid corn rapidly (Crabb, 1947).

Techniques for accelerating mutation rates with radiation and chemicals and through tissue culture were developed and widely applied in the genetic improvement of crops during the twentieth century (Maluszynski et al., 1995). These methods introduce mutations indiscriminately and require the growth of large numbers of seeds, cuttings, or regenerants to detect desirable changes. Nonetheless, all of these approaches have proven valuable in crop improvement. By the end of the twentieth century, more than 2,300 different crop varieties, ranging from varieties of cultivated wheat (*Triticum* ssp.) to strains of grapefruit (*Citrus paradisi*), had been developed using radiation and chemical mutagenesis (Maluszynski et al., 1995).

Mechanization of Agriculture

A major development with impact Malthus could not have envisioned is the mechanization of agriculture. Human and animal labor provided the motive force for agriculture throughout most of history and continues to do so in many less-developed countries. The invention of the internal combustion engine at the turn of the twentieth century eventually led to the development of small, maneuverable tractors. The mechanization of plowing, seed planting, cultivation, fertilizer and pesticide distribution, and harvesting accelerated in the U.S., Europe, and Asia following World War II (Binswanger, 1986). Major demographic changes followed the mechanization of agriculture virtually everywhere. In the U.S., 21 percent of the workforce was employed in agriculture in 1900 (Dimitri et al., 2005). By 1945, the fraction had declined to 16 percent and by 2000 the fraction of the population employed in agriculture had fallen to 1.9 percent. At the same time, the average size of farms increased and farms increasingly specialized in fewer crops. This profound demographic shift from agrarian to urban underlies the development of today's attitudes about food and farming in developed countries. Today the vast majority of the developed world's population is urban and far removed from primary food production.

The Green Revolution

Malthus penned his essay when the human population of the world stood at less than a billion. The population tripled over the next century and a half. As the second half of the twentieth century began, there were neo-Malthusian predictions of mass famines in developing countries that had not yet experienced science- and technology-based advances in agriculture. Perhaps the best known of the mid-century catastrophists was Paul Ehrlich, author of *The Population Bomb* (Ehrlich, 1968).

Remarkably, the extraordinary work of just a handful of scientists and their teams, principally plant breeders Borlaug, Swaminathan, Khush, and Rajaram, averted widely predicted Asian famines (Khush, 2001). The Green Revolution was based on the development of dwarf rice (*Oryza sativa*) and wheat varieties that responded to fertilizer application without falling over (lodging). Subsequent breeding for increased yield has continued to improve the productivity of these crops by as much as 1 percent per year. As noted earlier and perhaps most remarkably, the Green Revolution and other technological advances reduced the fraction of the world's hungry from half to less than a sixth, even as the population doubled from 3 to 6 billion. These accomplishments earned Borlaug a well-deserved Nobel Peace Prize in 1970, while all of the others have been World Food Prize laureates. Curiously, the Green Revolution is often vilified today.

Genetic Modification of Crops

The equally revolutionary molecular genetic advances that began in the 1960s led to the development of new methods of crop improvement. The basic methodology lies in the construction of hybrid DNA molecules designated "recombinant DNA" (R-DNA) because they consist of a piece of bacterial or viral DNA combined with a piece of DNA from a different kind of organism, plant, or animal (Fedoroff and Brown, 2004). The ability to multiply such hybrid DNA molecules in bacteria made it possible to develop the DNA sequencing techniques that underlie today's genomic revolution.

As well, techniques were developed to introduce genes into plants using either the soil bacterium *Agrobacterium tumefaciens*, which naturally transfers a segment of DNA into a plant cell, or mechanical penetration of plant cells using tiny DNA-coated particles (Birch, 1997). This combination of methods and knowledge made it possible to transfer a well-understood segment of genetic material from either the same or a related plant or from a completely unrelated organism into virtually any crop plant, creating what is known as a "transgenic" plant. Because genes work the same way in all organisms, this made it possible to introduce a desirable trait, such as disease- or pest-resistance, without the extensive genetic disturbance attending what we now consider to be the "conventional" crop improvement techniques of breeding and mutagenesis. Indeed, recent comparisons have revealed plant modification by molecular techniques has less impact on the genome and gene expression levels than do conventional genetic crosses (Richroch, 2013).

Several crop modifications achieved using these methods are now in widespread use. Perhaps the best known of these are crop plants containing a gene from the soil bacterium, *Bacillus thuringiensis*, long used as a biological pesticide. The gene encodes a protein that is toxic to the larvae of certain kinds of insects, but not to animals or humans (Hilder and Boulter, 1999). Such a toxin gene is often called the "Bt gene," but is actually a family of related toxin genes from a group of closely related bacteria and these are increasingly used in combinations to decrease the probability of resistance developing in the target insects, an approach that has been dubbed gene "stacking."

Herbicide tolerance is another widely accepted GM crop modification. Among the most common herbicides in use today are compounds that interfere with the production of certain amino acids that plants synthesize, but animals do not (Tan et al., 2006). Such herbicides therefore kill plants, but have low or no toxicity for animals or humans. Herbicide-tolerant crops make it possible to control weeds without damaging the crop and without tilling the soil. Such crops have been derived through natural mutations and induced mutations, as well as by introduction of genes from either bacterial sources or plant sources. Today, herbicide-tolerant varieties of many crops, most importantly soybeans (*Glycine max*) and canola (*Brassica napus*), are widely grown (Lee et al., 2014).

Papayas (*Carica papaya*) resistant to papaya ringspot virus (PRSV) saved the Hawaiian papaya industry and comprise the only such GM crop to emerge from public-sector GM research to date, although several others are being commercialized. Papaya ringspot virus is a devastating insect-borne viral disease that wiped out the papaya industry on Oahu in the 1950s, forcing its relocation to the Puna district of the Big Island, Hawaii. PRSV was first detected in the Puna district in 1992; by 1994 it was widespread and threatening the industry. A project initiated in 1987 introduced a gene from the PRSV into papayas based on reports that introducing a viral gene could make a plant resistant to the virus from which the gene came (Abel et al., 1986; Gonsalves, 1998). Transgenic seeds were released in 1998; by 2000, the papaya industry was returning to pre-1995 levels. This remarkable achievement of disease resistance enhanced a virus protection mechanism already present in the plant, much as vaccination protects people and animals from infection by pathogens (Tennant et al., 2001).

New methods are rapidly being developed that promise to further increase the specificity and precision of genetic modification. These techniques capitalize on growing knowledge of the dynamic processes underlying genome maintenance, particularly the repair of breaks in the genetic material, DNA. Known under the general rubric of "site-directed nuclease" (SDN) technology, this approach uses proteins (or protein-nucleic acid complexes) that seek out, bind to, and cut specific DNA sequences, introducing breaks in the DNA at one or a small set of sequences targeted for modification (Podevin et al., 2013). Repair of such DNA cuts by natural cellular processes results in precisely targeted genetic changes rather than the random ones introduced by older methods of mutagenesis.

The more recently developed system known by the acronym CRISPR/Cas is based on a bacterial defense system against invading viruses and promises extraordinary versatility in the kinds of very precise genome modifications that it can make (Zhang et al., 2016). The method can be used to edit multiple genes simultaneously, as well as to introduce a gene or genes at a preidentified site in the genome, something that could not be done with pinpoint specificity and precision by R-DNA methods. As well, such genetic changes can be made without creating a transgenic plant. The changes are the same at the molecular level as those that occur in nature or can be induced by older mutagenic techniques. What's new is that the genetic changes introduced by SDN techniques are not random, but confined precisely to the gene or genes selected by the breeder.

Adoption of GM Crops

GM crops have been adopted at unprecedented rates since their commercial introduction in 1996. In 2017, GM crops were grown in 24 countries on about 190 million hectares (ISAAA, 2017). Most importantly, more than 90 percent of the roughly 17 million farmers growing biotech crops today are smallholder, resource poor farmers (James, 2015). The simple reasons that farmers migrate to GM crops are that crop yields increase and farmer production costs decrease. A meta-analysis of 147 crop studies conducted over a period of 20 years concluded that the use of GM crops had reduced pesticide use by 37 percent, increased crop yields by 22 percent and increased farmers' profits by 68 percent (Klümper and Qaim, 2014). The vast majority of GM hectarage is devoted to the growing of GM corn, soybeans, cotton (*Gossypium* spp.), and canola with either Bt toxin-based pest resistance or herbicide tolerance traits. The reasons for the narrow GM crop and trait base to date lie in a combination of the economic, regulatory, and legal issues, discussed below.

While some resistance to the Bt toxin has developed, it has not been as rapid as initially feared and second-generation, two-Bt gene strategies to decrease the probability of resistance are already being implemented (Lemaux, 2009). Predicted deleterious effects on non-target organisms, such as monarch butterflies (*Danaus plexippus*) and soil microorganisms have either not been detected at all or are insignificant (Sears et al., 2001). The better cropping practices supported by GM crops have decreased the availability of the milkweed (*Asclepias* spp.) on which monarch larvae feed (Flockhart et al., 2015); hence efforts are being directed to the establishment of milkweed preserves (http://monarchjointventure.org/get-involved/create-habitat-for-monarchs/).

The development of herbicide tolerance in previously susceptible weeds, while not unique to GM crops, is becoming an increasing problem because of the widespread use of glyphosate with glyphosate-tolerant GM crops (Green, 2014). Moreover, the use of glyphosate has begun to be challenged by a contentious designation of glyphosate as a "probable" carcinogen by the

World Health Organization's International Agency for Research on Cancer (IARC), despite repeated rulings of the chemical as safe by the U.S. Environmental Protection Agency (EPA) (https://en.wikipedia.org/wiki/Glyphosate), and by recent lawsuits filed by cancer victims. Although the pace of herbicide discovery has slowed markedly since the 1980s, new combinations of herbicide-tolerant crops and older herbicides are coming on the market (Duke, 2014; Lombardo et al., 2016).

Safety Considerations

In 2016, the U.S. National Academy of Sciences issued a four-hundred-page report based on a review of hundreds of studies on the human and animal health effects of GM crops and on the economic and environmental impacts of their introduction (National Academies of Sciences & Medicine, 2016). On the question of human health and safety, the report carefully concludes: "On the basis of its detailed examination of comparisons between currently commercialized GE and non-GE foods in compositional analysis, acute and chronic animal toxicity tests, long-term data on health of livestock fed GE foods and epidemiological data, the committee concluded that no differences have been found that implicate a higher risk to human health safety from these GE foods than from their non-GE counterparts."

Moreover, there is no evidence that the use of GM techniques to modify organisms is associated with unique hazards. The European Union alone has invested more than €300 million in GMO biosafety research. Quoting from its recent report: "The main conclusion to be drawn from the efforts of more than 130 research projects, covering a period of more than 25 years of research and involving more than 500 independent research groups, is that biotechnology, and in particular GMOs, are not *per se* more risky than e.g. conventional plant breeding technologies." (http://ec.europa.eu/research/biosociety/pdf/a_decade_of_eu-funded_gmo_research.pdf) Every credible scientific body that has examined the evidence has come to the same conclusion (http://gmopundit.blogspot.com/p/450-published-safety-assessments.html).

Despite occasional one-of-a-kind, often sensationalized reports, feeding studies have identified no meaningful nutritional differences between GM and non-GM food and feed (Van Eenennaam and Young, 2014). Indeed, and perhaps unsurprisingly, comparative molecular analyses show that GM techniques have less impact on the genetic and molecular constitution of crop plants than conventional plant breeding techniques (Richroch, 2013). This is because conventional breeding mixes whole genomes comprising tens of thousands of genes that have previously existed in isolation, while GM methods generally add just a gene or two to an otherwise compatible genome. As well, crops modified by gene-editing techniques are less likely to have unexpected genetic effects than crops modified by the more conventional techniques of chemical and radiation mutagenesis methods simply because of the greater precision and predictability of molecular modification. Taken together with the closer scrutiny paid during product development to the potential for toxicity and allergenicity of novel proteins expressed by GM methods, GM crops are arguably the safest new crops ever introduced into the human and animal food chains.

Indeed, to date, the only unexpected effects of GM crops have been beneficial. Many grains and nuts, including corn, are commonly contaminated by mycotoxins, which are toxic and carcinogenic compounds made by fungi that follow boring insects into the plants. Bt corn, however, shows as much as a 90 percent reduction in mycotoxin levels because the fungi that follow the boring insects into the plants cannot get into the Bt plants (Munkvold, 2003). There is also evidence that planting Bt crops reduces insect pressure in non-GM crops growing nearby. The widespread adoption of Bt corn in the U.S. Midwest has resulted in an area-wide suppression of the European corn borer (*Ostrinia nubilalis*) (Hutchison et al., 2010).

Future Challenges in Agriculture

The human population has expanded more than seven-fold since Malthus's time. Agriculture in developed nations has become far less labor-intensive and has kept pace with population growth worldwide through science and technology. Today, less than one in fifty people living in developed countries grows crops or raises animals for food. But after a half-century's progress in decreasing the fraction of humanity experiencing chronic hunger, the food price and financial crises commencing in 2008 have begun to swell the ranks of the hungry once more (Lagi et al., 2011; Wu et al., 2014). Population experts anticipate the addition of another 2 to 4 billion people to the planet's population within the next three to four decades (Cohen, 2003; DESA, 2013; Gerland et al., 2014). But, as noted earlier, the amount of arable land has not changed appreciably in more than half a century (AG, 2009). Even as farmland is added through the destruction of tropical forests, arable land continues to be lost to urbanization, salinization, and desertification.

Supplies of fresh water for agriculture are under pressure as well. Today, about a third of the global population lives in arid and semi-arid areas, which cover roughly 40 percent of the land area. Climate scientists agree that over the coming decades, average temperatures will increase and dryland area will expand. Inhabitants of arid and semi-arid regions of all continents are extracting groundwater faster than aquifers can recharge—and often from fossil aquifers that do not recharge (Giordano and Villholth, 2007). Yet the major crops that now feed the world—corn, wheat, rice, and soy—require substantial amounts of water. Moreover, while it takes 500 to 2,000 liters of water to produce a kilogram of wheat, the amount of water required to produce a kilogram of animal protein can be as much as ten times greater, since the amount required to raise the grain to feed the animal must be included in the calculation (IWMI).

Increasing average temperatures and decreasing fresh water availability present critical challenges to agricultural researchers to increase crop performance under suboptimal conditions. Rapid advances in our knowledge of plant stress responses and improving molecular knowledge and tools for plant breeding have already resulted in the introduction of new drought-tolerant crop varieties, both GM and non-GM (Graff et al., 2013). New varieties of drought-tolerant maize produced using modern breeding approaches that employ molecular markers, but do not generate transgenic plants, have been released in the North American market by Syngenta and DuPont Pioneer, while Monsanto (now Bayer) and BASF have jointly developed MON87460 (aka Genuity DroughtGard Hybrids), a drought-tolerant maize variety expressing a cold-shock protein from the bacterium *Bacillus subtilis*, introducing it in the U.S. in 2013 (http://www.monsanto.com/products/pages/droughtgard-hybrids.aspx).

However, it should be kept in mind that suboptimal "stress" conditions necessarily move plants away from their peak ability to use sunlight to convert carbon dioxide, water, and other simple compounds into the carbohydrates and proteins that feed people and animals. Stress-tolerant varieties do not generally outperform less stress-tolerant varieties by much or at all under optimal conditions, but simply survive better under suboptimal conditions, losing less of their yield potential. Thus improvements in the stress tolerance of our existing crops can offset losses caused by climate warming, but will contribute little to increasing yields.

The Imperative to Do More with Less

The FAO has estimated that we will need a 70 percent increase the amount of food produced by 2050 (FAO, 2009). We will need more food, feed, and fiber not only because there will be more people, but they will be richer. Among the things that people demand as they become more affluent is more meat in their diet (Godfray et al., 2010). Because much of our grain goes to feed animals, the demand for more meat disproportionately increases the requirement for grain. But increasing the supply by expanding the land under cultivation cannot be sustained. All the best land is already under cultivation, and preserving what remains of our planet's rich biological heritage by leaving the remainder unplowed is a growing priority. Indeed, modeling exercises reveal that within just a few decades, the planet's natural resources will be insufficient to support developed-world consumption patterns (Odegard and van der Voet, 2014).

As well, the negative impact of climate change on agriculture is becoming increasingly apparent and is predicted to worsen (Challinor et al., 2014; IPCC, 2014). While more agriculturally suitable land may become available at greater distances from the equator as the climate warms, there is no guarantee that the productivity of these lands will compensate for productivity losses in the more populous equatorial regions. Whether our current highly productive food and feed crops can be modified and adapted to be substantially more productive at the higher temperatures expected or at more northern latitudes with shorter growing seasons is not known yet. Substantial research will be required not just on the salt, drought, and temperature tolerance of existing crop plants, but also for the domestication of plants that are not now used in agriculture, but that are capable of growing at higher temperatures and on saline water.

In today's highly productive developed-world agriculture, fertilizers and other chemicals are applied and used inefficiently, themselves becoming pollutants in our air, land, and water. As well, many of the chemicals used in both conventional and organic agriculture to control pests and diseases are toxic to people and to wildlife. While yield gains based on earlier scientific advances can still increase food production in many countries, particularly in Africa, such gains appear to have peaked in most developed countries, and recent productivity gains have been achieved largely through adoption of GM crops (Godfray et al., 2010; Grassini et al., 2013). For all of these reasons, transitioning to more sustainable agricultural practices while doubling the food and feed supply in the face of a warming climate is likely to be the greatest challenge of the twenty-first century (Kershen, 2013; Tilman et al., 2011).

The Innovation Pipeline

Despite the sheer magnitude of the food security challenge, there are advances coming on line and in the pipeline that have the potential to significantly increase both the amount of food the world can produce and the sustainability of its production. These all involve interrelated advances in science, particularly biology, and technology. Producing more food from existing land and water resources will necessitate greater reliance on a variety of technologies, ranging from improving and automating farm management to designing and scaling up covered agriculture, aquaculture, and animal husbandry facilities that have more sustainable resource use profiles and environmental footprints than today's. Both changing diet preferences and the ability to bypass the requirement for animals in producing edible animal protein also have the potential to make substantial contributions to sustainable food security.

The integration of remote sensing with GPS guidance makes it possible to take farming to a new level of precision (Lal and Stewart, 2015; Oliver et al., 2013; Thenkabail et al., 2016). Combining multi-year hyperspectral image collection with topographical information facilitates optimization of virtually all aspects of agriculture through programming of GPS-guided farm machinery. This extends from planting to irrigation and spraying to harvesting. Together with the availability of pest- and pathogen-resistant and herbicide-tolerant crops, precision farming can reduce the use of water, fertilizer, and chemicals to ameliorate agriculture's negative environmental impacts. Such precision farming systems can be enabled to operate over the internet with the potential for fully automated farming (Kaloxylos et al., 2012; Ryu et al., 2015). But implementation of highly automated farming systems will require a new level of sophistication in farmers and agricultural workers and is likely to operate best on a large scale, often today culturally resisted and derided as "factory farming."

Moving high-value crops, both plant and animal, into covered environments is happening around the world, reducing water use and increasing sustainability. Automation, robotics, and wireless sensor systems are well advanced in horticulture and, increasingly, in aquaculture and animal husbandry facilities (Grift et al., 2008; Malone, 2013; Saberioon et al., 2016; Straten et al., 2010). While open-cage aquaculture still predominates in both fresh and saline waters, recirculating land-based aquaculture systems decrease water requirements and implement waste-collection systems, all of which are increasingly automated (Lee, 1995; Mahalik and Kim, 2014; Soto-Zarazúa et al., 2011).

Integrating aquaculture and horticulture, often termed aquaponics, optimizes water efficiency and minimizes nutrient requirements by using the wastewater from aquaculture to grow plants hydroponically (Rakocy, 2012). To date, aquaponics systems have been small scale and not highly automated, although the Urban Organics facility (http://pentairaes.com/learn-about-aquaculture/urban-organics/urban-organics-brought-to-life-by-pentair) in East St. Paul, Minnesota is an example of the scale-up potential, particularly in urban settings. While aquaponic systems are most easily implemented in freshwater recirculating systems, the same principle has been developed for marine systems under the rubric of "integrated multitrophic aquaculture" and can be further adapted for land-based, recirculating systems (Granada et al., 2015; Waller et al., 2015).

Perhaps the greatest challenge in achieving food security is that the production of animal protein is inherently resource intensive. As noted earlier, it takes 500 to 2,000 liters of water to produce a kilogram of grain. Roughly half of global grain is now used to feed animals, including fish. As much as 90 percent of the feed powers the animal's own metabolism. Thus the consumption of animal protein rather than the grain itself is expensive in terms of water and agricultural resources and is a significant source of greenhouse gases (Godfray et al., 2010; IWMI; Steinfeld et al., 2006). Some animals are more efficient than others at converting grain to animal protein, with beef requiring roughly ten kilograms of grain to produce a kilogram of meat, while fish and shellfish require only about one to two kilograms (Godfray et al., 2010).

The optimal solution might seem to be a shift toward vegetarian diets, but people emerging from poverty understandably seek to add animal protein to their diet for both cultural and nutritional reasons (Steinfeld et al., 2006). Simply decreasing beef consumption and increasing fish and chicken consumption, both of which are emerging as positive for health reasons, can increase the amount of animal protein obtained per unit of agricultural resources because they require only 10 to 20 percent of the feed grain required to produce beef. An area of increasing contemporary interest and culinary innovation is entomophagy, the consumption of edible insects. While insects, sometimes now called "minilivestock," have provided human food since prehistoric times, they largely fell out of favor in historical times for cultural reasons and are today perceived as unacceptable food sources in many cultures (Abbasi and Abbasi, 2016). However, insects are as water- and feed-efficient sources of animal protein as fish, hence are of interest to efforts to provide more animal protein at less resource and environmental cost (Van Huis, 2013).

An example of an approach to sustainable production of more animal protein based on advances in both biology and technology is provided by the development of a GM salmon (*Salmo salar*) capable of growing continuously, rather than seasonally (https://aquabounty.com). The ability to grow continuously reduces the time and therefore the feed requirement for producing marketable salmon simply because wild salmon feed continuously, but grow only seasonally. The GM salmon produced by the company AquaBounty, a subsidiary of Intrexon (https://www.dna.com), has also been structured genetically so that the fish produced for market are sterile females, hence incapable of interbreeding with wild salmon. The GM salmon are grown in land-based recirculating systems, reducing the water, feed, and environmental footprint of offshore, open cage salmon farming, as well as the burden of pollutants and pathogens. Fish in general are already among the most efficient converters of grain to animal protein and the further feed reduction resulting from the genetic modification makes the AquaBounty GM salmon even more so.

Long-term, however, perhaps the potential for the most dramatic reduction in the resource and environmental footprints of animal protein production resides in the direct culturing of animal cells and tissues on an industrial scale for the production of synthetic meat products (Mattick et al., 2015; Post, 2012). A 2011 estimate suggests that cultured meat production would require 7 to 45 percent less energy, lower greenhouse gas emissions by 78 to 96 percent, and require roughly 1 percent of the land and 4 to 18 percent of the water required by conventional, whole animal meat production systems (Tuomisto and Teixeira de Mattos, 2011). Although palatable products are only now being perfected, cultured meat production offers major environmental benefits (http://www.cnn.com/2014/04/30/tech/innovation/cultured-meat/). Whether industrial scale cultured meat production will provide a substantial fraction of animal protein in the future depends on both economic and cultural factors.

Cultural Impediments to Achieving Food Security: The GMO Wars

To live sustainably within planetary constraints, we must produce more food on the same amount of land using less water, energy, and chemicals. This will necessitate the continuous invention and application to agriculture of ever more sophisticated methods and materials for producing foodstuffs. Having noted the importance of advances in remote sensing and automation in the previous section, it is the development of precise methods of genetic modification that emerges as perhaps the most critical technology, simply because we are moving out of the climate regime in which our major crops were domesticated and into climatic territory likely to have a deleterious impact on food production. We do not have the luxury of the thousands of years it has taken us to perfect the agricultural system that now feeds our population. Rapid improvements in protecting crop plants from plants, pests, pathogens, and environmental stress are not only possible through rapidly developing molecular technologies, but also increasingly critical. And yet, paradoxically, even though the use of GM technology has been accepted in medicine and for the production of such foods such as beer, bread, and cheese, its use in crop plants has evoked an unprecedented level of societal controversy. This has resulted in the proliferation of regulatory and legal constraints that threaten to cripple their use in achieving a more sustainable existence for humanity on planet Earth. The barriers to deploying modern technology in the service of agriculture are at once political, cultural, and economic.

Despite a global consensus among scientific societies that GM technology is safe, the political systems of Japan and most European and African countries remain opposed to growing GM crops. Many countries lack GM regulatory systems or have regulations that prohibit growing and, in some countries, even importing GM food and feed (Herring and Paarlberg, 2016). Even in countries such as the U.S. that have a GM regulatory framework (OSTP, 1986), the process remains complex, slow, and expensive. U.S. developers must often obtain the approval of three different agencies, the Environmental Protection Agency (EPA), the U. S. Department of Agriculture (USDA), and the Food and Drug Administration (FDA), to introduce a new GM crop into the food supply. Bringing a GM crop to market, including complying with the regulatory requirements, was estimated to cost $135 million in 2011 (McDougall, 2011). The effort, time, and cost for regulatory approval dramatically contracted the pipeline of GM innovations over the past two decades, including those that would directly benefit consumers (Graff et al., 2009). As discussed below, there are growing efforts to reduce the regulatory burdens in the light of the newest gene modification technologies.

In Europe, the regulatory framework is practically nonfunctional. Only one GM crop is currently being grown, and only two others have gained approval since 1990 when the EU first adopted a regulatory system (Kershen, 2014). The EU recently agreed to allow member countries decide individually whether to permit cultivation of an EU-approved GM crop (http://ec.europa.eu/food/plant/gmo/legislation/future_rules_en.htm). The impact of this decision will not be known for some time, but it is likely to further complicate trade and food aid as crops approved in one country await regulatory approval in others (Abrahams, 2014). Moreover, the increasing politicization of risk assessment makes it unlikely that uniform global safety standards for GM crops and animals will emerge in the foreseeable future (Smyth and Phillips, 2014). European influence has been especially detrimental in Africa, causing African leaders to be excessively precautionary in approving GM crops and even to ban the import of GM grain to alleviate famine (Paarlberg, 2009).

It is the case of Golden Rice, which has been genetically modified to produce the vitamin A precursor β-carotene, that provides the paradigmatic example of opportunity foregone to use GM technology to address a major global nutritional issue (Alberts et al., 2013). Severe vitamin A deficiency results in blindness, and half of the roughly half-million children who are blinded by it annually die within a year. Vitamin A deficiency also compromises immune system function, exacerbating many kinds of illnesses. It is a disease of poverty and poor diet, responsible for 1.9 to 2.8 million preventable deaths annually, mostly of children under five years old and women (Mayo-Wilson et al., 2011; Tanumihardjo, 2011).

Two scientists, Ingo Potrykus and Peter Beyer, and their teams developed a rice variety whose grains accumulate β-carotene, which our bodies convert to vitamin A. Collaborating with the International Rice Research Institute (IRRI) in the Philippines over a period of a quarter century, they developed and tested a transgenic rice variety that expresses sufficient quantities of β-carotene so that a few ounces of cooked rice can provide enough to eliminate the morbidity and mortality of vitamin A deficiency (Tang et al., 2012). Yet Golden Rice remains mired in controversy and has suffered repeated setbacks (Potrykus, 2010). In 2013, activists uprooted experimental plots of Golden Rice growing at IRRI, provoking more than 7000 scientists to sign a petition protesting such wanton destruction of potentially life-saving advances (Alberts et al., 2013). Although Golden Rice remains mainly in test plots in the countries where they could be most beneficial (Wesseler and Zilberman, 2014), recent regulatory approvals in the US, Australia and New Zealand raise the possibility that it could begin to move toward cultural acceptance (Coghlan, 2018). Such acceptance, however, is not a foregone conclusion in view of the extensive efforts of anti-GMO activists to turn the public against GMOs.

The increasing politicization of risk determination raises questions about the underlying motivations (Smyth and Phillips, 2014). NGOs, most vocally Greenpeace and Friends of the Earth, have conducted vigorous and successful campaigns of misinformation about GMOs first in Europe, then around the world (Paarlberg, 2014). Greenpeace remains adamantly opposed to even the most benign and beneficial uses of GM technology in agriculture, such as the development and distribution of Golden Rice. In 2016, more than a hundred Nobel Prize laureates signed a letter to Greenpeace protesting its stance on GMOs (http://supportprecisionagriculture.org/nobel-laureate-gmo-letter_rjr.html); it was met with stony silence (R. Roberts, personal communication). There is little doubt that such continued opposition to a harmless and beneficial technology is about the ability of such organizations to raise funds through fear-mongering (Apel, 2010; Paarlberg, 2014; Wesseler and Zilberman, 2014).

Despite the outsized influence of anti-GMO activists, the advent of gene- and genome-editing techniques has triggered a reassessment of regulatory approaches in the U.S. and, to some extent, in Europe and other countries. In particular, while earlier techniques added genetic material to the GM organism, the newer molecular techniques, generally referred to as gene-editing or genome-editing, can change DNA sequences in ways that are indistinguishable from the changes that occur spontaneously in nature and from those that can be induced through chemical and radiation mutagenesis. Previously, a GM organism could be identified by the presence of the foreign DNA. Organisms modified by gene-editing techniques commonly contain no foreign DNA and are indistinguishable at the DNA sequence level from spontaneous mutants or those induced by mutagenesis.

As of this writing, the political climate in the U.S. is favorable to radical regulatory reform, and the results may well have a salutary impact on the ability to use modern molecular techniques without undue regulatory oversight. In particular, the USDA has already announced that it will not regulate organisms, largely plants and some microorganisms, that have genetic changes that could have been achieved using older, unregulated mutagenic methods (https://www.usda.gov/media/press-releases/2018/03/28/secretary-perdue-issues-usda-statement-plant-breeding-innovation). Moreover, it proposes to confine regulation to organisms with traits that fall under the Plant Protection Act (https://www.aphis.usda.gov/plant_health/downloads/plant-protect-act.pdf). These are plant pests and weeds that have the potential to have a deleterious impact on agriculture. Moreover, the focus of regulation will be on organismal traits, not the method by which the trait was introduced (S. Abel and D. McKalip, USDA, personal communication). This radical regulatory reform, if achieved, would make it possible for scientists in all kinds of institutions and not just big biotech companies to use the most advanced molecular techniques to solve the many local agricultural problems that are becoming ever more acute as the climate warms and agriculture intensifies.

Urbanization and Its Consequences

Industrialization and large-scale agriculture underlie the migration of people into cities over the past two centuries. Most people live in cities today and obtain their food through a complex processing and distribution network that delivers the largely packaged food either to stores where people buy it or directly to their homes. The result is that city dwellers are far removed from the basics of agriculture and get their information primarily through advertising and the internet. A nostalgic view of pre-industrial "organic" agriculture as more "natural" has taken hold among urban elites largely through the marketing campaigns of the "organic food" industry. Support for local, organic, small-holder agriculture and opposition to technologically sophisticated, large-scale agriculture, pejoratively referred to as "industrial" agriculture, has developed into a deeply held belief system in the primarily urban populations of both developed and developing countries.

The most basic tenet of organic farming is a prohibition on the use of synthetic nitrogen fertilizers and a focus on the use of "natural" fertilizers, largely animal and green manures. The procedures that define organic farming in the U.S., commonly referred to as the "Organic Rule," forbid the growing of GM crops, as well as their use as animal feed (https://www.ams.usda.gov/grades-standards/organic-standards). To be profitable, however, marketers of "organic" produce must charge higher prices to cover the higher costs of this essentially nineteenth-century labor- and land-intensive system of agriculture (http://www.fao.org/organicag/oa-faq/oa-faq5/en/). Organic marketers have therefore promoted organically grown produce as more healthful for people, despite the absence of evidence that there are meaningful nutritional differences between organically and conventionally produced foods (Dangour et al., 2009).

Today the belief that organically grown foods are nutritionally superior is so deeply entrenched that it remains unaffected by exposure to empirical evidence that organically and conventionally produced foods are nutritionally indistinguishable (Olson, 2016; Smith-Spangler et al., 2012). Perhaps the most counterproductive—even sinister—development is the increasing vilification of GM foods as a marketing tool by the organic food industry and NGOs purportedly devoted to public welfare (Paarlberg, 2014; Schroeder, 2014). Just why we are particularly vulnerable to making and sticking to the negative views about GMOs promulgated by organic marketers and anti-GMO activists despite all evidence to the contrary is not entirely clear, but is likely to be rooted in intuitive fears and expectations about food (Blancke et al., 2015).

The organic food industry had its origins in rural India, where Sir Albert Howard, arguably the father of "organic" agriculture, developed composting methods capable of killing the pathogens that abound in animal manures and human wastes so that these could be used safely as fertilizers in agriculture (Fedoroff and Brown, 2004). Even as synthetic fertilizers were increasingly being used around the world, the organic movement grew in the UK and Europe, eventually finding an American champion in Jerome Rodale,

Founder of the Rodale Press, and pesticide crusader Rachel Carson, author of *Silent Spring*, the book that has been credited with starting the environmental movement (Carson, 2002). Rodale firmly believed that organically grown food was more "healthful" than conventionally grown food; his organization today remains dedicated to promoting the health and environmental benefits of organic farming (http://rodaleinstitute.org). Carson's concern was with the overuse of toxic chemicals, particularly the pesticide DDT, in conventional farming and she looked forward to the day the plants could be protected from pests biologically (Ronald, 2012). Paradoxically, today's environmental activists are among the most vociferous opponents of GM technology, despite its proven environmental benefits (Brookes and Barfoot, 2015).

With the establishment of organic retailers, such as Whole Foods and Wild Oats, the organic food business grew rapidly and certification organizations proliferated. To bring some uniformity to what was being certified as "organic," Congress established the National Organic Standards Board (NOSB) under the USDA through the Organic Food Production Act and charged it with developing national standards (Fedoroff and Brown, 2004). These were eventually published in 2000 and are generally referred to as the Organic Rule (https://www.ams.usda.gov/grades-standards/organic-standards). According to the NOSB, organic agriculture is a production system that makes minimal use of off-farm inputs and seeks to enhance "ecological harmony." The Organic Rule expressly forbids the use of GM crops, antibiotics, and synthetic nitrogen fertilizers in crop production and animal husbandry, as well as food additives and ionizing radiation in food processing.

Beliefs Matter

Organic food is food produced in compliance with the Organic Rule. The USDA's Organic Seal is a marketing tool that makes no claims about food safety or nutritional quality (https://www.ams.usda.gov/rules-regulations/organic/organic-seal). But the organic food industry has systematically used false and misleading marketing about the health benefits and relative safety of organic foods compared with what are now called "conventionally grown" foods (Schroeder, 2014). Indeed, organic marketers represent conventionally grown foods as swimming in pesticide residues, GM foods as dangerous, and the biotechnology companies that produce GM seeds as evil, while portraying organically grown foods as both safer and more healthful than conventionally grown food.

In recent years, organic marketers embarked on a deliberate strategy of stigmatizing GM foods by demanding that they be labeled (Schroeder, 2014).

Until recently, the Food and Drug Administration maintained its position that labeling is required only if the genetic modification makes a material difference relevant to health or the environment (http://www.fda.gov/Food/FoodScienceResearch/GEPlants/ucm346858.htm). The labeling campaigns have appealed to consumers emotionally through the creation of a new right, the "right to know" what's in their food. Since modern genetic modification is a technique, not a product, labeling a food as GM provides no information about either the safety or the nutritional content of the food. Because the GM foods on the market today are nutritionally indistinguishable from their non-GM counterparts (National Academies of Sciences & Medicine, 2016), attaching a GM label tells consumers nothing about what's "in their food." Yet objections to such a seemingly innocent demand have created the suspicion that deleterious health effects are being concealed. Creating such a suspicion is precisely the organic marketers' objective as a first step in driving GM crops out of the market and increasing the organic food industry's market share (Schroeder, 2014).

The U.S. Congress eventually passed a bill that imposes mandatory labeling of foods that contain ingredients from GM organisms, with just a few exceptions (http://www.agriculture.senate.gov/imo/media/doc/Ag%20biotech%20compromise%20proposal.pdf). However, the legislation left open and up to the USDA to determine precisely how such a requirement would be implemented. Among the alternatives proposed are on-product labels, 800 numbers, and websites where further information can be obtained about ingredients. While the legislative objective, supported by the USDA and organizations such as the Grocery Manufacturers Association (http://www.gmaonline.org), has been to satisfy the demands of GMO opponents for transparency without stigmatizing foods, this legislation could prove self-defeating. Many food companies have already added on-package labels or have begun to reformulate their products to exclude GM ingredients in anticipation of the backlash from anti-GMO activists. And as of this writing, the USDA has passed its 2018 deadline to complete its task of devising widely acceptable means of conveying information about GM ingredients to customers.

Because it is consumer preferences that ultimately drive the food industry, belief systems about organically produced food and modern genetic modification have the potential to drive advances in agricultural technology out of the market globally. A group of investigators from Purdue University is exploring the economic and environmental consequences of both a complete global ban on GMOs and an acceleration of GM penetration globally to match U.S. levels today (Mahaffey et al., 2016). Projecting the results of a complete GMO ban into the future, they conclude that yields would decline

(7 to 10 percent) and that the impact on welfare would be felt disproportionately by importing countries with the lowest GM crop penetration. As importantly, more land would be converted into agricultural production, resulting in an estimated increase in greenhouse gas emissions of 13.8 percent (Mahaffey et al., 2016). Farmers' costs would increase, as would their reliance on pesticides, reversing the environmental gains of rapid GM crop adoption since their commercial introduction (Brookes and Barfoot, 2015). This is a step in the wrong direction, negatively affecting both agricultural productivity and sustainability. Hence it may be our own belief systems that constitute the most serious impediment to achieving sustainable food security for the soon-to-be 10 billion people on planet Earth.

Education Matters

Finally, it would be irresponsible to finish a discussion about achieving food security through scientific and technological advances without mentioning that both resources and education determine how quickly new technologies are adopted. While some technologies are relatively easy to promulgate among farmers with little education, as judged by the rapid gallop of cell phones around the planet, others are not. As well, farmers, particularly those in less developed countries, are subject to the anti-technology rhetoric of NGOs, most notably Greenpeace, that in turn influence national policies (Paarlberg, 2014).

Nonetheless, technology adoption is positively correlated with educational level (Foster and Rosenzweig, 2010). But the educational gulf between the most advanced and least advanced nations, between urban and rural populations, remains deep and wide, particularly in the agricultural sector. As well, there are large disparities in the availability of credit and crop insurance between the farmers of developed and developing countries. The spread of cell phones and internet connectivity around the globe makes it possible to deploy innovative approaches to both resources and education, in turn opening prospects for accelerating the modernization of agricultural everywhere (Duncombe, 2016; Mugwisi et al., 2015).

Conclusions

In sum, it is not difficult to imagine a world of 10 billion that is able to feed itself sustainably, were it possible to focus on rapid development and implementation of sophisticated farming and other types of food production systems that use water, land, and energy efficiently and are already in the innovation pipeline. These will all, however, require both cultural changes in how we view food and substantial increases in educational levels worldwide, both of which present challenges as great as that posed by the necessary technological advances.

References

Abbasi, T., and Abbasi, S. (2016). Reducing the global environmental impact of livestock production: the minilivestock option. *J Cleaner Production, 112*, 1754–1766.

Abel, P. P., Nelson, R. S., De, B., Hoffmann, N., Rogers, S. G., Fraley, R. T., and Beachy, R. N. (1986). Delay of disease development in transgenic plants that express the tobacco mosaic virus coat protein gene. *Science, 232*, 738–743.

Abrahams, D. (2014). Legal considerations related to the authorization, import and cultivation of GM crops in the European Union: A precedent for other regulated industries? *BioSciences Law Rev, 13*, 155–160.

AG, L.C.A.M. (2009). *The land commodities global agriculture & farmland investment report 2009.*

Alberts, B., Beachy, R., Baulcombe, D., Blobel, G., Datta, S., Fedoroff, N., Kennedy, D., Khush, G. S., Peacock, J., and Rees, M. (2013). Standing up for GMOs. *Science, 341*, 1320–1320.

Apel, A. (2010). The costly benefits of opposing agricultural biotechnology. *New Biotechnol, 27*, 635–640.

Ayensu, E., van Claasen, D. R., Collins, M., Dearing, A., Fresco, L., Gadgil, M., Gitay, H., Glaser, G., Juma, C., and Krebs, J. (1999). International ecosystem assessment. *Science, 286*, 685–686.

Bear, F. E. (1938). *Theory and practice in the use of fertilizers.*, Vol. 46. New York: John Wiley & Sons.

Binswanger, H. (1986). Agricultural mechanization: a comparative historical perspective. *World Bank Res Obs, 1*, 27–56.

Birch, R. G. (1997). Plant transformation: problems and strategies for practical application. *Annu Rev Plant Biol, 48*, 297–326.

Bittman, M. (2014, November 11). Don't ask how to feed the 9 billion. In *The New York Times*, p. A27.

Blancke, S., Van Breusegem, F., De Jaeger, G., Braeckman, J., and Van Montagu, M. (2015). Fatal attraction: the intuitive appeal of GMO opposition. *Trends Plant Sci, 20*, 414–418.

Boyer, C. and Shannon, J. (1983) The use of endosperm genes for sweet corn improvement. In J. Janik (ed.) *Plant Breeding Reviews*, pp. 139–161. The Avi Publishing Company, Inc.

Brookes, G. and Barfoot, P. (2015). *GM crops: global socio-economic and environmental impacts 1996–2013.* PG Economics Ltd.

Cardinale, B. J., Duffy, J. E., Gonzalez, A., Hooper, D. U., Perrings, C., Venail, P., Narwani, A., Mace, G. M., Tilman, D., and Wardle, D.A. (2012). Biodiversity loss and its impact on humanity. *Nature, 486*, 59–67.

Carlson, E. A. (1966). *The gene: a critical history*. Philadelphia: W. B . Saunders

Carson, R. (2002) *Silent Spring*. Houghton Mifflin Harcourt.

Challinor, A., Watson, J., Lobell, D., Howden, S., Smith, D., and Chhetri, N. (2014). A meta-analysis of crop yield under climate change and adaptation. *Nature Climate Change, 4*, 287–291.

Cohen, J. E. (2003). Human population: the next half century. *Science, 302*, 1172–1175.

Coghlan, A. (2018, May 30). GM golden rice gets approval from food regulators in the US. *New Scientist*. Retrieved from https://www.newscientist.com/article/mg23831802-500-gm-golden-rice-gets-approval-from-food-regulators-in-the-us/.

Crabb, A. R. (1947). *The hybrid-corn makers. Prophets of plenty*. New Brunswick: Rutgers University Press.

Crow, J. F. (1998). 90 years ago: the beginning of hybrid maize. *Genetics, 148*, 923–928.

Dangour, A. D., Dodhia, S. K., Hayter, A., Allen, E., Lock, K., and Uauy, R. (2009). Nutritional quality of organic foods: a systematic review. *American J Clinical Nutrition, 90*, 680–685.

de Saussure, N.-T. (1804) *Recherches chimiques sur la vegetation.*

DESA (2013). *World population prospects: the 2012 revision.* United Nations, Department of Economic and Social Affairs.

Diamond, J. (1987). The worst mistake in the history of the human race. *Discover, 8*, 64–66.

Dimitri, C., Effland, A., and Conklin, N. (2005). *The 20th century transformation of US agriculture and farm policy.* USDA Economic Information Bulletin No. 3.

Doebley, J. F., Gaut, B. S. and Smith, B. D. (2006). The molecular genetics of crop domestication. *Cell, 127*, 1309–1321.

Duke, S. O. (2015). Perspectives on transgenic, herbicide-resistant crops in the USA almost 20 years after introduction. *Pest Manag Sci., 71*(5), 652–657.

Duncombe, R. (2016). Mobile phones for agricultural and rural development: a literature review and suggestions for future research. *Europ J Dev Res, 28*, 213–235.

Ehrlich, P. (1968). *The population bomb.* New York: Random House.

FAO (2009). *How to feed the World in 2050.*

Fedoroff, N. V., 2003, Prehistoric GM corn. *Science, 302*, 1158–1159.

Fedoroff, N. V. and Brown, N. M. (2004) *Mendel in the kitchen: a scientist's view of genetically modified food.* Joseph Henry Press.

Flockhart, D., Pichancourt, J. B., Norris, D. R., and Martin, T. G. (2015). Unravelling the annual cycle in a migratory animal: breeding-season habitat loss drives population declines of monarch butterflies. *J Animal Ecol, 84*, 155–165.

Foster, A. D. and Rosenzweig, M. R. (2010). Microeconomics of technology adoption. *Annu Rev Econ, 2*, 395-424.

Fuller, D. Q. and Allaby, R. (2009). Seed dispersal and crop domestication: shattering, germination and seasonality in evolution under cultivation. In *Annu Plant Rev (*pp. 238–295). Wiley-Blackwell.

Gerland, P., Raftery, A. E., Ševčíková, H., Li, N., Gu, D., Spoorenberg, T., Alkema, L., Fosdick, B. K., Chunn, J., and Lalic, N. (2014). World population stabilization unlikely this century. *Science, 346*, 234–237.

Giordano, M., and Villholth, K. G. (2007). *The agricultural groundwater revolution: opportunities and threats to development*, Vol. 3: CABI.

Godfray, H. C. J., Beddington, J. R., Crute, I. R., Haddad, L., Lawrence, D., Muir, J. F., Pretty, J., Robinson, S., Thomas, S. M., and Toulmin, C. (2010). Food security: the challenge of feeding 9 billion people. *Science, 327*, 812–818.

Gonsalves, D. (1998). Control of papaya ringspot virus in papaya: a case study. *Annu Rev Phytopathol, 36*, 415–437.

Graff, G., Hochman, G., and Zilberman, D. (2013). The research, development, commercialization, and adoption of drought and stress-tolerant crops. In Tuteja, N. and Gill, S. S. (eds.), *Crop improvement under adverse conditions*, pp. 1–33. New York: Springer.

Graff, G. D., Zilberman, D., and Bennett, A. B. (2009). The contraction of agbiotech product quality innovation. *Nature Biotechnology, 27*, 702–704.

Granada, L., Sousa, N., Lopes, S., and Lemos, M. F. (2015). Is integrated multitrophic aquaculture the solution to the sectors' major challenges? *Reviews in Aquaculture.* https://doi.org/10.1111/raq.12093

Grassini, P., Eskridge, K. M., and Cassman, K. G. (2013). Distinguishing between yield advances and yield plateaus in historical crop production trends. *Nature Communications, 4.*

Green, J. M. (2014). Current state of herbicides in herbicide-resistant crops. *Pest Manat Sci, 70*, 1351–1357.

Grift, T., Zhang, Q., Kondo, N., and Ting, K. (2008). A review of automation and robotics for the bioindustry. *J Biomechatronics Engi, 1*, 37–54.

Harris, D. R. (1996). *The origins and spread of agriculture and pastoralism in Eurasia: crops, fields, flocks and herds.* London: University College London Press.

Herring, R., and Paarlberg, R. (2016). The political economy of biotechnology. *Annu Rev Res Economics*, 8, 397-416.

Hilder, V. A., and Boulter, D. (1999). Genetic engineering of crop plants for insect resistance—a critical review. *Crop Protection, 18*, 177–191.

Hutchison, W., Burkness, E., Mitchell, P., Moon, R., Leslie, T., Fleischer, S. J., Abrahamson, M., Hamilton, K., Steffey, K. L., and Gray, M. (2010). Areawide suppression of European corn borer with Bt maize reaps savings to non-Bt maize growers. *Science, 330*, 222–225.

IPCC (2014). *Summary for policymakers.*

ISAAA (2017). Global status of commercialized biotech/GM crops in 2017: Biotech crop adoption surges as economic benefits accumulate in 22 years. ISAAA Brief No. 53. ISAAA: Ithaca, NY.

IWMI (2007). *Does food trade save water?*

Jaenicke-Despres, V., Buckler, E. S., Smith, B. D., Gilbert, M. T. P., Cooper, A., Doebley, J., and Pääbo, S. (2003). Early allelic selection in maize as revealed by ancient DNA. *Science, 302*, 1206–1208.

James, C. (2015). 20th anniversary (1996 to 2015) of the global commercialization of biotech crops and biotech crop highlights in 2015. *ISAAA Brief.*

Kaloxylos, A., Eigenmann, R., Teye, F., Politopoulou, Z., Wolfert, S., Shrank, C., Dillinger, M., Lampropoulou, I., Antoniou, E., and Pesonen, L. (2012). Farm management systems and the future internet era. *Computers and electronics in agriculture, 89*, 130–144.

Kershen, D. L. (2013). The contested vision for agriculture's future: sustainable intensive agriculture and agroecology. *Creighton Law Review, 46, No.4, 591-618.*

Kershen, D. L. (2014). European decisions about the "Whack-a-mole" game. *GM Crops & Food,* 5: 4–7.

Khush, G. S. (2001). Green revolution: the way forward. *Nature Rev Genet, 2*, 815–822.

Klümper, W., and Qaim, M. (2014). A meta-analysis of the impacts of genetically modified crops. *PLoS ONE, 9*, e111629.

Lagi, M., Bertrand, K. Z., and Bar-Yam, Y. (2011). The food crises and political instability in North Africa and the Middle East. *Available at SSRN 1910031.*

Lal, R., and Stewart, B. A. (eds.) (2015). *Soil-Specific Farming: Precision Agriculture*, Vol. 22. Boca Raton, Florida: CRC Press.

Lee, P. G. (1995). A review of automated control systems for aquaculture and design criteria for their implementation. *Aquacult Engineering,* 14: 205–227.

Lee, S., Clay, D., and Clay, S. (2014). Impact of herbicide tolerant crops on soil health and sustainable agriculture crop production. In Songstad, D. D., Hatfield, J. L., and Tomes, D. T. (eds), *Convergence of food security, energy security and sustainable agriculture.* Biotechnology in Agriculture and Forestry, pp. 211–236. Springer.

Lemaux, P. G. (2008). Genetically engineered plants and foods: a scientist's analysis of the issues (Part I). *Annu Rev Plant Biol, 59*, 771.

Lemaux, P. G. (2009). Genetically engineered plants and foods: a scientist's analysis of the issues (Part II). *Plant Biology, 60*, 511.

Lombardo, L., Coppola, G., and Zelasco, S. (2016). New technologies for insect-resistant and herbicide-tolerant plants. *Trends Biotechnol, 34*, 49–57.

Mahaffey, H., Taheripour, F., and Tyner, W. E. (2016). Evaluating the economic and environmental impacts of a global GMO ban. In *2016 Annual Meeting, Agricultural and Applied Economics Association.*

Mahalik, N. P., and Kim, K. (2014). Aquaculture monitoring and control systems for seaweed and fish farming. *World J Agricultl Res, 2*, 176–182.

Malone, R. (2013). Recirculating aquaculture tank production systems: a review of current design practice. *SRAC Publication-Southern Regional Aquaculture Center.*

Malthus, T. (1798). An essay on the principle of population. St. Paul's Church-Yard, London: J. Johnson.

Maluszynski, M., Ahloowalia, B. S., and Sigurbjörnsson, B. (1995). Application of in vivo and in vitro mutation techniques for crop improvement. In A. C. Cassells and P. W. Jones (eds.), *The methodology of plant genetic manipulation: Criteria for decision making* (pp. 303–315). Netherlands: Springer.

Mattick, C. S., Landis, A. E., Allenby, B. R., and Genovese, N. J. (2015). Anticipatory life cycle analysis of in vitro biomass cultivation for cultured meat production in the United States. *Env Sci Technol, 49*, 11941–11949.

Mayo-Wilson, E., Imdad, A., Herzer, K., Yakoob, M. Y., and Bhutta, Z. A. (2011). Vitamin A supplements for preventing mortality, illness, and blindness in children aged under 5: systematic review and meta-analysis. *BMJ, 343*, d5094.

McDougall, P. (2011). *The cost and time involved in the discovery, development and authorisation of a new plant biotechnology derived trait.* Crop Life International.

Mugwisi, T., Mostert, J., and Ocholla, D. (2015). Access to and utilization of information and communication technologies by agricultural researchers and extension workers in Zimbabwe. *Information Technology for Development, 21*, 67–84.

Munkvold, G. P. (2003). Cultural and genetic approaches to managing mycotoxins in maize. *Annu Rev Phytopathol, 41*, 99–116.

National Academies of Sciences, E. and Medicine (2016). *Genetically Engineered Crops: Experiences and Prospects.* The National Academies Press, Washington, D.C.

Odegard, I., and van der Voet, E. (2014). The future of food—scenarios and the effect on natural resource use in agriculture in 2050. *Ecol Economics, 97*, 51–59.

Oliver, M., Bishop, T., and Marchant, B. (2013). *Precision agriculture for sustainability and environmental protection.* Routledge.

Olson, E. L. (2016). The rationalization and persistence of organic food beliefs in the face of contrary evidence. *J Cleaner Production*, 140. doi: 10.1016.

OSTP (1986). Coordinated framework for regulation of biotechnology. *Federal Register, 51*, 23,302–323,393.

Paarlberg, R. (2009). *Starved for science: how biotechnology is being kept out of Africa*: Harvard University Press.

Paarlberg, R. (2014). Consequences of the anti-GMO campaigns. In *Bread and Brain, Education and Poverty.* Vatican: Pontifical Acad Sci.

Podevin, N., Davies, H. V., Hartung, F., Nogue, F., and Casacuberta, J. M. (2013). Site-directed nucleases: a paradigm shift in predictable, knowledge-based plant breeding. *Trends Biotechnol, 31*, 375–383.

Post, M. J. (2012). Cultured meat from stem cells: Challenges and prospects. *Meat Sci, 92*, 297–301.

Potrykus, I. (2010). Regulation must be revolutionized. *Nature, 466*, 561–561.

Priestly, J. (2013). *Experiments and observations on different kinds of air*, Vol. 2: Cambridge University Press.

Rakocy, J. E. (2012). Aquaponics—integrating fish and plant culture. *Aquacult Production Sys*, 344–386.

Ray, D. K., Mueller, N. D., West, P. C. and Foley, J. A. (2013). Yield trends are insufficient to double global crop production by 2050. *PLoS ONE, 8*, e66428.

Richroch, A. E. (2013). Assessment of GE food safety using "-omics" techniques and long-term animal feeding studies. *New Biotechnol, 30*, 351–354.

Rimas, A., and Fraser, E. (2010). *Empires of food: feast, famine, and the rise and fall of civilizations.* Simon and Schuster.

Ronald, P. (2012, August 12). Would Rachel Carson embrace "frankenfoods"?—this scientist believes "yes." Forbes. Retrieved from https://www.forbes.com/sites/realspin/2012/08/12/would-rachel-carson-embrace-frankenfoods-this-scientist-believes-yes/#43b8e7eb4586.

Russel, D. A., and Williams, G. G. (1977). History of chemical fertilizer development. *Soil Sci Soc America J, 41*, 260–265.

Ryu, M., Yun, J., Miao, T., Ahn, I.-Y., Choi, S.-C., and Kim, J. (2015). Design and implementation of a connected farm for smart farming system. Paper presented at the SENSORS, 2015 IEEE, 2015.

Saberioon, M., Gholizadeh, A., Cisar, P., Pautsina, A., and Urban, J. (2016). Application of machine vision systems in aquaculture with emphasis on fish: state-of-the-art and key issues. *Rev Aquacult.*

Schroeder, J. (2014). *Organic Marketing Report.* Academics Review.

Sears, M. K., Hellmich, R. L., Stanley-Horn, D. E., Oberhauser, K. S., Pleasants, J. M., Mattila, H. R., Siegfried, B. D., and Dively, G. P. (2001). Impact of Bt corn pollen on monarch butterfly populations: a risk assessment. *Proceedings of the National Academy of Sciences, 98*, 11937–11942.

Smil, V. (2004). *Enriching the earth: Fritz Haber, Carl Bosch, and the transformation of world food production.* MIT Press.

Smith-Spangler, C., Brandeau, M. L., Hunter, G. E., Bavinger, J. C., Pearson, M., Eschbach, P. J., Sundaram, V., Liu, H., Schirmer, P., and Stave, C. (2012). Are organic foods safer or healthier than conventional alternatives?: a systematic review. *Ann Int Med, 157*, 348–366.

Smyth, S. J. and Phillips, P. W. (2014). Risk, regulation and biotechnology: The case of GM crops. *GM Crops & Food, 5*, 170–177.

Soto-Zarazúa, G. M., Peniche-Vera, R., Rico-García, E., Toledano-Ayala, M., Ocampo-Velázquez, R., and Herrera-Ruiz, G. (2011). An automated recirculation aquaculture system based on fuzzy logic control for aquaculture production of tilapia (Oreochromis niloticus). *Aquacult Int, 19*, 797–808.

Steinfeld, H., Gerber, P., Wassenaar, T., Castel, V., and de Haan, C. (2006). *Livestock's long shadow: environmental issues and options.* FAO.

Tan, S., Evans, R., and Singh, B. (2006). Herbicidal inhibitors of amino acid biosynthesis and herbicide-tolerant crops. *Amino Acids, 30*, 195–204.

Tang, G., Hu, Y. Y., S, Wang, Y., Dallal, G. E., Grusak, M. A., and Russell, R. M. (2012). beta-Carotene in Golden Rice is as good as beta-carotene in oil at providing vitamin A to children. *Am J Clin Nutr, 96*, 658–664.

Tanumihardjo, S. A. (2011). Vitamin A: biomarkers of nutrition for development. *Am J Clin Nutrition, 94*, 658S–665S.

Tennant, P., Fermin, G., Fitch, M., Manshardt, R., Slightom, J., and Gonsalves, D. (2001). Papaya ringspot virus resistance of transgenic Rainbow and SunUp is affected by gene dosage, plant development, and coat protein homology. *Eur J Plant Pathol, 107*, 645–653.

Thenkabail, P. S., Lyon, J. G., and Huete, A. (2016). *Hyperspectral remote sensing of vegetation.* CRC Press.

Tilman, D., Balzer, C., Hill, J., and Befort, B. L. (2011). Global food demand and the sustainable intensification of agriculture. *Proc Natl Acad Sci USA, 108*, 20260–20264.

Tuomisto, H. L. and Teixeira de Mattos, M. J. (2011). Environmental impacts of cultured meat production. *Env Sci Technol, 45*, 6117–6123.

Van Eenennaam, A., and Young, A. (2014). Prevalence and impacts of genetically engineered feedstuffs on livestock populations. *J Animal Sci, 92*, 4255–4278.

Van Huis, A. (2013). Potential of insects as food and feed in assuring food security. *Annu Rev Entomol, 58*, 563–583.

Van Straten, G., van Henten, E., van Willigenburg, L., and Van Ooteghem, R. (2010). *Optimal control of greenhouse cultivation.* CRC Press.

Waller, U., Buhmann, A. K., Ernst, A., Hanke, V., Kulakowski, A., Wecker, B., Orellana, J., and Papenbrock, J. (2015) Integrated multi-trophic aquaculture in a zero-exchange recirculation aquaculture system for marine fish and hydroponic halophyte production. *Aquacult Int, 23*, 1473–1489.

Werrell, C. E., and Femia, F. (2013). *The Arab Spring and climate change: a climate and security correlations series.* Center for American Progress.

Wesseler, J., and Zilberman, D. (2014). The economic power of the Golden Rice opposition. *Env Development Economics, 19*, 724–742.

Wu, S.-H., Ho, C.-T., Nah, S.-L., and Chau, C.-F. (2014). Global hunger: a challenge to agricultural, food, and nutritional sciences. *Crit Rev Food Sci Nutrition, 54*, 151–162.

Zhang, D., Li, Z., and Li, J.-F. (2016). Targeted gene manipulation in plants using the CRISPR/Cas technology. *J Genet Genomics, 43*, 251–262.

TRADE'S CONTRIBUTION TO GLOBAL FOOD SECURITY

Kym Anderson
University of Adelaide and Australian National University

FEEDING THE WORLD AS NATIONAL populations and per capita incomes grow is a perennial concern. That concern is heightened when international food prices spike, as in 2008–2012. Yet even when food prices slump, too many people remain hungry: The world has around 700 million extremely poor and 820 million hungry and undernourished people (FAO, 2017), who deserve more. Unfortunately, there has been a slowdown since late last century in farm productivity growth (Fuglie, Wang, and Ball, 2012), and climate change is expected to slow it further in the tropics. That is worrying, because two-thirds of the poor are developing country farmers (World Bank, 2016).

More investment in agricultural research and development (R&D) would eventually lead to expanded production and lower food prices, but growth in public investments in such R&D has slowed in recent decades. As well, increasing shares of those funds are being directed toward off-farm issues such as food safety, health and nutrition, biofuels technology, and the environment (Pardey et al., 2016). True, food and agricultural R&D by the private sector is substituting for public spending to some extent, but nearly half of that is spent on food processing rather than farm productivity (Alston and Pardey, 2017). Meanwhile, regulatory restrictions or bans on modern biotechnology are discouraging investment by life science firms in many countries (Qaim, 2016).

More investments in farm R&D would be useful, however their impact would take decades to be fully felt. Moreover, it is less useful the more markets are distorted by government policies such as trade restrictions. A complementary way to reduce the plight of those 700–820 million poor, hungry, and undernourished people, therefore, is to increase the trade openness of markets for farm products.

During much of the twentieth century, national border restrictions slowed the growth in international food trade and thereby reduced the efficiency of global resource use in agriculture. Policy reforms in the two decades from the mid-1980s eased the situation considerably (Anderson, 2009), but that rosy period ended with the financial crisis of 2008. In the decade since then there have been outbreaks of anti-globalization sentiment in numerous western countries that are putting at risk the policy reform progress made late last century. Hence considerable scope remains for further reforming trade-related farm policies that are still distorting food prices and limiting the role international trade can play in strengthening global food security. Coincidentally, that role is becoming ever easier as real transport costs keep falling, and it is becoming ever more important as climate changes bring more/larger extreme weather events that are increasing year-to-year fluctuations in national food production.

The purpose of this chapter is to lay out the case for further opening up to trade, to document the current extent of trade in farm products and of barriers to that trade, to report on projections of trade under various assumptions to 2030, and to offer ways to increase the openness of agricultural markets. It concludes with a review of prospects for stronger WTO disciplines, for freeing up farm trade with bilateral and regional preferential free trade agreements, and especially for unilateral market liberalization as more efficient instruments to assist the most food-insecure households become more feasible.

Trade Openness Boosts Food Security

Numerous developing countries place long-run food self-sufficiency high on the list of their policy priorities. If that were to lead governments to reduce the current underinvestment in public agricultural R&D, national and global food security would be enhanced. But so often it leads governments to raise food import barriers, which undermines rather than boosts their national food security. It undermines by raising domestic food prices and hence the cost of food consumption to net buyers of food. Simultaneously, it reduces prospects for export growth by more-competitive farmers in more-open developing countries. By contrast, international and intercontinental trade multiplies the global gains from production specialization and market exchange, and reduces the extent of global food insecurity, malnutrition, and risk of famine.

Trade and investment openness also expands the scope for raising diet diversity and food safety and quality. By speeding the development of supermarkets in developing countries, the range and safety of available food products increases for those who find it more convenient to shop there, even if it reduces the number of small shops available in nearby neighborhoods (Reardon, Henson, and Gulati, 2010). If all countries were open to international trade and investment, the use of resources devoted to producing, marketing, distributing and retailing the world's food would be optimized, and fluctuations in trade volumes and international food prices would be minimized.[1] Openness thus contributes to all four key components of food security: availability, access, utilization, and market stability.

The Ups and Downs of Trade-Distorting Policies

Poor countries tend to be net importers of farm products, even though a large proportion of their populations are still engaged in farming. One reason for the lack of alignment between agriculture's shares of national employment and exports has to do with the distortions to agricultural incentives that have been rife in most nations. From the 1950s to the mid-1980s, agricultural protectionism and farm subsidies in high-income countries encouraged growth in their food production but dampened their consumption. Meanwhile, governments of many newly independent developing countries directly taxed farm exports from the 1960s, and also harmed farmers indirectly with an industrialization strategy that involved restrictions on imports of manufactured goods and an overvalued currency. The anti-trade bias in the policies of both groups of countries depressed international food prices and the share of global agricultural production that was traded. In 2007, that production share averaged just 11 percent for farm products compared with 31 percent for manufactured goods and 42 percent for minerals and energy products (Narayanan, Aguiar, and McDougall, 2012).

Had there been less of an anti-trade bias in policies affecting farmer incentives globally, international food prices would have been less volatile, a larger number of countries would have emerged as significant food traders, and all countries would be enjoying a more diversified (and potentially more nutritious) diet.

---

1 Some observers worry about the advertising strength of multinational processing or retail companies selling low-priced processed foods, some of which may be considered less nutritious than the more-expensive traditional foods they replace (e.g., Hawkes, Grace, and Thow, 2015). In such circumstances, there are more-efficient policy responses than preventing such trade or foreign direct investment, such as ensuring there is adequate nutritional information available (e.g., on product labels) and education campaigns to guide consumers to healthy food choices, regardless of whether the firms processing or retailing the available foods are local- or foreign-owned.

Since the mid-1980s, many developing country governments have been reforming their agricultural, trade, and exchange rate policies, thereby reducing their anti-agricultural and anti-trade bias. Some high-income countries have reduced their farm price supports too, most notably in Western Europe. Associated with those policy reforms are reductions in the distortions to consumer prices of food products. When placed in historical perspective, the reforms since the mid-1980s were as dramatic as the opposite policy changes in the preceding three decades (Anderson, 2009).

The extent and speed of those reforms varied across regions though: Among developing countries it has been least for Africa and greatest for Asia. There is still much variation across developing countries in both the level and rate of change in price distortions. There is also a wide range of interventions across products within each country's farm sector, and the anti-trade bias persists. Phasing out such measures would boost farm trade, "thicken" international food markets, and thus not only raise the average level but also lower the volatility of prices in those markets. It would also ensure that the world's productive resources in the farm sector would be put to their best use and so make global agricultural production more sustainable from an environmental viewpoint.

Because food price volatility can cause political unrest, it leads governments in many countries to insulate domestic food markets from gyrations in international prices. It may reduce the volatility of domestic prices in countries that insulate heavily, but the collective impact of such interventions by a large number of countries is to increase the volatility of international food prices, and thereby of domestic prices in more-open countries. Yet if a similar proportion of the world's food exporters insulate to the same degree as a group of food importers, each group will fully offset the other's attempt to prevent their domestic price from moving as much as the international price (Martin and Anderson, 2012; Jensen and Anderson, 2016). Moreover, developing countries as a group would probably see fewer of their people fall into poverty when international food prices spike if they and all other countries agreed to abstain from altering trade restrictions in the hope of insulating their domestic markets from such spikes (Anderson, Ivanic, and Martin, 2014).

One other area of food trade policy concern has to do with products containing genetically modified organisms (GMO). Some groups fear such products may be unsafe as food or animal feed, or that their production may have adverse effects on non-GMO crops or the natural environment. This has led numerous countries to procrastinate on approving GMO crop production or use. Those policies have persisted for two decades in spite of the fact that there is no evidence that GMO crops have greater adverse impacts than non-GMO crops on health or (with simple safeguards in place) on other crops or the environment. On the contrary, GMO crop varieties are reducing the need for agrochemicals that harm the environment and farmer health, and have the potential to greatly enhance the micronutrient content of food staples (Qaim, 2015). Nonetheless, import barriers to food markets in Western Europe have led many African and Asian governments to ban GMO crop production, for fear their country's agricultural exports, even of non-GMO products, may otherwise be rejected by concerned governments abroad. Such bans generate little if any net benefit to the developing countries imposing them, as the domestic consumer loss net of any protectionism boost to African and Asian farmers is far more than the small gain in terms of greater market access to the European Union (EU) (Anderson, 2010; 2016, ch. 7). Removing those moratoria would speed poverty alleviation and boost food security directly, substantially, and relatively rapidly in those developing countries willing to do so. Moreover, prospective gains from this new technology will increase as climate change proceeds and requires adaptation by farmers to warming and to increased weather volatility and higher costs of irrigation water.

Projections of Market Developments

The world had 1.00 hectare per capita of cultivated, grass, and woodlands in 2000, a quarter of which was cultivated. Both those numbers will be about one-third less by 2050, according to Fischer et al. (2010) and IFPRI (2017). The world's farm resources, and especially its cultivated cropland, has to become more productive if the world is to be at least as well fed as it is today.

The two regions where poverty, hunger, and malnutrition are worst now, and will also be in 2050, are Sub-Saharan Africa and South Asia. They also happen to be the regions that are most vulnerable to adverse impacts of climate change on farm productivity. Both will become more food import-dependent unless their farm productivity development is exceptionally fast (IFPRI, 2017). But in an interconnected world of nations, what happens elsewhere in the world also will influence the comparative advantages of these two poorest regions. Rapid economic growth in natural resource-poor economies in East Asia is shifting the global economic and industrial center of gravity away from the north Atlantic, and increasing the demand for exports from natural resource–rich economies. This is a continuation of the process begun in Japan in the 1950s and followed by Hong Kong, Korea, and Taiwan from the late 1960s, then by Southeast Asian countries and China. The earlier

Northeast Asian group represents just 3 percent of the world's population, so its rapid industrial growth was accommodated by the rest of the world without much difficulty, including in markets for primary products. China, by contrast, accounts for one-sixth of humanity. Its rapid and persistent industrialization therefore is having far greater global significance for food and other primary product markets.

To get a sense of how food markets in various parts of the world might develop over coming decades, the results of one set of global economywide modeling projections to 2030 are summarized here, with and without potential changes in trade-related policies (and ignoring climate change over this relatively short period). Anderson and Strutt (2016) project China to continue to have a rapidly rising share of world production, raising the developing countries' GDP share from 27 percent in 2007 to 40 percent by 2030 and their share of world agricultural GDP from one-half to two-thirds. Yet that group's share of global farm exports is projected to remain unchanged at 35 percent, while their share of farm imports is projected to nearly double, from 32 percent to 56 percent (and similarly for their share of other primary product imports). Asia is the main contributor to those changes in world agricultural import shares. Both Latin America and Sub-Saharan Africa, by contrast, are projected to see their shares of global exports of farm products rise, and by more than the small rise in their share of global imports of primary products.

The structural changes associated with global growth over this period to 2030 alter sectoral shares of national exports and imports in disparate ways. Farm products become a smaller part of exports of China, South Asia, and Sub-Saharan Africa, but a larger part of Latin American and Southeast Asian exports. Manufactures will continue to dominate Asia's exports but less so for India, where exports of services are projected to become even more important. Meanwhile, the share of food in national imports is projected to triple for China and nearly double for South Asia while remaining much the same in other developing country regions in the baseline scenario. Even so, national agricultural self-sufficiency percentages do not change greatly between 2007 and the baseline projection scenario for 2030: They rise for Brazil and South Africa plus Europe and North America, and fall for China and India (compare columns 1 and 2 of Table 1). Prices of food relative to other goods in international markets in 2030 are just 11 percent higher in this baseline scenario than those in 2007 (just before the recent food price spike), and the share of food production that is internationally traded is only slightly higher in 2030 than in 2007.

TABLE 1. Agricultural self-sufficiency ratio,[a] 2007, 2030 baseline and 2030 alternative scenarios (%).

	2007	2030 conservative Asian growth	2030 faster Asian growth	2030, with faster LA & SSA agricultural productivity	2030, with increased agric. protection in China and India
Argentina	170	169	177	181	168
Brazil	119	135	141	151	133
Chile	117	113	115	128	112
Peru	100	94	97	102	93
Rest of L. America	104	101	105	112	101
M. East+ N. Africa	84	86	88	83	86
South Africa	101	124	119	117	123
Rest of Africa	100	100	103	110	99
Europe	97	105	105	102	103
NAFTA	105	116	120	113	111
Australia+ NZ	131	132	138	129	130
China	97	88	87	87	94
Rest of East Asia	93	95	100	93	94
South Asia	100	94	95	93	94
All high-income countries	100	109	111	106	106
All developing countries	100	96	96	97	97
World	100	100	100	100	100

[a] Agricultural self-sufficiency ratio excludes "Other (processed) food, beverages and tobacco products."

Source: Anderson and Strutt (2016).

Of course, these projections alter when the baseline simulation assumptions are changed. The farm trade outcomes are particularly sensitive to the assumed rates of overall national economic growth and of farm productivity growth, and to agricultural trade policies by 2030 (which are assumed unchanged from those of 2007 in the baseline projection), particularly of populous China and India. To show how those assumptions affect national self-sufficiency in agricultural and food products, the model was re-run with a change to each of them in turn. Their impacts on the degrees of agricultural self-sufficiency of various countries and regions are shown in Table 1.

For example, what if GDP, skilled labor, and capital in China and India were to grow one-third faster than in the baseline, which is still below their actual rates of growth during the past decade? In this alternative scenario, it is also assumed that annual multi-factor productivity growth in primary sectors globally is one percentage point faster, in response to faster growth in demand for those sectors' output by China and India. These amendments to Asian GDP growth and to global primary sector productivity growth lead to real global export prices for farm products in 2030 being only 3 percent instead of 11 percent above those in 2007. Not surprisingly, the faster Asian growth scenario leads to a substantial increase in the importance of exports to Asia of resource-rich economies: The share of total Latin American exports in 2030 going to Asia increase from 30 percent to 38 percent, while for Sub-Saharan Africa the increase is from 50 percent to 57 percent of its total 2030 exports. Although overall developing country agricultural self-sufficiency stays constant in this faster-growth scenario, the agricultural self-sufficiency rates increase for those two regions by 3 to 5 percentage points (compare columns 2 and 3 of Table 1). The share of developing Asia in world imports of food and agricultural products increases from 34 percent to 40 percent.

Trade-Related Policy Prospects

The above scenarios to 2030 assume trade and subsidy policies do not change over the projection period. Yet in the course of each nation's economic development, its policies have gradually moved away from discouraging farm relative to non-farm production (e.g., no longer taxing agricultural exports), and some countries have begun to explicitly encourage farming. Policies that subsidize farmers or protect them from import competition can, however, be equally damaging to national economic welfare. Such policies also harm countries that export farm products. The latter group of countries were successful during the GATT's Uruguay Round of multilateral trade negotiations in bringing discipline to such policies and securing policy reform commitments from high-income countries that reduced their price- and trade-distorting effects. But developing countries have yet to make legal bindings in the WTO of their agricultural import tariffs and domestic support to the same extent as high-income countries, leaving open the possibility of them raising those barriers to food trade in the future.

Given also that considerable distortions remain in agricultural markets of high-income countries, and that bilateral, regional and other preferential trade agreements in recent years have done little to discipline farm policies (Bureau and Jean, 2013; Fulponi, 2015; Grant, 2013), another round of multilateral trade negotiations could bring large benefits and greater food security to the world. The WTO's Doha Development Agenda (DDA) was launched in 2001, with further agricultural trade liberalization high on its agenda. Despite ample modeling evidence suggesting the economic welfare gains from such reform could be substantial (Anderson and Martin, 2006; Bouët and Laborde, 2017; Laborde, Martin, and van der Mensbrugghe, 2017), no agreement is in sight. Those gains would be even greater if, in the absence of a DDA agreement, agricultural protectionism were to grow in developing countries, as it did in earlier decades in more-advanced economies (Anderson et al. 2016).

The risk of rising costs from protectionism increases with the size of a nation's food-market economy. The above simulations suggest that both China and India—the two biggest national food markets by far among developing countries—will face declining self-sufficiency in farm products between 2007 and 2030: for China from 97 percent to 87–88 percent, and for South Asia from 100 percent to 94–95 percent in the baseline and faster growth scenarios (Table 1). This may prompt China and India to follow the earlier-industrializing Northeast Asian countries in imposing import restrictions on key food grains. As well, in the interest of boosting farm incomes to reduce the yawning urban-rural income gap, they may impose import restrictions also on meat and milk products (but not on coarse grains and oilseed products required for animal feedstuffs). Indeed there are signs already of such a rise in agricultural supports for farmers in these two countries—and in Indonesia, which has the next-largest national food market (Figure 1).

If such restrictions were in the form of tariff equivalents severe enough to eliminate imports of those selected products in 2030, then, according to the same modelers (Anderson and Strutt, 2016), such a trade policy response by China and India would raise substantially the share of their imports of agricultural products that are not protected (Table 2). As their domestic resources move toward rice, wheat, and livestock production, self-sufficiency

FIGURE 1: Agricultural Nominal Rates of Assistance[a] in China, India, Indonesia, and EU28, 1985 to 2018 (%).

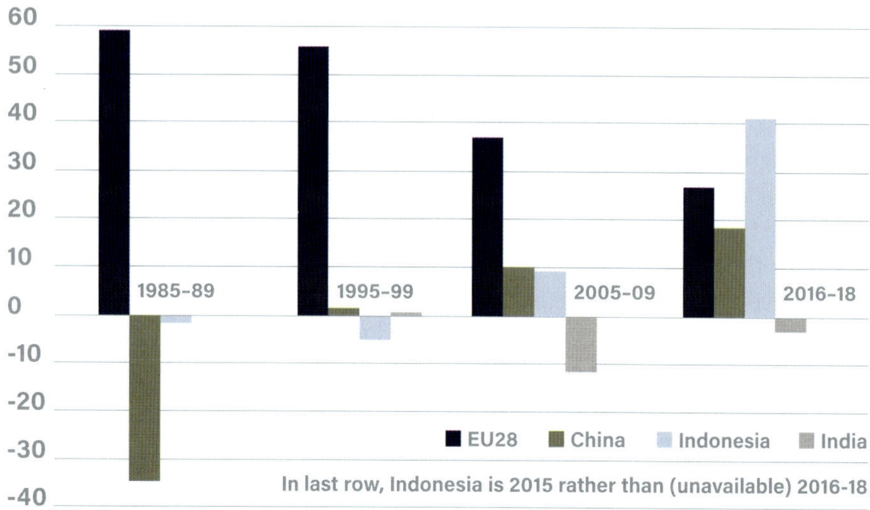

[a] The Nominal Rate of Assistance is the percentage by which gross returns to farmers have been raised by national farm policies (predominantly import restrictions). The final bar for Indonesia is 2015, as 2016–18 is unavailable. Source: Compiled from estimates in Anderson and Nelgen (2013), updated from OECD (2019).

would fall further for crops that provide inputs into livestock feedstuffs, and also for other crops. The tariff equivalents of such import restrictions would range from 115 percent for wheat to 255 percent for red meats for China, and between 136 percent and 326 percent for those products in India. These are well above bound out-of-quota tariffs in numerous cases (compare the last two columns for China and for India in Table 2) and so would be inconsistent with WTO commitments under international law. Moreover, such a policy response would impose a burden on those Chinese and Indian households that are net buyers of grain, meat, and milk products, because the tariffs would cause domestic consumer prices for those products to rise along with producer prices. This may substantially undermine national food security in China and India by reducing real household income and food prices. In addition, it would reduce sales opportunities for food-exporting developing countries and thus undermine their economic access to food as well.

TABLE 2: Shares of agricultural imports and agricultural tariff rates before and after increased agricultural protection, China and India, 2030 (%).

	China					India				
	Share of agric. imports, core sim., 2030	Share of ag. imports, 2030 with selected food import bans	2030 tariff rates	2030 tariff rates, with selected import bans	China's out-of-quota bound tariffs at WTO	Share of agric. imports, core sim., 2030	Share of ag. imports, 2030 with selected food import bans	2030 tariff rates	2030 tariff rates, with selected import bans	India's out-of-quota bound tariffs at WTO
*Rice	1	0	2	196	65	0	0	43	256	80
*Wheat	0	0	2	115	65	7	0	100	326	80
Coarse grains	0	1	2	2	65	0	0	24	25	60-80
Fruit & vegetable	8	16	7	8	11	23	26	35	35	25-50
Oilseeds	11	15	2	2	3	1	1	41	41	75
Vegetable oils	18	30	2	2	3	28	30	82	81	75
Sugar	1	2	0	0	50	1	1	96	96	na
Cotton	3	4	4	4	40	7	8	10	10	na
Other crops	1	2	8	8	na	17	21	48	48	na
*Beef & sheepmeat	1	0	11	255	12	0	0	17	136	na
*Other meats	26	0	8	164	12	3	0	17	156	na
*Dairy products	4	0	8	159	11	1	0	31	153	60
Other+ processed food	25	30				13	13			
TOTAL	100	100				100	100			
Prop'n of total imports	13	10				9	8			

* Indicates sectors subject to the self-sufficiency policy.

Source: Anderson and Strutt (2016).

The above projections do not attempt to take into account possible disruptive new technologies, since their impacts are too difficult to predict. Nor do they allow for an acceleration of technological progress in renewable energy production (possibly stimulated by widespread adoption of taxes on carbon emissions), which may reduce or increase the assumed growth in demand for farm products as an input into biofuels. Taxes on other greenhouse gas emissions, if imposed on farmers, also would impact markets for food (e.g., beef and dairy).

Implications for Markets, Institutions, and Policies

Openness of each national economy to international trade and investment in goods, services, capital, and technologies will optimize the use of resources devoted to producing the world's food, and will maximize real incomes globally. Openness also minimizes fluctuations in international prices and quantities of products traded across borders. Hence trade opening should be among the food policy options national governments consider when seeking to reduce income inequality, poverty, malnutrition, and hunger, to boost diet diversity and food safety, and to raise food quality. Those national and global gains from trade, and trade's positive impact on food security and nutrition, would be even greater if land and water property rights and markets for those key farm resources were better established in each country, and any negative production externalities (e.g., from applying farm chemicals) were appropriately taxed.

If climate change will become more damaging to food production via larger and more-frequent extreme weather events, then all the more reason for nations to be open to international food markets and allow trade to buffer seasonal fluctuations in each country's domestic production: The more countries that do so, the less volatile will be international food prices. Currently, national trade restrictions add non-trivially to international food price volatility in two ways: through "thinning" international food markets, and through "insulating" domestic food markets from international price fluctuations.

The solution to the first ("thinning") problem is for countries to agree to simultaneously open further their markets to food trade. That was what happened in the Uruguay Round, and it is what has been aspired to by members of the WTO via their Doha Development Agenda. Meanwhile, various bilateral and multilateral negotiations on options for regional integration and free-trade areas have concluded successfully and others are under discussion. However, the benefits from such preferential agreements are always far smaller than those from a multilateral agreement—and, in those agreements too, agriculture often is the sector liberalized least.

The optimal solution to the second ("insulating") problem also involves the WTO. In a many-country world, it is clear from recent analyses that the trade policy actions of individual countries can be offset by those of other countries to the point that the interventions become ineffective in achieving their stated aim of reducing domestic food price volatility. This is a classic international public good problem that could be reduced by a multilateral agreement to restrain the variability of trade restrictions, e.g., by agreeing at the WTO to convert specific tariffs into *ad valorem* ones (percent instead of per ton).

Regardless of whether WTO member countries liberalize their food trade and bind their trade taxes on exports as well as imports at low or zero levels, there would still be occasions when international food prices spike. This raises the question as to what alternative instruments governments could use at such times to avert short-term losses for significant groups in their societies. A standard answer from economists is that food security for consumers, most notably food affordability for the poor, is best dealt with using generic social "safety nets"—or, better still, "trampolines." Such measures can be targeted to the most vulnerable households to offset the adverse impacts of a wide range of different shocks on poor people—net sellers as well as net buyers of food—without imposing the costly by-product distortions that necessarily accompany the use of nth-best trade policy instruments for social protection. The measures might take the form of conditional targeted income supplements to only the most vulnerable households, and only while the price spike lasts.

This standard answer has far greater power now than just a few years ago, thanks to the digital information and communication technology (ICT) revolution. In the past, critics often claimed such payments were unaffordable in poor countries because of the fiscal outlay involved and the high cost of administering such handouts. However, the ICT revolution has made it possible for cash transfers to be provided electronically at very low cost as direct assistance to even remote households, bypassing intermediaries such as village chiefs.

What if countries are still unsatisfied with the contribution of their farmers to national food security, as reflected in food self-sufficiency ratios, or feel their farmers are missing out on the benefits of rapid economic growth and industrialization? Again, agricultural import protection measures are far from first-best ways of dealing with these socio-political concerns. Alternative measures include subsidizing investments in agricultural R&D, in rural education and health, and in roads and other rural infrastructure improvements. If the social rates of return from those investments are currently high and above rates of return on other investments, as is typically the case in developing countries, expanding such investments will boost economic growth. Almost certainly they would also reduce poverty and boost national food security, including through raising net farm incomes while lowering the consumer price of food.

References

Alston, J. M., and Pardey, P. G. (2017), Transforming traditional agriculture redux. In Monga, C., and Lin, J. Y. (eds.), *Handbook of structural transformation*, London: Oxford University Press.

Anderson, K. (ed.) (2009). *Distortions to agricultural incentives: a global perspective, 1955–2007*, London: Palgrave Macmillan and Washington DC: World Bank. Retrieved from https://openknowledge.worldbank.org/handle/10986/9436.

Anderson, K. (2010). Economic impacts of policies affecting biotechnology and trade. *New Biotechnology, 27*(5), 558–64, November.

Anderson, K. (2016). *Agricultural trade, policy reforms, and global food security*. London: Palgrave Macmillan.

Anderson, K., Ivanic, M., and Martin, W. (2014). Food price spikes, price insulation, and poverty. In Chavas, J.-P., Hummels, D., and Wright, B. D. (eds.), *The economics of food price volatility*, pp. 311–339. Chicago: University of Chicago Press for NBER.

Anderson, K., and Martin, W. (2006). *Agricultural trade reform and the Doha development agenda*, London: Palgrave Macmillan and Washington, D.C.: World Bank. Retrieved from https://openknowledge.worldbank.org/handle/10986/6889

Anderson, K., and Nelgen, S. (2013). *Updated national and global estimates of distortions to agricultural incentives, 1955 to 2011*. Retrieved from www.worldbank.org/agdistortions.

Anderson, K., and Strutt, A. (2016). Impacts of Asia's rise on African and Latin American trade: projections to 2030. *The World Economy, 39*(2), 172–194.

Anderson, K., Strutt, A., Nelgen, S., and Jensen, H. G. (2016). What is the appropriate counterfactual when estimating effects of multilateral trade policy reform? *Journal of Agricultural Economics, 67*(3), 764–778.

Bouët, A., and Laborde, D. (eds.) (2017). *Agriculture, development, and the global trading system, 2000–2015*. Washington, D.C.: International Food Policy Research Institute.

Bureau, J. C., and Jean, S. (2013). The impact of regional trade agreements on trade in agricultural products. OECD Food, Agriculture and Fisheries Papers, No. 65. Paris: OECD.

FAO (2017). *The future of food and agriculture: trends and challenges*, Rome: FAO.

Fischer, G., Hizsnyik, E., Prieler, S., and Wiberg, D. (2010). *Scarcity and abundance of land resources: competing uses and the shrinking land resource base*, SOLAW Background Thematic Report TR02, Rome: FAO. Retrieved from http://www.fao.org/nr/solaw.

Fuglie, K. O., Wang, S. L., and Ball, V. E. (eds.) (2012). *Productivity growth in agriculture: an international perspective*, London: CABI.

Fulponi, L. (2015). Regional trade agreements and agriculture, OECD Food, Agriculture and Fisheries Papers, No. 79, Paris: OECD.

Grant, J. H. (2013). Is the growth of regionalism as significant as the headlines suggest? lessons from agricultural trade, *Agricultural Economics, 44*(1), 93–109.

Hawkes, C., Grace, D., and Thow, A. M. (2015). Trade liberalization, food, nutrition and health. In Smith, R., Blouin, C., Mirza, Z., Beyer, P., and Drager, N. (eds.) *Trade and health: towards a national strategy*, pp. 92–116. Geneva: World Health Organization.

IFPRI (2017). *Global food policy report 2017*. Washington, D.C.: IFPRI.

Jensen, H. G., and Anderson, K. (2017). Grain price spikes and beggar-thy-neighbor policy responses: a global CGE analysis, *World Bank Economic Review, 31*(1), 158–275.

Laborde, D., Martin, W., and van der Mensbrugghe, D. (2017). Measuring the impacts of global trade reform with optimal aggregators of distortions. *Review of International Economics, 25*(2), 403–425.

Martin, W., and Anderson, K. (2012). Export restrictions and price insulation during commodity price booms. *American Journal of Agricultural Economics, 94*(2), 422–427.

Narayanan, B. G., Aguiar, A., and McDougall, R. A. (eds.) (2012). *Global trade, assistance, and production: the GTAP 8 data base.* Center for Global Trade Analysis, Purdue University, West Lafayette.

OECD (2017). *Producer and consumer support estimates.* Online database retrieved from www.oecd.org.

Pardey, P. G., Chan-Kang, C., Dehmer, S. P., and Beddow, J. M. (2016). Agricultural R&D is on the move. *Nature, 15*(537), 301–303.

Qaim, M. (2016). *Genetically modified crops and agricultural development.* London: Palgrave Macmillan.

Reardon, T., Henson, S., and Gulati, A. (2010). Links between supermarkets and food prices, diet diversity and food safety in developing countries. In Hawkes, C., Blouin, C., Henson, S., Drager, N., and Dubé, L. (eds.), *Trade, food, diet and health: perspectives and policy options*, pp. 111–130. Oxford: Wiley-Blackwell.

World Bank (2016). *Poverty and shared prosperity, 2016,* Washington, D.C.: World Bank.

FEEDING A WORLD IN THE WAKE OF CLIMATE CHANGE AND RESOURCE CONSTRAINTS

Selecting an Appropriate Farm-Level Response Strategy

James N. Blignaut, School of Public Leadership, Stellenbosch University, South Africa, South African Environmental Observation Network

Avi Perevolotsky, Department of Natural Resources, Institute of Plant Sciences, the Volcani Center, Israel

Introduction

One of the great challenges humanity currently faces is the production of enough quality food to meet the needs of the growing global population (Godfray et al., 2010). It is projected that by the mid-twenty-first century, global food demand will be double the current figures (Tilman et al., 2002). Demand for food continues to increase not only due to population increase, but also because of change in diets, improvement of incomes, and cultural shifts (Garnett et al., 2013).

The influence of the so-called green revolution of the 1960s—that is, intensive use of resources such as chemicals, water, and energy—combined with cultivar breeding, helped to meet the increasing food demand up to the present (Matson et al., 1997). Increased production is the goal of agricultural research and practice. The fact that global food production has doubled from 1960 to 2000 emphasizes the contribution of the green revolution (Tomich et al., 2011). But it has not been without a price: A significant burden was imposed on the environment by intensive agricultural practices (Skinner et al., 1997; Tilman et al., 2001; Robertson and Swinton, 2005). Until the 1980s, little concern was raised about the environmental impact of agriculture even though some of the warning signs have been blinking for years (Park, 1988).

In addition to cultivar improvement, principal agents in the advancement of the green revolution have been chemicals such as fertilizers, herbicides, pesticides, fungicides, and others. Their contribution to the increase in yields and food security is beyond doubt. However, the consequences of their use have been ignored for a long time. Agricultural fertilization is not a very efficient process, seeing as only 4 percent of the nitrogen in the fertilizer actually reaches the human diet (14 percent in the vegetarian diet). It is estimated that 50 percent of the fertilizer applied in the field is lost to the environment, either in a gaseous form or dissolved in water (Tilman et al., 2002; Vitousek, 2009). Water with fertilizer residues is known to contaminate underground water and causes eutrophication in wet habitats. Getting rid of, or neutralizing, the impacts of these undesired chemicals is an expensive undertaking. Residues of pesticides and other control materials are deposited in the soil, where they accumulate over time, as well as in the air and in living organisms. These residues may induce an array of negative consequences, called *externalities* in economics. This inefficient linear system of providing high quantities of external inputs with relatively low levels of uptake, and its associated polluting impacts, yet still demanding more external inputs (also with its external impacts such as mining), is clearly unsustainable. There is therefore no doubt that the growing demand for food and the attempt to improve the standards of living, place much stress on resources and on the environment.

The total global area allocated for cropping increased by 466 percent from 1700 to 1980 (Meyer and Turner, 1992), with associated loss in habitats and extinction of native species. At the beginning of the current millennium 10 percent of the global land area was devoted to intensive cultivation and 17 percent to extensive cultivation; 40 percent was rangelands and pastures. In other words, 67 percent of the land surface was under agricultural use in one way or another (Mooney et al., 2005). The dramatic changes in land use and, in particular, the conversion of natural habitats to cropland has become very evident in recent decades. Such land use cascades are expected to impose negative consequences on local biodiversity (Daily, 2001; Phalan et al., 2011) and on the provision of ecological services (DeFries et al., 2004).

The conflict between agriculture and nature is not only with regard to land area. Farming management affects biodiversity, in the agricultural area and in its proximate environment. For example, agricultural intensification (heavy machinery, fertilization, limited crop rotation, and enlargement of monoculture fields), especially of cereals, has led to a significant decline (30 percent) in species richness of farmland birds all over Europe (Donald et al., 2001).

Water for irrigation has been an essential component of agricultural intensification. Agriculture is, by far, the principal user of fresh water in the world. In the U.S. alone, 87 percent of the water retrieved from natural sources is used for irrigation (Postel, 1992). Climate change is expected to worsen the delicate water balance in many regions of the world (Turral et al., 2011). Pimentel et al. (1997) analyzed the global use of water and concluded that conservation and recycling together with an improvement in the efficiency of water use are the future responses to water scarcity, rather than large development projects.

Climate change (i.e., alterations in quantities and distribution of precipitation, temperature changes, and/or the intensities of climatological events) may affect agriculture significantly, mostly negatively. Crop productivity may decrease, thus threatening food availability in certain regions, especially those that are already suffering shortage of food, and interrupting the stability of food systems (Wheeler and Von Braun, 2013). The vulnerability of agricultural productivity to climate variations is well demonstrated by the impact of the recent El Niño climatic phenomenon on global yields of grains (Howden et al., 2007). Climate change is not only about water scarcity. In many regions of the world, temperatures are expected to rise considerably, thus significantly affecting crop yields. A recent research study showed that the yield of the four most important staple crops (wheat, rice, maize, and soybeans) would decrease globally by 3 to 7 percent due to forecasted climate change (Zhao et al., 2017).

The combination of the above-mentioned factors with respect to the increasing demand for food in the wake of limited and even declining resources, both in terms of quality and quantity, not to mention global financial and political instability, give rise to a wicked problem (Rittel and Webber, 1973; Pacanowsky, 1995; Bueren et al., 2003; Batie, 2008). Batie (2008, p. 1176) defines wicked problems as follows:

> Wicked problems, which are sometimes called social messes or untamed problems, are dynamically complex, ill-structured, public problems. The causes and effects of the problem are extremely difficult to identify and model; wicked problems tend to be intractable and elusive because they are influenced by many dynamic social and political factors as well as biophysical complexities. Also, most wicked problems are connected to, or are symptoms of, other problems. As a result, there is no consensus on what exactly the problem is. Indeed, a wicked problem is not well understood until after formulation of a potential

> solution, and therefore, the problem definition tends to change over time. However, because of their complex interdependences, wicked problems are never solved, but rather they become better or worse.
>
> Wicked problems always occur in a social context, and there can be radically different views and understanding of the problem by different stakeholders, with no unique "correct" view. Thus, their wicked nature stems not only from their biophysical complexity but also from multiple stakeholders' perceptions of them and of potential trade-offs associated with problem solving. Identification of solutions becomes as much a social and political process as it is a scientific endeavor. Also, wicked problems are characterized as having high uncertainty associated with them, not only with outcomes but also with the potential causes and effects underlying the problems.

From the above it is evident that a wicked problem does not have an absolute "right" or "wrong," or "yes" or "no" answer. The outcomes of interventions can generally be described as either "better" or "worse" (i.e., improving or worsening a situation). The question, therefore, for agriculture, given the above-stated challenges of increasing food production for more people in the face of climate change and increasing resource scarcity, is which method(s) would either improve or worsen production in the long run considering sustainability-related aspects.

Decision Making in a Complex and Dynamic World: A Scenario-Based Framework

As indicated above, there is an urgent need for increased agricultural output with cognizance of both climate change and environmental concerns. These concerns, and the biophysical, social, and institutional contexts within which a farming operation exists, vary greatly. A farming practice that could be considered appropriate in one region is not necessarily appropriate in another. The same is true of the socio-economic context of each agro-ecosystem. Within the context of wicked problems, there is therefore no "one-size-fits-all" solution that is universally better irrespective of the situation. Figure 1 is a scenario-based framework that illustrates this.

Based on Figure 1, the following observations can be made:

1. The horizontal axis reflects uncertainty with respect to climate change and resource constraints (lumped together for simplicity). It also implies the consequences of negative externalities influencing the quality of the resource (e.g., water). A continuum

exists between high/very high impact as a worst-case option and low/moderate impact as a best-case scenario.

2. The vertical axis reflects the range of possible farming practices ranging from appropriate to inappropriate within a specific biophysical, social and institutional context or region.
3. The four quadrants represent a range of possible outcomes: The most likely outcomes of choosing appropriate *vis-á-vis* inappropriate farming practices under conditions of uncertainty with respect to future climate change impacts and resource constraints, or a continuum thereof, across a range of different contexts and landscapes, can vary considerably.

FIGURE 1: A scenario-based framework illustrating the plausible outcomes between selecting the appropriate or inappropriate farming techniques under various degrees of uncertain resource and climate change constraints.

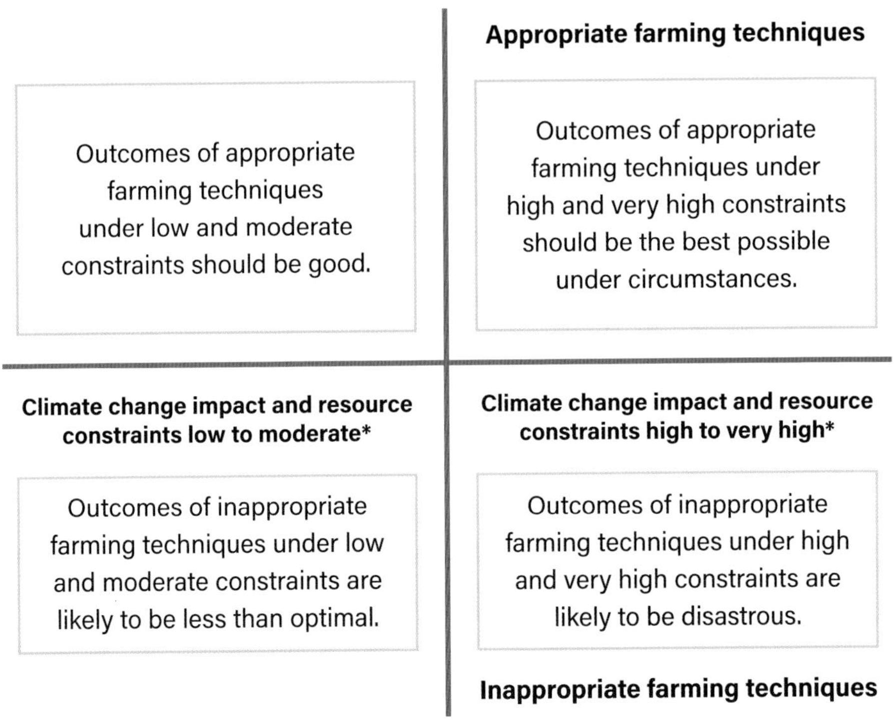

*It should be noted that resource constraints and climate change are lumped together here for illustrative purposes and simplicity, but that might not be the case.

This analysis leads to two questions:

1. What is the range of farming management practices that a farmer can choose from?
2. How should a farmer decide which farming practice to follow within a specific context or site?

We firstly consider a range of plausible farming practices that can be considered, and secondly propose a technique to select the most appropriate farming system within each situation.

A Range of Plausible Farming Systems: A Brief Review

Many authors have defined the great challenge faced by humanity: doubling the amount of food produced globally but doing it in a sustainable manner while confronting significant changes in climate (Tilman et al., 2002; Godfray et al., 2010; Lal, 2013; Smith and Gregory, 2013). Some experts believe that conventional intensive agriculture is the only solution for the forecasted shortage of food in the world (Loomis and Connor, 1992; Connor et al., 2011). Although many practitioners in the agricultural world, including many farmers, share this opinion, we could not find much support for it in the scientific literature. There is a claim that the world produces, and will continue to produce, enough food to sustain the growing human population but the factors that perpetuate poverty and malnutrition—lack of means to purchase the required food, malfunctioning distribution of food, etc.—perpetuate misery (e.g., Alexandratos, 1999). Consequently, a more equal, efficient, and just food distribution in the world is proposed as a key solution to future shortage (Alexandratos and Bruinsma, 2012).

Many experts, however, believe that food will be at a shortage in the future and support a strategy of "sustainable intensification" based on principles of agro-ecology (Garnett et al., 2013; Godfray and Garnet, 2014; Pretty and Bharucha, 2014). This strategy is not well defined in most cases and lacks operational details. However, it is supposed to facilitate higher production of vegetative food while minimizing the environmental impact of the agricultural practice and withstanding the expected global climatic changes. This seems to be a tremendous challenge to pursue.

The scientific literature contains many attempts to define the agro-ecology discipline; therefore, we follow Francis et al. (2003, p. 100) who see agro-ecology as "the integrative study of the ecology of the entire food system, encompassing ecological, economic and social dimensions." Agro-ecology attempts to modify all facets of the agricultural practice into a more environmentally friendly activity without losing productivity.

Fertile soil is an essential condition for successful cropping everywhere. Even though this is known to most farmers, in the U.S., for example, almost 20 percent of their farmlands have been degraded and lost productivity to some extent since 1945 as an outcome of human mismanagement (Tilman et al., 2002). All over the world cultivable soils have been degraded and depleted due to erosion, compaction, loss of organic matter and nutrients, and contamination by pesticides and heavy metals (Lal, 1990; Pimentel, 2006).

One of the pillars of agro-ecology is "conservation agriculture," which aims to cope with these destructive processes. Conservation agriculture (CA) is actually a combination of three principal management activities intending to minimize soil disturbance and maximize crop yields over time: no- or shallow-till, application of cover crop or mulch, and a well-orchestrated crop rotation (Hobbs, 2007; Hobbs et al., 2008). A review of more than two hundred studies from all over the world indicated that under conventional tillage soil loss due to erosion is one to two times higher than the rate of new soil creation in natural habitats (Montgomery, 2007). In other words, conventional tillage is not sustainable. On the other hand, the rate of erosion from fields with no ploughing is like the rate of soil production (Montgomery, 2007).

Conservation agriculture management is supposed to improve the soil in its three dimensions (chemical, physical, and biological) as well as improve soil moisture by increasing infiltration, improving recycling of nutrients, increasing organic matter and carbon content, and decreasing weed presence. If applied correctly, CA is also supposed to lower production costs (in energy and chemicals) and increase crop yields, thereby increasing the profitability of the farms (Huggins and Reganold, 2008). Moreover, CA can help to lower pest and disease damage due to higher biodiversity that includes natural enemies of pests and disease vectors (Hobbs, 2007; Hobbs et al., 2008). Greenhouse gas emissions and water contamination are also expected to be reduced.

At the beginning of the 1970s, on a world scale, CA was applied to less than 1 million hectares; but in 2003, 45 million hectares were under this sustainable management practice, mostly in North and South America but also in Australia and South Africa (Kassam et al., 2015). In 2013, 11 percent of the field crops in the world (1.5 billion hectares) were cultivated through CA management (Hobbs, 2007; Kassam and Brammer, 2013). CA is also strongly recommended by the FAO as the correct way of cropping (FAO, 2014).

The fact that many farmers in North America, South America, and Australia are adopting CA management is evidence that this practice is rewarding through significant yield increase and/or lower production costs. But the question is whether this practice is relevant to small-scale farmers, peasants, and small-holder agriculturalists that cultivate a significant part of the worldwide agricultural land. It appears that these groups, dominant in Africa and Southeast Asia, are not overly enthusiastic about adopting this new practice. A survey among small-holder farmers demonstrated that no-till management produced smaller yields than conventional management. However, the findings are not conclusive due to the low number of case studies, different crops, different environmental conditions and changing methodology (Brouder and Gomez-Macpherson, 2014). Similar findings concerning the impact of no-till compared to conventional management on the yield of 48 crops were summed up in a review relating to 610 studies in controlled experiments in more than 60 countries (Pittelkow et al., 2014). When cover crops or mulch were added to the experiments, the results improved.

Most of the peasants operate a diverse production system and, in many cases, livestock are integrated with the cropping system. Under such circumstances agricultural residues and waste serve as supplement to the livestock diet and farmers cannot use these for mulching the fields (Palm et al., 2014). It has been established that as the environment becomes poorer in resources, farmers become keener on taking full advantage of the crop residues (Pittelkow et al., 2014).

Lal (2013) defined the challenge faced by modern agriculture as follows: "It must produce more from less per capita land and water resources and under changing and harsh climate." One may also add that at present the environmental footprint is required to be minimal. The fact that per capita arable land decreased globally from 0.42 hectares in 1960 to 0.22 hectares in 2004 (Lal, 2013) implies that the expected production goal is going to be very difficult to reach. In developing countries, increasing food production is strongly connected with the existence of irrigation infrastructure. Kurukulasuriya et al. (2006) surveyed nine thousand farmers in eleven African countries and revealed that the revenues from cropping of most farmers are declining unless they have access to irrigation. On the other hand, it is proposed that post-harvest food losses and food waste in developed countries can be reduced by 10 to 50 percent, thus doubling food availability (Lal, 2013).

All in all, there are convincing arguments for the integration of agroecological measures in the present intensive world agriculture. An ecological approach to agricultural soils, as advocated by conservation agriculture, is an essential starting point. However, the successful application of such a management practice depends on an array of factors, mostly non-agronomic, and in some cases, it may not be a feasible option. We now turn toward defining a decisionmaking framework that could be applied at farm, regional, and national scales to assist in determining the best land use options considering a multitude of aspects.

Determining the Most Appropriate Response: A Conceptual Framework

From the above discussion it should be evident that there is a range of possible agronomic responses to the challenges resulting from the need to feed more people under the impact of climate change and various resource constraints. It is therefore exceedingly difficult, if not impossible, to select one response that will perform best in all cases. This does contaminate the scientific search for the best option. And, perhaps, it is not a single response that will suit and serve society and the environment best under all circumstances. In this regard, please consider Figure 2.

FIGURE 2: An elementary multi-factor framework depicting the farming technology used, the geo-climatic and bio-physical context as well as social and economic contexts within which the farm operates, and the suite of outputs and consequences as a result thereof, and its impact, with a delay over time indicated by the two vertical lines, feeding back into the system.

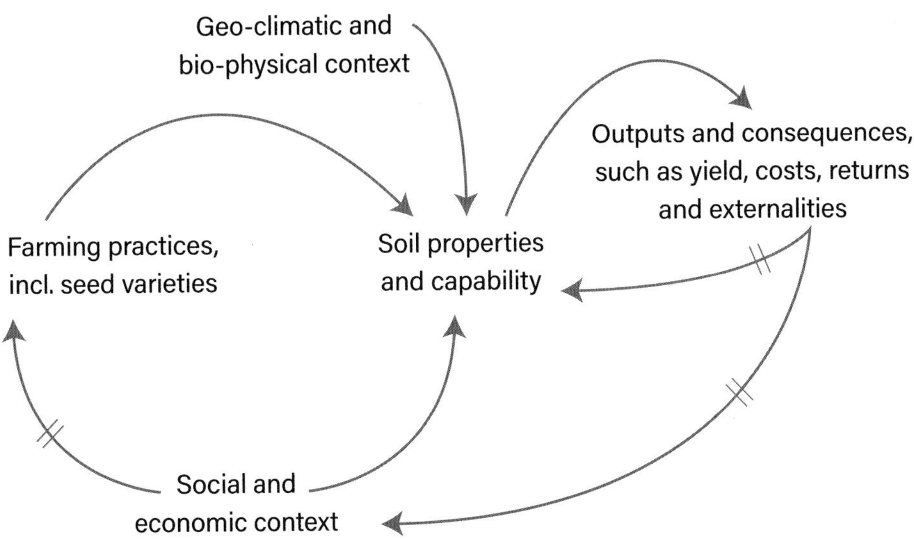

Figure 2 provides a basic and elementary framework of the various aspects that have an impact on the output and the consequences of farming a particular land parcel in a particular way. First, there is the farming practice, including the selection of seed varieties, management techniques, and the degree and type of mechanization discussed above. The agricultural activity takes place within a specific geo-climatic and bio-physical context. The geographic locality of a land parcel plays a major role in determining the soil type and the site productivity. The exact location will also influence the quality of the groundwater, and the vulnerability and susceptibility of the aquifer. It will also have a major impact on the erodibility of the soils, whether of water, sun, or wind or a combination of these factors. In addition, the farming activity takes place within specific social and economic contexts. Those are derived from the values and perceptions of people, their objectives, their type and level of training, their financial status, and their ability to raise finance and access to capital. Not only are these factors interrelated, but they have a meaningful impact on the soil properties and capability over time. The soil and its capability can either improve or deteriorate substantially. All the factors mentioned above combine to determine the outcomes and consequences of the farming activity, be it the yield, the costs to achieve such yield, return on investment, or environmental externalities such as impacts on soil, water, habitat, species, etc. There is therefore a feedback, over time, whereby the outcomes and the consequences have an impact on the future management practices.

Governmental policies are also important to this discussion. The government can help to distribute agricultural knowledge (extension), support commerce in agricultural produce, subsidize environmentally friendly practices, or develop resources. It may also do nothing. All these interventions affect the farmer's decision making on which cropping strategy to pursue.

This multiplicity of factors introduces complexities, which are ever-changing and impacting on each other; it also introduces dynamics and makes the decision regarding both current and future farming practices extremely difficult. Optimizing the outcomes of a complex and dynamic system is not achieved through elementary single function equations.

The real challenge, however, is the fact that not only are the outcomes and consequences over time linked with the type of farming practice and various other aspects, but there is an array of outcomes and consequences to consider. This implies that there are multiple input factors leading to multiple outputs and consequences, and these operate within a dynamic context. Optimizing such a system is the quest.

Determining the Most Appropriate Response: From Concept to Practice

In seeking to optimize the system described above we turn to the Nobel laureate Jan Tinbergen's Golden Rule (Tinbergen, 1952). While there is a range of interpretations of his rule, the most common is found in Acocella et al. (2013, p. 2):

> ...a policymaker can reach any given (fixed) set of independent target values if the number of independent instruments equals or exceeds the number of targets.

Following from this, if you have one instrument, namely the selection of the farming practice, for each individual crop, within a specific geo-climatic and bio-physical as well as socio-economic context, you can only have one output target (target in this context should be read as an objective). Seeking to achieve multiple targets (or objectives) with one instrument inevitably leads to failure in one or more of the targets or can even lead to less desirable and suboptimal outcomes in all. For example, we cannot seek to achieve both maximum profit and maximum environmental sustainability simultaneously as two independent targets/objectives. Based on Tinbergen's Golden Rule, we either have to increase the complexity and increase the number of instruments, or we need to try and reduce the complexity by redefining the multiple targets/objectives into a single one. Increasing complexity is rarely possible or desirable. A farmer is restricted to the selection of one farming practice at a time; also, he has a multiple number of input factors to select from already and is not keen on adding complexity. Furthermore, selecting one farming practice, by definition, excludes other options on the same land parcel at the same time. This leaves a change in the target definition as the only viable option through which Tinbergen's Golden Rule can be satisfied. This implies reducing the number of objectives from a plethora of options such as profit maximization, cost minimization, externality reduction, and job creation, to one. This implies setting a single, yet multi-dimensional, objective, and then seeking the best farming practice to achieve it.

One tool that could assist in this regard is a multi-criteria decision making analysis (MCDA) framework. Using what is called the analytical hierarchy process (Saaty, 1987; Jagtap and Bewoor, 2017) it is possible for decision makers to select the most appropriate option by modeling "a problem in a hierarchical structure determining the relationships between the goals, objectives (criteria), sub-objectives, and alternatives" (Jagtap and Bewoor, 2017, p.

1929). Within the context of selecting the most appropriate farming system, for the particular crop or crops involved, this implies determining:

1. The range of possible farming practices or options that could be applied
2. The range of indicators to represent a multiple number of objectives
3. The criteria for determining high (good) or low (bad) scores
4. A weight for each of the number of indicators—the default value will be an equal weight if that is not specified
5. A score for the various farming practices/options at the hand of the indicators
6. A rank for each of the options to identify which option performed best

Despite its obvious strengths (i.e., namely being coherent, consistent, replicable, easy to understand and to execute using Excel, and transparent, to mention but a few), there are a few caveats in this approach as well. Some of these are:

1. The selection of the indicators to represent the objectives is important. An indicator not selected, or not part of the evaluation process, will not be scored even if it is significant.
2. The determination of the weight of the various indicators could have an overbearing impact on the outcome. This raises the question about who should determine the weights and how that process should unfold.
3. The people involved, the information used, and the process followed to determine the scoring could also have an impact on the result, complicating the interpretation.
4. This type of MCDA tends to be static (i.e., done at a specific point in time). It is indeed possible to add a dynamic component to it using, among others, system dynamics modeling software such as Vensim or Stella.

Despite these caveats, MCDA can assist greatly in determining an appropriate response strategy making use of the best available information within a range of contexts. It does not propagate one ideology above another, and it does not make broad-brush suggestions that cannot function within site-specific realities. Rather, it embraces site-specific realities, and the fact that there are a multiple number of possible farming practices that could be applied on a specific land parcel to a specific crop. It further embraces the fact that the selection of the most desired/appropriate farming practice is rarely, if ever, the result of only one selection criteria. It, therefore, embraces complexity, makes it explicit and does not shy away from it.

As mentioned above, one of the caveats is the selection of indicators to represent objectives. While the indicators to be used could be subjective and open to debate, a very useful framework that could avoid much of the subjectivity is to use the Five Capitals approach. These can be defined as follows (Aronson et al., 2007):

1. Financial capital (i.e., money or its substitutes)
2. Manufactured capital (i.e., buildings, roads and other human-produced fixed assets)
3. Social capital (i.e., institutions, relationships, social networks, and shared cultural beliefs and traditions that promote mutual trust)
4. Human capital (i.e., individual or collective efforts and intellectual skills)
5. Natural capital (i.e., an economic metaphor for the stock of physical and biological natural resources)

Using this approach will go a long way in being broad and encompassing of many variables, yet structured enough to avoid double counting. Under each of these five main headings various sub-objectives can be proposed in which the specifics should be defined. See Table 1 for a generic example. Examples of indicators that could be considered under each of the five main forms of capital are:

1. Financial capital
    - Internal rate of return
    - Net present value
    - Benefit-cost ratio
    - Pay-back period
    - Debt service cost to net return ratio
    - Cost of potential environmental offset or mitigation action or possible environmental penalty
2. Manufactured capital
    - Capital enhancement or betterment of the farm
    - Operating cost of capital equipment
    - Water use efficiency impact
    - Energy use efficiency impact

3. Social capital
   - Contribution to food security
   - Promotion of nation building
   - Formation of social networks and regional co-operation
   - Contribution to community safety risk reduction
   - Adaption to climate change
   - Contribution to tax revenue
   - Contribution foreign exchange
4. Human capital
   - Employment opportunities/job creation
   - Up-skilling, education, awareness building
   - Contribution to human health risk reduction
5. Natural capital
   - Impact on fresh water resources
   - Impact on flooding risk
   - Impact on soil erosion risk
   - Impact on indigenous species
   - Transformation of natural areas/change in habitat
   - Impact on veld and/or forest fire risks
   - Impact on problem species occurrence and risk

TABLE 1. A generic MCDA framework for assisting in the selection of the most desirable farming practice.

Multiplication X. Each option is being scored using, for example, a Likert scale of 1–5, for each of the indicators at the hand of the criteria provided.

Capital:	Financial capital			Manufactured capital			Social capital			Human capital			Natural capital			
Sub-objective	Indicator 1 (user define)	Ind 2 (user define)	Ind n, (user define)	Ind 1 (user define)	Ind 2 (user define)	Ind n, (user define)	Ind 1 (user define)	Ind 2 (user define)	Ind n, (user define)	Ind1 (user define)	Ind 2 (user define)	Ind n, (user define)	Indr 1 (user define)	Ind 2 (user define)	Ind n, (user define)	
Criteria: 1-5	1 = bad; 5 = good (user to define the range)	1 = bad; 5 = good (user to define the range)	1 = bad; 5 = good (user to define the range)	1 = bad; 5 = good (user to define the range)	1 = bad; 5 = good (user to define the range)	1 = bad; 5 = good (user to define the range)	1 = bad; 5 = good (user to define the range)	1 = bad; 5 = good (user to define the range)	1 = bad; 5 = good (user to define the range)	1 = bad; 5 = good (user to define the range)	1 = bad; 5 = good (user to define the range)	1 = bad; 5 = good (user to define the range)	1 = bad; 5 = good (user to define the range)	1 = bad; 5 = good (user to define the range)	1 = bad; 5 = good (user to define the range)	
Weight: 1-5	1 – 5 (user define a)	1 – 5 (user define b)	1 – 5 (user define c)	1 – 5 (user define d)	1 – 5 (user define e)	1 – 5 (user define f)	1 – 5 (user define g)	1 – 5 (user define h)	1 – 5 (user define i)	1 – 5 (user define j)	1 – 5 (user define k)	1 – 5 (user define l)	1 – 5 (user define m)	1 – 5 (user define n)	1 – 5 (user define o)	Sum of weights (a – o = z)
Normalized weight	a / z * 100	b / z * 100	c / z * 100	d / z * 100	e / z * 100	f / z * 100	g / z * 100	h / z * 100	i / z * 100	j / z * 100	k / z * 100	l / z * 100	m/z * 100	n / z * 100	o / z * 100	Sum = 100
Option 1	X	X	X	X	X	X	X	X	X	X	X	X	X	X	X	Weighed score
Option 2	X	X	X	X	X	X	X	X	X	X	X	X	X	X	X	Weighed score
...	X	X	X	X	X	X	X	X	X	X	X	X	X	X	X	Weighed score
Option n	X	X	X	X	X	X	X	X	X	X	X	X	X	X	X	Weighed score

Conclusion

At a global level, food production is, generally speaking, still within the surplus zone. This means that enough food is being produced for the world's people. This is as a result of the recent rapid increases in production due to land conversion and the use of chemicals and improved technologies. The current challenge is not food shortage per se, but its distribution (i.e., getting the agricultural produce from the surplus to the deficit zones in an affordable, fast, and efficient manner while maintaining food quality). Access is thus the issue, not production. This scenario, however, is likely to come to an end sooner rather than later. We're facing an era where society will be faced with the double challenge of distribution and shortage. Under such a scenario the distribution of food is likely to become a game of political, institutional, and economic power to the point of potential outright conflict. Why? If society cannot meet the challenge of appropriate and adequate food distribution within times of surplus, imagine a time of shortage.

The mentioned shortage is as a result of both supply and demand side factors. From the supply side, changes in climatic patterns and increasing resource constraints are the dominant factors. Some of the resource constraints are self-induced as a result of the wasteful way in which agriculture has been practiced over the last few decades leading to erosion, to unsustainable linear supply system of nutrients and chemicals, and to concomitant contamination of soils and water bodies. These negative externalities and their consequences are tied directly to the past success within the agricultural sector. From the demand side it should come as no surprise that the demand for food is rising as a result of, among other factors, a growth in income, changing dietary patterns, and an increase in population.

Given both the supply and demand factors, the pressures on agriculture are rapidly rising—and that within an ever-increasing cauldron of vested interest and power. This leaves the decisionmaker, be it a farmer, a regional institution, a national government or the global society, with, broadly speaking, four possible response options leading to four scenarios or "future trajectories." Within each context, the decisionmaker has to select the most appropriate agricultural practice within the constraints of climate change and resource limitations. While some factors are beyond the scope of influence of an individual decisionmaker, this is not so regarding the decision with respect to the farming method applied. That is within the decisionmaker's domain of influence.

Selecting the most appropriate farming practice is, however, not based only on a single set of inputs. It involves simultaneously considering a range of factors varying from financial, social, human, infrastructural, and environmental. No decision maker can, however, seek to reach multiple objectives using one instrument. That would violate Tinbergen's Golden Rule. It is thus proposed that decision makers, at any level, use a multi-criteria decisionmaking framework based on the five capitals, or a surrogate thereof, and implement the land use option and farming practice with the highest score. This method is replicable, consistent, and coherent, easy to implement, and science-informed.

Today's generation has to farm with the future in mind. Decisions have to be made today to ensure that farming will be sustainable in all respects and not only financially beneficial to the individual farmer in the short-run. This is a matter of national security concern as failing to sustainably manage the resources from which the future's food is to be derived from, is to plan for increased conflict. It is therefore better to plan for sustainability while embracing the current challenges.

References

Acocella, N., Bartolomeo, G. D., and Hallett, A. H. (2013). *The theory of economic policy in a strategic context*. Cambridge: Cambridge University Press.

Alexandratos, N. (1999). World food and agriculture: outlook for the medium and longer term. *PNAS, 96*, 5908–5914. doi: 10.1073/pnas.96.11.5908.

Alexandratos, N., and Bruinsma, J. (2012). World agriculture towards 2030/2050: the 2012 revision. ESA Working paper No. 12-03. Rome: FAO.

Aronson, J., Milton, S., and Blignaut, J. N. (eds.) 2007. *Restoring natural capital: science, business and practice*. Washington: Island press.

Batie, S. S. (2008). Wicked problems and applied economics. *American Journal of Agricultural Economics, 90*(5), 1176–1191.

Bueren, E. M., Klijn, E. H., and Koppenjan, J. F. M. (2003). Dealing with wicked problems in networks: Analyzing an environmental debate from a network perspective. *Journal of Public Administration Research and Theory, 13*(2), 193–212.

Brouder, S. M., and Gomez-Macpherson, H. (2014). The impact of conservation agriculture on small-holder agricultural yields: a scoping review of the evidence. *Agriculture, Ecosystems and Environment, 187*, 11–32.

Connor, D. J., Loomis, R. S., and Cassman, K. G. (2011). *Crop ecology: Productivity and management in agricultural systems*, 2nd ed. New York: Cambridge University Press.

Daily, G. (2001). Ecological forecasts. *Nature, 411*, 245.

DeFries, R. S., Foley, J. A., and Asner, G. A. (2004). Land-use choices: balancing human needs and ecosystem function. *Frontiers in Ecology and Environment, 2*, 249–257.

Donald, P. F., Green, R. E., and Heath, M. F. (2001). Agricultural intensification and the collapse of Europe's farmland bird populations. *Proceedings of the Royal Society of London, Biological Science B, 268*, 25–29.

FAO (2014). What is conservation agriculture? Retrieved from http://www.fao.org/ag/ca/1a.html.

Francis, C., Lieblein, G., Gliessman, S., et al. (2003). Agroecology: the ecology of food systems. *Journal of Sustainable Agriculture, 22*, 99–118.

Garnett, T., Appleby, M. C., and Balmford, A. (2013). Sustainable intensification in agriculture: premises and policies. *Science, 341*, 33–34.

Godfray, H. C. J., Beddington, J. R., Crute, I. R., et al. (2010). Food security: the challenge of feeding 9 billion people. *Science, 327,* 812–818.

Godfray, H. J. C., and Garnett, T. (2014). Food security and sustainable intensification. *Philosophical Transactions B of the Royal Society of London, Biological Science, 369.* doi: 10.1098/rstb.2012.0273.

Hobbs, P. R. (2007). Conservation agriculture: what is it and why is it important for future sustainable food production?*The Journal of Agricultural Science,145,* 127–137.

Hobbs, P. R., Sayre, K., and Gupta, R. (2008). The role of conservation agriculture in sustainable agriculture. *Philosophical Transaction of the Royal Society Series B, 363,* 543–555.

Howden, S. M., Soussana, J-F., Tubiello, F. N., et al. (2007). Adapting agriculture to climate change. *PNAS, 104,* 19691–19696

Huggins, D. R., and Reganold, J.P. (2008). No-till: the quiet revolution. *Scientific American, 299,* 70–77.

Jagtap, H. P., and Bewoor, A. K. (2017). Use of analytic hierarchy process methodology for criticality analysis of thermal power plant equipments. *Materials Today: Proceedings, 4,* 1927–1936.

Kassam, A., and Brammer, H. (2013). Combining sustainable agricultural production with economic and environmental benefits. *The Geographical Journal, 179,* 11–18. doi: 10.1111/j.1475-4959.2012.00465.x.

Kassam, A., Friedrich, T., Derpsch, R., et al. (2015). Overview of the worldwide spread of conservation agriculture. *Field Actions Science Reports, 8* [online].

Kurukulasuriya, P., Mendelsohn, R., Hassan, R., et al. (2006). Will African agriculture survive climate change? *The World Bank Economic Review, 20,* 367–388.

Lal, R. (1990). Soil erosion and land degradation: The global risks. In Lal, R., and Stewart, B. A. (eds.), *Advances in soil science: soil degradation.* New York: Springer-Verlag.

Lal, R. (2013). Food security in a changing climate. *Ecohydrology & Hydrobiology, 13,* 8–21.

Loomis, R. S., and Connor, D. J. (1992). *Crop ecology, productivity and management in agricultural systems.* New York: Cambridge University Press.

Matson, P. A., Parton, A. G., Power, M., and Swift, J. (1997). Agricultural intensification and ecosystem properties. *Science, 277,* 504–509.

Meyer, W. B., and Turner II, B.L. 1992. Human population growth and global land-use/cover change. *Annual Review of Ecology and Systematics, 23,* 39–61.

Montgomery, D. R. (2007). Soil erosion and agricultural sustainability. *PNAS, 104,* 13268–13272.

Mooney, H. A., Cooper, A., and Reid, W. (2005). Confronting the human dilemma. *Science, 434,* 561–562.

Pacanowsky, M. (1995). Team tools for wicked problems. *Organizational Dynamics, 23*(3), 36–51.

Palm, C., Blanco-Canqui, H., DeClerck, F., et al. (2014). Conservation agriculture and ecosystem services: an overview. *Agriculture, Ecosystems and Environment, 187,* 87–105.

Park, J. R. (ed.) (1988). *Environmental management in agriculture: European perspectives.* London: Belhaven Press.

Phalan, B., Balmford, A., Green, R. E., and Scharlemann, J. P. W. (2011). Minimising the harm to biodiversity of producing more food globally. *Food Policy, 36,* s62–s71.

Pimentel, D. (2006). Soil erosion: a food and environmental threat. *Environment, Development and Sustainability, 8,* 119–137.

Pimentel, D., Houser, J., Preiss, E., et al. (1997). Water resources: agriculture, the environment, and society. *BioScience, 47*(2), 97–106.

Pittelkow, C. M., Liang, X., Linquis, B. A., et al. (2014). Productivity limits and potentials of the principles of conservation agriculture. *Nature, 517,* 365–368.

Postel, S. (1992). *Last oasis: facing water scarcity.* New York: W. W. Norton and Co.

Pretty, J., and Bharucha, Z. P. (2014). Sustainable intensification in agricultural systems. *Annals of Botany, 114,* 1571–1596. doi: 10.1093/aob/mcu205.

Rittel, H., and Webber, M. 1973. Dilemmas in a general theory of planning. *Elsevier Policy Sciences, 4,* 155–169.

Robertson, G. P., and Swinton, S. M. (2005). Reconciling agricultural productivity and environmental integrity: a grand challenge for agriculture. *Frontiers in Ecology and the Environment,* 3(1), 38–46.

Skinner, J. A., Lewis, K. A., Bardon, K. S., et al. (1997). An overview of the environmental impact of agriculture in the UK. *Journal of Environmental Management, 50,* 111–128.

Saaty, R. W. (1987). The analytic hierarchy process—what it is and how it is used? *Mathematical Modelling, 9*(3–5), 161–176.

Smith, P., and Gregory, P. J. (2013). Climate change and sustainable food production. *Proceedings of the Nutrition Society, 72,* 21–28.

Tilman, D., Fargione, J., Wolff, B., et al. (2001). Forecasting agriculturally driven global environmental change. *Science, 292,* 281–284.

Tilman, D., Cassman, K. G., Matson, P. A., et al. (2002). Agricultural sustainability and intensive production practices. *Nature, 418,* 671–677.

Tinbergen, J. (1952). *On the theory of economic policy.* Amsterdam: North-Holland Publishing Company.

Tomich, T. P., Brodt, S., Ferris, H., et al. (2011). Agroecology: a review from global-change perspective. *Annual Review of Environment and Resources, 36,* 193–222.

Turral, H., Burke, J. and Faurès, J.-M. (2011). *Climate change, water and food security.* Rome: FAO.

Vitousek, P. M., Naylor, R., Crews, T., et al. (2009). Nutrient imbalances in agricultural development. *Science, 324,* 1519–1520.

Wheeler, T. and Von Braun, J. (2013). Climate change impacts on global food security. *Science, 341,* 508–513.

Zhao, C., Liu, B., Piao, S., et al. (2017). Temperature increase reduces global yields of major crops in four independent estimates. *PNAS, 114,* 9326–9331.

CLIMATE CHANGE AND FOOD SECURITY

Mathew Abraham and Prabhu Pingali
Tata-Cornell Institute, Charles H. Dyson School of Applied Economics and Management, Cornell University

THE DEPENDENCE OF AGRICULTURAL PRODUCTION on weather conditions makes it one of the most vulnerable sectors to climate change. Evidence that anthropogenic activity is leading to the rise in global temperature and has increased variations in average weather conditions globally is consistent and strong (Böhm et al., 2010; Hansen et al., 2010; Mann et al., 2009; Screen and Simmonds, 2010). Flora and fauna in all ecosystems adapt to prevailing climatic conditions, and changes will have an impact on their physiology leading them to be less productive or perish. Global food security is defined as a) the availability of quality food in required quantity adequate to feed the global population b) ability of individuals and households to access nutritious foods and utilize it to nutritional and physiological well-being, and c) assured stability in securing food at all times (Wheeler and von Braun, 2013). Rising temperatures, extreme climatic events such as droughts, floods, cyclones, and storms accentuate production risks and unpredictability, impacting agricultural production and consequently global food security.

Making the agricultural sector resilient to climate change is essential to eradicate global hunger in the existing population and safeguard future generations from it. The FAO (2017) currently puts the global hunger figures at around 11 percent or 815 million and as the population is expected to grow to 9.7 billion by 2050, feeding the increase in numbers and eradicating existing hunger would require at least a 70 percent increase in global food production. The green revolution in the late 1960s ushered in an era of productivity growth in wheat (*Triticum aestivum*), rice (*Oryza sativa*), and maize (*Zea mays*) using research and development in high yielding technologies and introducing new farm management practices through increased input intensity. While increasing food availability, it also led to robust declines in poverty in many developing countries (Pingali, 2012). The challenges for twenty-first-century agriculture are different when we consider the potential impact of climate change. As one of the largest emitters of greenhouse gases (GHG), agriculture affects and is influenced by climate change. Increasing production without expanding the cropped area, reducing environmental externalities such as emissions and land and water degradation, while dealing with increased agroclimatic risks resulting from climate change are new challenges.

In this chapter, we try to unravel the main characteristics of climate change challenges in meeting the food security needs of the future and the institutional approaches needed to design adaptation strategies to reduce impacts. This chapter has three parts. In the first part, we look at how the agricultural sector influences and is influenced by climate change. In the second part, we assess the role of technology, management practices in agriculture, and international cooperation in dealing with food security and climate change. In the past technological interventions enabled yield increases, helping address the issue of food shortage, but in the wake of climate change, technology needs to be coupled with management practices and cooperation for countries and food systems to adapt to climate change: technology to improve yields, reduce externalities, and adapt to climatic risks such as floods and droughts; management to deal with environmental externalities and efficiently utilize land and water resources; cooperation to ensure effective redistribution of food and aid from areas least affected to areas most affected. The last part looks at adaptation and resilience-building strategies essential to ensure the availability, stability, access, and utilization of food.

Agricultural Production and Climate Change

Agricultural production influences and is influenced by changing weather patterns and therefore forms an essential part of any debate on climate change. Agriculture is a large emitter of carbon dioxide (CO2) based and non-CO2 based greenhouse gases (GHG) such as methane and (CH4) and nitrous oxide (N2O), contributing to changes in temperature. This, in turn, influences production conditions and the risks under which agricultural production takes place. Sustainable production and consumption, therefore, forms an essential part of climate change mitigation and adaptation strategies. In this section, we look at how agricultural production influences climate change resulting from emissions and how climate change affects production conditions and risks involved in food production. In understanding this dual role, we can make a case for mitigation and adaptation strategies that may help to reduce the impact of climate change in meeting the food security needs of the future.

Emissions from Production: How Agriculture Contributes to Climate Change

Agriculture, forestry, and other land use (AFOLU) contributes to 21 percent of the world GHG emissions, which consists of CO_2, CH_4, and N_2O. The main activities contributing to emissions are crop production, deforestation, soil and nutrient management, and livestock production. The main GHG affecting global temperatures are the non-CO_2 gases, CH_4 and N_2O. CO_2 emission from agriculture is usually considered neutral as it is associated with the annual cycles of carbon fixations and oxidation through photosynthesis. However, CO_2 fluxes resulting primarily from deforestation and depletion of carbon sinks remain a concern. Agriculture drives about 75 percent of global land clearing, as the most suitable land for agricultural production is also covered by forests. Forests store around 269 gigatons of carbon and over the period between 1990 and 2015, carbon stocks have been reduced by 11.2 gigatons. Around 442 million tons of carbon was lost per year from the conversion of forests for agriculture and settlements and through degradation (FAOSTAT, 2015). Despite a decline in forest resources, over the past decade, there has been a 50 percent reduction in net forest loss (FAO, 2016a; IPCC, 2014). While deforestation continues in the African continent and South America, there has been a rise in afforestation globally. Parallel to a decrease in natural forests, there has been a rise in planted forests (ibid).

Non-CO_2 emissions from the agricultural sector are 56 percent of the total global emissions, and China, the United States and India contribute 31 percent, 14 percent, and 11 percent, respectively. Agriculture contributes to about 72 percent of N_2O emissions and 39 percent of CH_4. Although currently agriculture is the largest contributor of CH_4, at the rate at which the energy sector is growing, it will soon replace agriculture as the biggest emitter.

• • •

FIGURE 1: Percentage share of anthropogenic GHG emissions in various sectors.

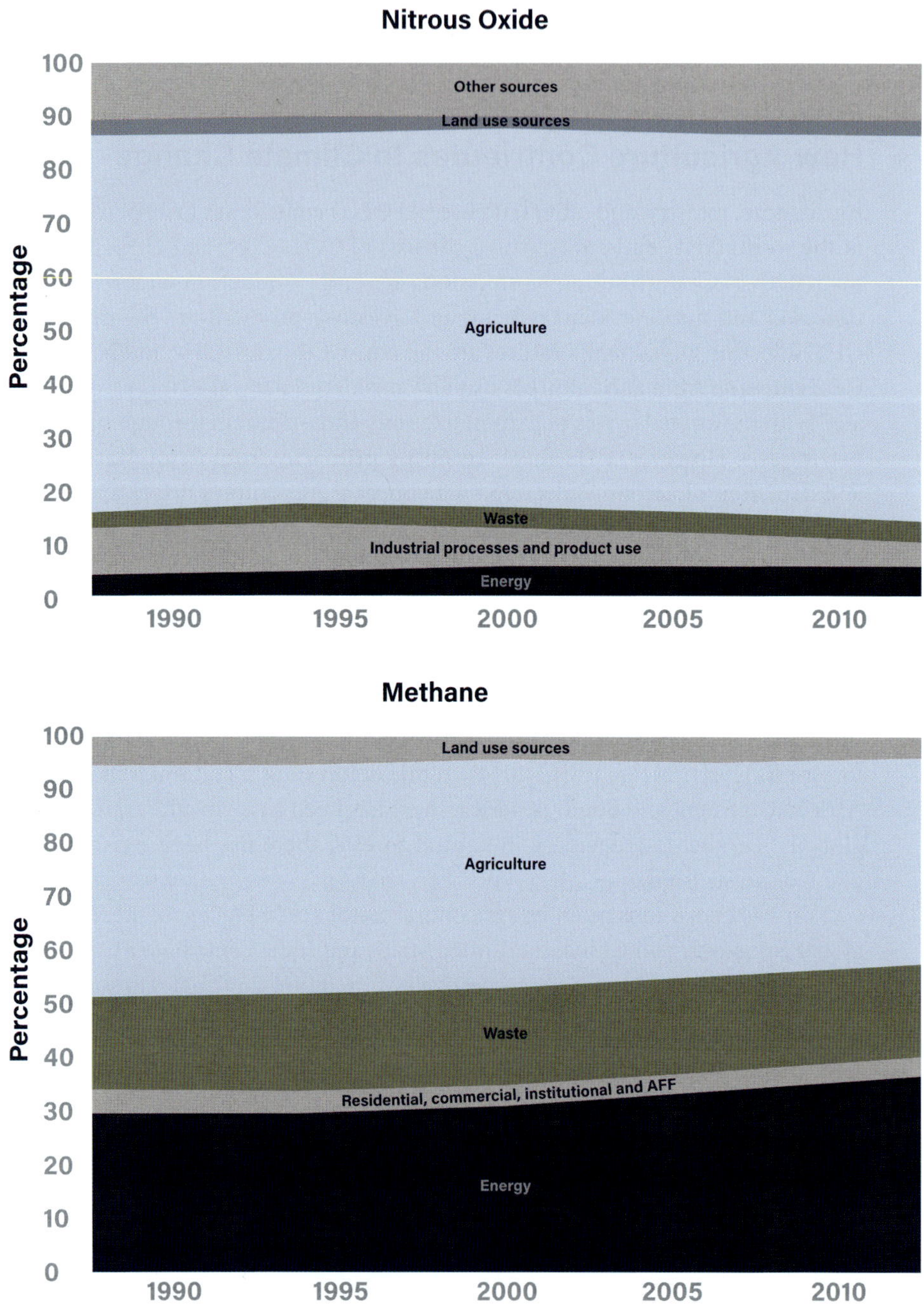

Source: FAOSTATS

Figure 2 shows the breakup of non-GHG emissions from the agricultural sector from 1961–2016. Enteric fermentation from the digestive processes of livestock, biomass burning, and rice cultivation are the largest contributors of CH4, while manure management, soils, and synthetic fertilizers are the largest contributors to N2O. Since 2002, a larger annual growth rate of N2O emissions especially from synthetic fertilizers (2.02 percent increase) and crop residue (2.06 percent increase) was observed. According to the IPCC (2014), in the last ten years, synthetic fertilizers overtook manure in its emission of N2O, and this is largely due to increasing input intensity for productivity gains. CH4 emissions from enteric fermentation grew at 0.75 percent per annum and rice cultivation at 0.5 percent per annum during the same period. About 75 percent of enteric fermentation emissions come from developing countries while 90 percent of rice cultivation emissions are from Asia.

FIGURE 2: Global greenhouse gas emissions from agriculture (1961–2016).

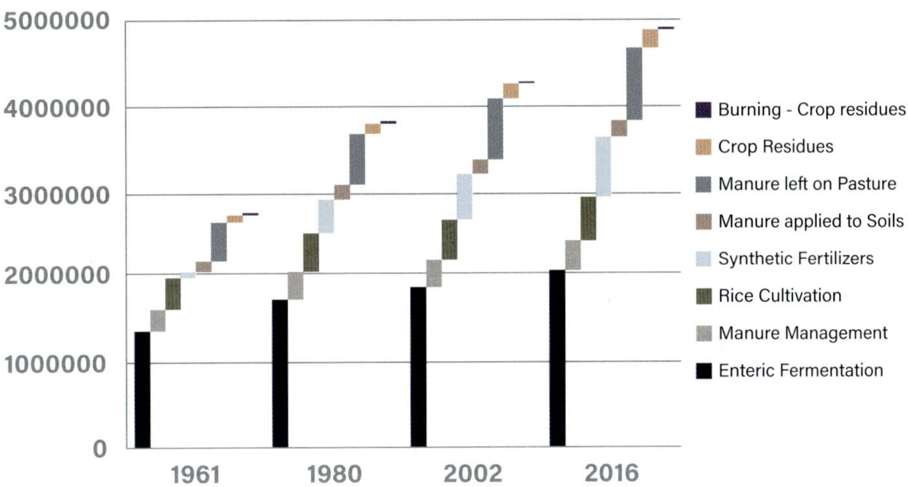

Source: FAOSTATS

The green revolution played a central role in improving yields of major staples in various parts of the developing world. Although there were welfare and resource-utilization benefits, there were also environmental costs. As the dominant technologies that were used were yield-increasing, they allowed for the release of marginal lands out of agricultural production and slowed expansion of agricultural land (Pingali, 2012). However, poor management also led to diminishing biodiversity, increased GHG emissions from injudicious increases in inputs, and land and water degradation from higher resource utilization, chemical runoffs, and other factors (ibid.). Increased productivity in crops and livestock, and higher use of inputs—especially synthetic fertilizers and intensive cropping systems—led to greater emissions from the agricultural sector. With rising populations and growing need for diversified food, this will continue to be on an upward trend.

How Changing Demand Influences Emissions from Agriculture

Population and income growths influence the demand for food and exert production pressures on the agricultural sector. As population increases, the leading global challenge is to provide sufficiently for the existing population, eradicate all forms of hunger, and safeguard future generations from it by producing enough, requiring a significant increase in production. Among the projected production growth of various food groups by the year 2050, cereal production is projected to grow slower than other commodities, primarily due to low-income elasticity, as consumption is not expected to grow with changing income. According to Alexandratos and Bruimsma's (2012) projections, meat, sugar and sweeteners, and oil crops that have higher income elasticity by which demand would grow with changing incomes are expected to increase by approximately 75 percent in the next 38 years.

FIGURE 3. Dietary transition by food group in selected countries (1961–2013).

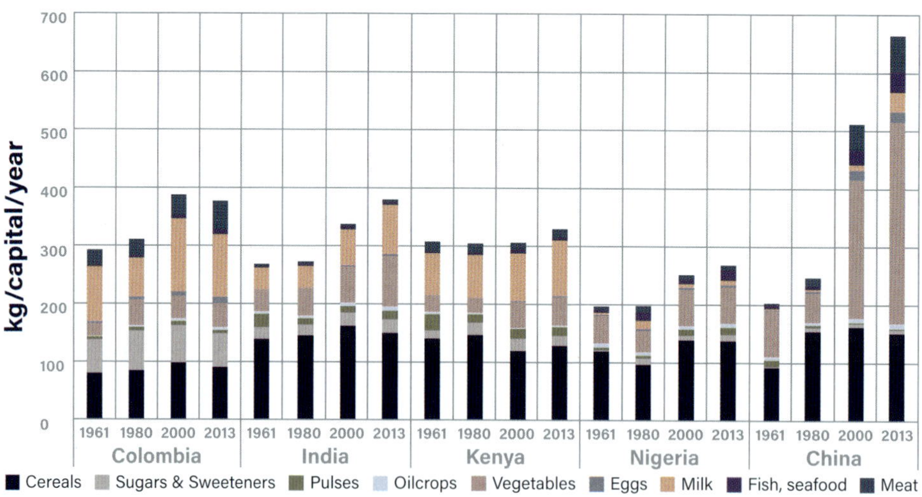

Source: FAOSTAT

Both population growth and income growth are expected to be higher in developing countries. Many of these countries, especially in Asia, benefited from the green revolution. The consequent agricultural development stimulated growth in the non-agricultural sectors. This led to rising incomes and urbanization that influenced rapid diversification of diets and boosted demand for higher value crops and livestock products (Pingali, 2007; Pingali and Khwaja, 2004; Reardon et al., 2009; Reardon and Minten, 2011). Figure 3 shows the dietary transition in selected developing countries on three continents between 1961 and 2013. In all countries, we see a decline in the availability of cereals and high growth in the availability of vegetable and milk-based products. Meat, seafood, and eggs along with oil crops also show an upward trend of different magnitudes in different countries. The dual challenge for the agricultural sector is to increase production and diversity of agricultural produce in line with both population and income growth. Diversification to higher-value animal-based products and input-intensive production practices in other crops is bound to increase emissions from the agricultural sector (Aleksandroweiz et al., 2016; He et al., 2018). However, productivity growth will also be influenced by the availability of expandable land and water resources and by climate change–related factors.

The Effects of Climate Change on Agricultural Production

The complexity of agriculture comes from the many risks associated with it as an economic activity. These risks emerge from imperfectly predictable biological, climatic, and price factors. Compared to other economic activity, outputs from agriculture are more uncertain (The World Bank, 2005). Climate change influences all these risks of agricultural production, playing an important role in achieving food security. In this section, we look at how climate change has influenced and is expected to influence agricultural production and the consequences this may have on the global food systems. Agriculture production affects food availability, food access, and food stability. The productivity of the agricultural sector determines food availability, while the ability to acquire and utilize various foods at the regional, national, and international levels determines food access. Unpredictable production conditions also make food prices extremely volatile. Stability in food availability is therefore central in determining food price and access. The effects of climate change will, however, be felt differently in different parts of the world. In the low-latitude regions and countries the effects of temperature increase on crop production will be consistent and negative, while in the northern latitudes, the impacts are uncertain and, in some cases, there may be advantages through yield increases resulting from warmer climate (Hoffman and Vogel, 2008; Intergovernmental Panel on Climate Change, 2014; Porter et al., 2014).

Production Conditions and Food Availability

Food availability is determined by the physical availability of food in a given context (regional, national, or international) and this is determined by the level of output from the agricultural sector after accounting for losses at the production, harvest, and post-harvest stages. Availability is determined by crop and livestock yield or production capacity, biophysical and climatic risks, management practices such as harvesting, storage, and transportation methods, and trade and commerce that determine the inflow and outflow of commodities. The two climatic factors shown to affect food availability are temperature and precipitation.

A major effect of higher temperatures on agricultural production is a significant drop in crop yields (Rosenzweig et al., 2014). There are a large number of studies looking at the effects of temperature increase on yields of major crops, especially food grains. Lobell, Schlenker, and Costa-Roberts (2011) showed that between 1980 and 2008 the yield of wheat fell by 5.5 percent and maize by 3.8 percent compared to a scenario where the temperature was constant. Nelson et al. (2014) showed that yields of coarse grains, oilseeds, wheat, and rice will decline 17 percent because of exogenous climate change shocks if global annual GHGs continue to rise throughout the twenty-first century. In a review of results from over 91 studies with 1,722 estimates, it was seen that the inter-annual variability of crop yields is likely to increase from 2030 (Challinor et al., 2014). Porter et al. (2014) in a meta-analysis of 1,090 studies showed that changes in temperature would adversely affect yields of wheat, maize, rice, and soybeans (*Glycine max*) in the long run. Higher temperatures could also mean a significant change in disease and pest ecologies, leading to potential increases in the risk of pest attacks and crop failures due to disease (O'Brien et al., 2004). Concerning livestock, heat stress will lower animal resistance to vectors, parasites, and pathogens (Niang et al., 2014; Thornton et al., 2009). Feed crop and forage quality and quantity, water availability, animal and milk production, and biodiversity are other factors that will affect animal husbandry in the context of climate change (Rojas-Downing et al., 2017).

Extreme weather events usually have a major effect on precipitation in the form of floods and droughts. Literature is still unclear about the link between anthropogenic GHG emissions and extreme weather events (Khandekar, 2013; Welsh, 1997), or whether such incidence has been on the rise in recent years. One observable event that has had a notable impact has been the El Niño-Southern Oscillation or ENSO, which is an increase in surface temperatures in the tropical Pacific Ocean. An event that happens every two to seven years, it is known to increase rainfall leading to floods, droughts, tropical cyclones, and forest fires, among other disasters. The last ENSO was one of the most intense and widespread of the past hundred years (FAO, 2016b). The impact of such events on agriculture is often large. An FAO study (2015) showed that between 2003 and 2013 agriculture absorbed 22 percent of the total economic impact resulting from natural disasters in forty-eight developing countries. In crop production, floods contributed the largest losses, while droughts are most damaging to livestock. IPCC (2007) states that increased frequency of extreme weather events such as heat waves, droughts, and floods will contribute more greatly to food insecurity than temperature change.

Food Stability: Prices, Supply Response, and Climate Change

Volatility in food prices is often high due to a number of risks in agricultural production and marketing. Volatility refers to the movement of prices of commodities over a period, and becomes a problem when their magnitude is high and unanticipated leading to suboptimal decision making at both the farm and national level. Price can be influenced by both endogenous and exogenous market shocks. In economies with segmented markets, local supply and demand conditions influence price, while in integrated markets international production conditions can also have an influence. Net food importing countries that are dependent on international markets are influenced by exogenous shocks, while other developing economies can be influenced by endogenous shocks (and exogenous shocks if they also import food) as markets may lack the capacity to absorb domestic shocks during periods of stable international prices.

The effects of climate change leading to rising extreme weather events can increase uncertainty in planting and harvesting (Rosenzweig et al., 2001) and crop loss influencing price volatility. As a majority of world agricultural production is carried out by small and marginal farms with less than two hectares of land, the ability to manage risks is often limited. Climate shocks exacerbate effects on small farms with poor access to technologies and poor soil and water management practices. A relatively small climatic incident in these conditions may have serious effects at the regional level (FAO et al., 2011). Small producer agriculture households along with urban households are net consumers of food (Demetriades and Esplen, 2008; Hertel et al., 2010; Poulton et al., 2010; Skoufias et al., 2011; Skoufias and Vinha, 2013); therefore the effects of price volatility affects agricultural producers. Higher risk may also result in less than optimal production and investment decisions as smallholders go back to a low-risk, low-return-based farming by reducing investments in technology and inputs (Antle and Crissman, 1990; Fafchamps, 1992; Sadoulet and De Janvry, 1995), further accentuating supply shock, especially in a scenario of increasing demand.

Access and Utilization of Food under Climate Change

Smallholder agricultural production is closely linked to access to nutrition and food security in three ways. First, it ensures availability through production. Second, it reduces the real cost of food, making it more affordable. Lastly, it improves incomes of farming households, enabling access to nutritious foods (Ivanic and Martin, 2008; Pingali et al., 2015; Swinnen and Squicciarini, 2012). We see that countries with pro-agricultural growth policies witness lower incidence in child stunting compared to countries that do not (Webb and Block, 2012). Analysis using FAO data shows that increases in agriculture growth correlate with decreases in hunger, stunting, and child mortality in SSA; countries with low agricultural productivity have consistently performed poorly in all three indicators (Abraham and Pingali, 2017).

Climate change affects food availability, incomes, and prices by increasing production risks influencing capita access to calories and child mortality, especially for poor households who are net buyers of food (Porter et al., 2014; Zezza et al., 2008). High food prices and lower incomes will be the main drivers of food insecurity. Ruel, Deitchler, and Arimond (2010) states that at the household level, both the quality and quantity of food consumed will be affected, while Klotz et al. (2008) argued that quality of food is reduced before quantity and therefore micronutrient deficiencies will precede weight loss in the wake of income shocks.

Various estimates predict different intensities of increased hunger; the general agreement is that there may be an increase of 34 to 600 million food-insecure people by 2050, depending on the estimate (IPCC, 2014; Parry et al., 2005; Yohe et al., 2007). As the largest number of poor reside in Sub-Saharan Africa and South Asia, the biggest impact will be felt in these regions. Along with quantity, emerging evidence shows that rising temperatures will also affect the quality of food, especially concerning its nutrition content. Myers et al. (2014) research using experimental data shows that high levels of CO2 in wheat and rice cultivation decreases the grains' micronutrients such as zinc (by 9.3 percent), iron (5.1 percent), and protein (7.8 percent in rice and 6.3 percent in wheat). Their research also shows that there was a small decrease in protein in field peas. Other research shows that protein concentrations may be lower, while carbohydrate concentrations increase (Dietterich et al., 2015; Taub, 2010).

The agricultural sector affects and is affected by climate change. As one of the largest emitter of GHGs, especially methane and nitrous oxide, and with these emissions expected to grow, the primary challenge is reducing and sequestering emissions to reduce their impact on temperature increase. At the same time, as climate change transforms the conditions under which agricultural production takes place because of accentuated production risks through unpredictable weather phenomena, the projections of food availability, access, and stability are adverse. Addressing these issues is central to ensuring food security. Can we do this? In the next section, we look at the various approaches and interventions that are needed to begin addressing concerns of climate change through proactive measures, anticipating climatic challenges in future agricultural production.

Can Agriculture Weather Climate Change? An Optimistic Look at Technology, Management, and Cooperation as Adaptation Strategies

About 90 percent of the 570 million farms globally are small or marginal in size (less than two hectares), cultivated by 1.5 billon of the world's poor (Rapsomanikis, 2015). About 80 percent of these farms are in Asia and Sub-Saharan Africa, where the problems of hunger and poverty are most severe (FAO, 2012). The development of these farms is crucial for income growth, poverty reduction, food security, gender empowerment, and environmental sustainability (Byerlee et al., 2009; Pingali, 2010). However, the capacity of smallholder agriculture to manage climatic risks, especially during adverse events is limited; it can trap them in chronic poverty (Carter and Barrett, 2006; Fafchamps, 2003; Kebede, 1992). As the impact of climate change would disproportionately fall on these socio-economic groups, adaptation strategies need to primarily focus on them.

Adaptation strategies, therefore, need to essentially focus on three aspects: 1) to increase production for a growing population, keeping in mind limitations of land and water resource availability; 2) to manage resource utilization and externalities from agricultural production to prevent accentuation of climate change; and 3) to nurture institutions that will enable vulnerable populations to cope in times of climate change–related events and extreme climatic incidences. The impacts of climate change will be felt disproportionately across the world and strategies need to be tailored to specific regions depending on the stage of structural transformation or dependence on the agricultural sector (Abraham and Pingali, 2017; Pingali et al., 2016). Countries where a major proportion of GDP comes from the agricultural sector are referred to as low-productivity agricultural systems. Most of the Sub-Saharan African countries are classified as such: Green revolution technologies did not benefit these regions initially as yield-increasing technology made a late entry. Hunger and poverty remain high, and because of poor infrastructures such as irrigation and a lack of good roads, vulnerabilities to climate change are high. In modernizing agricultural systems where per capita incomes range from $5,000 to $15,000 and GDP contributions from agriculture are between 10 and 30 percent, gained from green revolution technologies resulting in agricultural productivity increases and reduction of poverty and hunger since the 1970s. Latin American, Southeast Asian and South Asian countries fall into this classification. The remaining economies are commercialized agricultural economies with high per capita incomes where agriculture share in GDP is less than 4 percent.

A majority of low productivity and modernizing agricultural systems are in the lower latitudes, near the equator and the tropics, and it is in these regions that rising temperatures will affect agriculture most (Hoffman and Vogel, 2008; IPCC, 2014). As the dependence on agriculture in these economies is much higher, most of the strategies need to be focused in these regions. This section looks at how technology, management practices, and institutions to ensure effective distribution will play a key role in enabling food security and cushioning the effects of increasing climate change.

The Role of Technology: Going beyond Yield Increase

Green revolution technologies have played a crucial role in increasing yields of essential staples such as wheat, rice, and maize in many parts of the world, helping people become food secure in terms of per capita calorie availability at the national level. What made it successful was the fact that it was a scale-neutral technology and state and policy support to help in its adoption was high. Subsidies for inputs and their effective distribution along with extension services at the production stages and assured prices and procurement by the state to create buffers incentivized their adoption in a big way (Pingali, 2012). However, there were also limits to the uptake of green revolution technologies. First, they were limited to major staples only and did not include other nutritional sources such as dense, coarse grains and pulses, leading to shrinking of land under cultivation of the latter through a substitution effect (ibid.). Second, its adoption bypassed regions with agro-climatic, socio-economic, and technological or biophysical constraints, where input-intensive green revolution technologies could not be adopted (Pingali et al., 2014). While in Asia 82 percent of the area under staples were modern high-yielding varieties in 1998, Sub-Saharan Africa had only 27 percent (Evenson, 2003). Third, although the green revolution prevented expansion of agriculture into marginal lands, leading to conservation, other environmental impacts due to injudicious use of inputs included water and soil degradation and increased GHG emissions (Pingali, 2012).

Technology still plays a central role in assuring food security in the wake of changing environmental conditions and challenges. However, this needs to go beyond yield-increasing technologies in staples to other crops such as coarse grains and pulses, developing crops that are resilient to heat and excess precipitation and using crops and practices that reduce emissions. In low-productivity agricultural systems, in recent years yield increases have been largely through expansion of area under cultivation (Binswanger-Mkhize and McCalla, 2010). Expansion of yield-increasing R&D to coarse grains and pulses that grow in semi-arid regions of low productivity and modernizing agricultural systems will increase output and availability of these nutritious food groups. The development of heat-tolerant and drought-resistant crops will allow for the expansion of crops in not only regions that were bypassed by the green revolution but also temperate regions susceptible to temperature increases. This will also help reduce yield gaps within staples such as maize as compared to other regions in low-productive agricultural systems. IFPRI's IMPACT model that forecasts agricultural output projects heat tolerant varieties can contribute to the highest yield increase in maize by the year 2050 (Rosegrant et al., 2014).

Technology to reduce emissions from growing various crops is also an important intervention. Plants with traits such as Biological Nitrification Inhibition (BNI) that suppresses the loss of nitrogen from the soil can improve uptake efficiency and boost crop productivity (Subbarao et al., 2017). These technologies, coupled with management practices such as zero-tillage, can reduce emissions significantly. The IMPACT models forecast that 124 million people—or 12 percent of undernourishment in developing countries—can be reduced through nitrogen-efficient crops, 9 percent or 91 million people through zero tillage, and 8 percent or 80 million people if heat-tolerant and precision agriculture was adopted (FAO, 2016b; Rosegrant et al., 2014). Technological intervention needs to be coupled with management practices for efficient implementation and reducing environmental externalities, as witnessed under the green revolution.

Management Practices, Production, and Consumption Efficiency in Dealing with Climate Change

Agricultural intensification that allows for increasing productivity, lower costs of production, and limiting negative externalities from agricultural production such as diminishing biodiversity, increased greenhouse emissions, and land and water degradation (among others), is referred to as sustainable intensification (FAO, 2016b; Matson et al., 1997; Pretty et al., 2011). This, coupled with risk mitigation, forms the central strategy for food systems to adapt to future production conditions where climate change is an inevitable reality. The two-pronged aim of management in food systems is to limit the effects of agricultural production on the environment while increasing agricultural production. Limiting the negative effects of agricultural intensification entails the management of emissions and the judicious use and replenishment of resources while limiting the effects of climate change on production, which requires mitigation strategies. These strategies will be different in different locations and regions depending on production conditions and levels of agroclimatic risks.

Effective soil management by maintaining balances in the carbon and nitrogen cycles plays an important role in the reduction of GHG from the agricultural sector (Burney et al., 2010; Wollenberg et al., 2016). Soils hold the second-largest pool of carbon after the oceans and changes in the stock of organic carbon in them can affect atmospheric CO2 to a great extent (Chappell et al., 2015). The type of soil, nutrient, and water management practices adopted by the farmers during agricultural production influences these cycles. The amount of fertilizer applied to the soil, fallowing practices, nitrogen-fixing crop cover, and tilling practices are the main determinants of these cycles and of the amount of CO2 and NO2 that is released into the atmosphere. Agroforestry that allows for the cultivation of crops and tree cover and agroecology practices with sustainable interactions between plant, animal, and human environments that allows for proper tilling, fallowing, and crop cover practices can improve the capacity of farms to limit nitrogen emissions and sequester carbon.

Judicious use of nitrogen fertilizers is also critical to limit NO2 emissions. In East Asia regions where there are problems of mineral fertilizer overuse, environmental harm in the form of water contamination and GHG emissions are major concerns. In regions like Sub-Saharan Africa, where more nutrients are removed with harvest than are replenished, leading to soil depletion, increasing the level of mineral fertilizer use is essential to boost crop productivity (FAO, 2016b). According to IFPRI's impact model, in the business-as-usual scenario, NO2 emissions would increase to 7.5 million tons from 4.1 million tons in 2010. However, reduction strategies through effective land management can cut emissions by 3.3 million tons by 2050.

Increasing water use efficiency is essential to ensure the availability of water and to mitigate the effects of prolonged drought scenarios. This requires an institutional and technical change at the farm and national level that enables conservation, replenishment, and effective allocation. In regions like Sub-Saharan Africa where 95 percent of farms are rainfed and 43 percent is dryland prone to extreme weather events (Cooper et al., 2008; UNDP, 2009), investment in irrigation facilities and promotion of activities such as water harvesting will play an important role in improved water availability (HLPE, 2014). Watershed and aquifer management in semi-arid regions, along with informational services on climatic variability, need to be built into water management practices to help regions cope or manage water stresses that may increase in the years to come (Bates et al., 2008; FAO, 2013; Sadoff and Muller, 2009). Water and drought stresses are often exacerbated by land degradation in sub-humid and semi-arid conditions to a greater degree than in purely arid conditions (Adhikari, 2013). The sensitivity of a particular food system to climate change increases with scarcity and degradation of natural resources. Therefore, systems by which resources are judiciously used and replenished while limiting externalities such as emissions and degradation are central to managing the impacts of climate change.

Reducing food loss and wastage in harvest and post-harvest stages can also ease productivity pressures on the food system. In many developing countries a major part of food loss and waste happens at the harvest and marketing stages, while in developed countries waste takes place at the consumption stages (Foley et al., 2005; Parfitt et al., 2010). According to Godfray et al. (2010) approximately 30 to 40 percent of food produced is lost between harvest and consumption. Investments in value chains in the form of better connectivity and storage facilities are essential in many developing countries, while increased consumer awareness about food loss and behavior change aimed at reducing food waste will be essential in the long run.

Institutions, International Cooperation, and Trade: Reducing the Impacts of Climate Change

It is clear that the impact of climate change will be disproportionately felt across the world, with the food systems of lower latitude countries in the tropical and equatorial regions being more affected than higher latitude countries. These regions may not have the capacity (expertise, resources, and finances) to build adaptation strategies and resilience to climate change. While adaptation strategies involve technological interventions and management practices discussed in the previous sections, resilient livelihood according to the FAO (2016b) is ensuring adequate income and food security that will equip individuals and communities to withstand, recover, and adapt to climate-related circumstances. It involves identifying and implementing actions to deal with changing circumstances, both in response to climate events and foreseeing potential risks that may influence production and livelihood. The role of institutions such as social protection, aid, and trade is central to building adaptation strategies. Therefore, along with technology and management, institutional support and cooperation in the form of social protection, aid, and trade form the third pillar to building climate change resilience in food systems.

Aligning social protection with climate risk management and disaster risk reduction is important to reduce the impact of climate change (Asfaw et al., 2016). As the poor are most vulnerable, effective safety nets will lower the effects of climate change on livelihood and health. Social protection programs such as cash transfers, school meals, and programs to protect vulnerable social groups (especially women) during reduced agricultural activities are needed. Cash transfers in countries where currency and market functions are stable have the advantage of cost-effectiveness and flexibility, giving households better chances to access food or capital and to improve resilience (FAO, 2016b). Index-based insurance systems dependent on rainfall and yield indexes rather than a failure of crops can help vulnerable groups cope with income losses (FAO, 2016b; Ludi et al., 2007). To make them more accessible, state subsidies for these insurance programs may also play a crucial role in assuring most vulnerable groups are covered (GIZ, 2015; Tack et al., 2018).

Different organizations forecasting the cost of climate-proofing estimate that annual expenditure may be anywhere between \$9 and \$109 billion annually, depending on the estimate and region (Lipinski and McGray, 2010).

This may be an excessive burden for many low-productivity agricultural economies as well as modernizing economies. The UN Framework Convention on Climate Change (UNFCCC) stipulates that developing countries have the right to obtaining support from developed countries in climate change adaptation. Humanitarian aid to support and strengthen safety nets, as well as official development assistance, may play a crucial role here. Currently official development assistance to the agricultural sector of low- and middle-income countries amounts to 1.4 percent of total agricultural expenditures (Lowder, Carisma, and Skoet, 2015). In the long run, this may not be enough. The role of international donor organizations and developmental organizations such as the UN and the World Bank will be important in enabling access to international aid and providing technical support in utilizing it.

International trade in food will emerge to play an important role when climate-induced production changes lead to higher trade flows from high and mid-latitudes to low latitudes where yields will be affected (Huang et al., 2011). Price movements in domestic and international markets are different, and how domestic prices are affected by international prices depends on the level of integration to the world market. Countries that insulate themselves tend to transmit local instability to the international markets (FAO et al., 2011). The 2008 food crisis is an example of how insulating domestic markets in times of high prices can lead to greater price spikes on the international scale (Anderson and Nelgen, 2012), affecting net food importing countries. Demeke et al. (2009), in a study that surveyed eighty-one countries, show that forty-three reduced their import tariffs and twenty-five countries banned exports or increased taxes, leading to further spiking of global food prices. This rise also limited the capacity of international relief organizations to purchase food to support populations that could not access food, exposing the limited readiness of global relief responses in times of high price volatility (FAO et al., 2011). In the future, with rising climatic uncertainties, integrated global markets may allow for dispersing risk and enabling availability in times of local weather-related adversities. An integrated global market for food through the reduction of trade barriers and trade-distorting supports can allow for freer flow of information about production conditions that allows for better supply response. It can enable better participation in futures markets where price risks can be hedged and enable more transparent and realistic buffer stock systems that allow for realistic food reserves and discourage unrealistic protectionism—and also allow for international coordination at times of localized climate-related food crisis.

• • •

The Way Forward: Adapting and Building Resilience in the Food System

The agricultural sector is one of the biggest emitters of GHG, contributing substantially to climate change. It is also the sector that is the most affected by climate change, as the production and marketing of agricultural produce is wrought with production and marketing risks directly linked to climatic conditions. Therefore, the food security challenge of feeding the future hinges on ensuring availability, access, and utilization and stability in food. For this, enabling food systems to adapt to climate change, while building resilience to cope with extreme weather events as well as gradual temperature increases and corresponding change is essential.

Adaptation Strategies to Increase Production and Limit Emissions

Central to adaptation strategies is the need to increase food production to feed a growing population while ensuring the externalities from this increase remain at manageable levels—not only accentuation but also limiting the impact of climate change. Increasing yields is essential to improving food availability without an increase in cropped area and to safeguarding forest resources, which are often the most suited for agricultural expansion. In parts of Sub-Saharan Africa and semi-arid regions and unirrigated regions of Asia, the adoption of green revolution technologies was limited. Closing the yield gaps in already cultivated lands is an important strategy for improving food availability. The largest yield gaps are in Africa, China, and Eurasian plains for maize, parts of India for rice, and in India and Eurasia for wheat (Licker et al., 2010; Mueller et al., 2012). In areas of Asia and Africa where the majority of coarse grains and pulses are grown, yields are low due to lack of HYV resulting from poor R&D. To increase yields, there is a need for infrastructure development, especially irrigation and increase in inputs through fertilizer application. Nitrogenous fertilizer coupled with irrigation have contributed to 30–50 percent increases in production. In Sub-Saharan Africa and some parts of Asia, fertilizer application rates are low, and an increase in this will be essential for yield increase.

While there is a need to increase fertilizer application in some regions, effective management of farming practices, including managing and reducing over-application of nitrogenous application is essential. This is especially true in East and Southeast Asia. Reducing emissions through land and livestock management while creating resources and mechanisms to sequester

GHG will be a major contribution of the agricultural sector. Institutional support through extension services and capacity-building at the farm level is essential for managing emissions from land and livestock. In developing countries, land-management practices such as zero or reduced tillage, contour farming, mulches, and crop cover practices are essential to reduce nitrous oxide and increase stocks of soil carbon. Reduction of soil and water degradation is also an essential part of managing externalities from agriculture. In developing countries of Asia and Africa, community-based land and resource management programs such as agroforestry and watershed programs to sequester atmospheric carbon and replenish groundwater are necessary to reduce the impact of emissions and increase usage of resources in the process of increasing yields.

Low quality and low digestible feed, along with poor manure management in the form of improper storage and collection mechanisms, leads to higher emission from livestock. Managing feed to increase nutrients and digestibility and storage and biogas capture from manure will be important to increasing yields and reducing emissions. In crop-based meat and dairy production, especially in OECD countries and China, these systems can be more easily managed. In grazing communities, effective management to reduce overgrazing of common lands will require linkages with farms to manage feed and nutrient availability. Integrating crop and livestock production systems will enable reducing risks and improving incomes and nutrition at the household levels in developing economies (Ramankutty et al., 2008), where specialization in crop or animal production is low.

Improving Resilience of Food Systems to Climate Change

Improving resilience in food systems involves reducing the sensitivity to climate shocks and also limiting the impact on availability, access, and utilization of food. The two components of building resilience are biological and institutional. Biological resilience in plants and livestock is essential to reduce the impact of climate change on their production capacities, while institutions are essential to improve society's resilience to the effects of climate change, especially when the regional impacts of climate change will be different. Technology and practices are both essential to improve biological resilience of food systems. Technology, especially during the green revolution stages, concentrated on improving yields in major staple grains. In the wake of unpredictable weather and extreme climatic events, crops that can withstand this unpredictability become important. Heat-tolerant and drought-resistant varieties of crops, especially in maize and wheat, and flood-resistant rice in rainfall heavy regions would be essential safeguard yields and prevent crop failure. In regions where longer winters and shorter wet periods may occur, shorter-duration varieties of crops will be essential to adapt to the smaller growing seasons. Zoonotic and vector-borne diseases resulting from changing the interaction between climate variables and pathogens, vectors, and hosts can influence outputs and disease burdens in both livestock and humans. Here vaccines and good management practices in meat and dairy production are important to manage and limit these.

Agro-ecological practices where attention is given to variety and ecological compatibility in livestock and the promotion of polyculture and crop-livestock mixed systems will be important to improve agrobiodiversity, and resilience to disease and weather. Breeds that harbor disease resistance to local conditions could better weather changes. Institutions to develop technology and practice and making them through better access to credit and information are essential for wider adoption and implementation. The role of the private sector and the public sector will be essential for investments in R&D. In situations where returns on investments in research are low, government subsidies to improve private sector response will be crucial.

The effects of climate change will be felt disproportionately across the globe, with poorer economies where a majority of the population are involved in agriculture being the worst affected. Enabling safety nets to reduce the impacts of extreme events, provisioning of services such as insurance to account for livelihood loss, and assuring the availability of food through effective redistribution through trade and aid becomes important to build the resilience of different societies to climate change in relation to food. International trade free from protectionist tariffs will enable the transfer of food, especially to net food importing countries, and remedy the spatial imbalances caused by climate change. However, it is also important for regions most affected and prone to extreme weather events to have local institutions to cope. Index-based insurance systems to deal with different levels of income losses and cash transfers to compensate for the loss of livelihood will allow for better access to food in times of distress. Many countries may not have the capacity to invest in adaptation strategies and build resilience; here global cooperation for aid and the transfer of research, knowledge, and practices becomes important. International organizations will play an important role in coordinating the processes of knowledge and resource transfers to help the agricultural sector adapt and build resilience in tackling climate change.

References

Abraham, M., and Pingali, P. (2017). Transforming smallholder agriculture to achieve the SDGs. In Riesgo,L., Gomez-Y-Paloma, S., and Louhichi, K. (eds.), *The role of small farms in food and nutrition security*. Springer.

Adhikari, B. (2013). Poverty reduction through promoting alternative livelihoods: implications for marginal drylands. *Journal of International Development*, *25*(7), 947–967. doi: 10.1002/jid.1820.

Aleksandrowicz, L., Green, R., Joy, E. J. M., Smith, P., and Haines, A. (2016). The impacts of dietary change on greenhouse gas emissions, land use, water use, and health: a systematic review. *PLoS ONE*, *11*(11), e0165797. doi: 10.1371/journal.pone.0165797.

Alexandratos, N., and Bruinsma, J. (2012). *World Agriculture towards 2030/2050 the 2012 revision* (ESA Working Paper No. 12-03). Rome, Italy. Retrieved from http://www.fao.org/fileadmin/templates/esa/Global_persepctives/world_ag_2030_50_2012_rev.pdf.

Anderson, K., and Nelgen, S. (2012). Trade barrier volatility and agricultural price stabilization. *World Development*, *40*(1), 36–48. doi: 10.1016/j.worlddev.2011.05.018.

Antle, J. M., and Crissman, C. C. (1990). Risk, efficiency, and the adoption of modern crop varieties: evidence from the Philippines. *Economic Development and Cultural Change*, *38*(3), 517–537. doi: 10.1086/451813.

Asfaw, S., McCarthy, N., Lipper, L., Arslan, A., and Cattaneo, A. (2016). What determines farmers' adaptive capacity? Empirical evidence from Malawi. *Food Security*, *8*(3), 643–664. doi: 10.1007/s12571-016-0571-0.

Bates, B. C., Kundzewicz, Z. W., Wu, S., and Palutikof, J. P. (2008). *Climate change and water* (Technical Paper of the Intergovernmental Panel on Climate Change). Geneva, Switzerland.

Binswanger-Mkhize, H., and McCalla., A. F. (2010). The changing context and prospects for agricultural and rural development in Africa. In Pingali, P. L., and Evenson, R. E. (eds.), *Handbook of agricultural economics vol. 4* (4th ed.), pp. 3571–3712. Oxford: Elsevier B.V.

Böhm, R., Jones, P. D., Hiebl, J., Frank, D., Brunetti, M., and Maugeri, M. (2010). The early instrumental warm-bias: a solution for long central European temperature series 1760–2007. *Climatic Change*, *101*(1), 41–67. doi: 10.1007/s10584-009-9649-4.

Burney, J. A., Davis, S. J., and Lobell, D. B. (2010). Greenhouse gas mitigation by agricultural intensification. *Proceedings of the National Academy of Sciences*, *107*(26), 12052 LP-12057. Retrieved from http://www.pnas.org/content/107/26/12052.abstract.

Byerlee, D., Janvry, A. de, and Sadoulet, E. (2009). Agriculture for development: toward a new paradigm. *Annual Review of Resource Economics*, *1*, 15–31.

Carter, M. R., and Barrett, C. B. (2006). The economics of poverty traps and persistent poverty: An asset-based approach. *The Journal of Development Studies*, *42*(2), 178–199. doi: 10.1080/00220380500405261.

Challinor, A. J., Watson, J., Lobell, D. B., Howden, S. M., Smith, D. R., and Chhetri, N. (2014). A meta-analysis of crop yield under climate change and adaptation. *Nature Climate Change*, *4*(4), 287.

Chappell, A., Baldock, J., and Sanderman, J. (2015). The global significance of omitting soil erosion from soil organic carbon cycling schemes. *Nature Climate Change*, *6*, 187. doi: 10.1038/nclimate2829.

Cooper, P. J. M., Dimes, J., Rao, K. P. C., Shapiro, B., Shiferaw, B., and Twomlow, S. (2008). Coping better with current climatic variability in the rain-fed farming systems of sub-Saharan Africa: an essential first step in adapting to future climate change? *Agriculture, Ecosystems & Environment*, *126*(1–2), 24–35. doi: 10.1016/j.agee.2008.01.007.

Demeke, M., Maetz, G., and Materne, P. (2009). *Country responses to the food security crisis: nature and preliminary implications of the policies pursued*. Rome, Italy. FAO.

Demetriades, J., and Esplen, E. (2008). The gender dimensions of poverty and climate change adaptation. *Ids Bulletin*, *39*(4), 24–31.

Dietterich, L. H., Zanobetti, A., Kloog, I., Huybers, P., Leakey, A. D. B., Bloom, A. J., … Myers, S. S. (2015). Impacts of elevated atmospheric CO2 on nutrient content of important food crops. *Scientific Data*, *2*, 150036. doi: 10.1038/sdata.2015.36

Evenson, R. E., and Gollin, D. (2003). Crop variety improvement and its effect on productivity: the impact of international agricultural research, pp. 447–472. Cambridge, MA: CABI.

Fafchamps, M. (1992). Solidarity networks in preindustrial societies: rational peasants with a moral economy. *Economic Development and Cultural Change*, *41*(1), 147–174. Retrieved from http://www.jstor.org/stable/1154225.

Fafchamps, M. (2003). *Rural poverty, risk and development*. Cheltenham, UK: Edward Elgar Publishing.

FAO (2012). Smallholders and family farmers. Retrieved from http://www.fao.org/fileadmin/templates/nr/sustainability_pathways/docs/Factsheet_SMALLHOLDERS.pdf

FAO (2013). *Climate-smart agriculture source book*. Rome, Italy: Food and Agriculture Organization of the United Nations.

FAO (2015). *The impact of natural hazards and disasters on agriculture and food security and nutrition: A call for action to build resilient livelihoods*. Retrieved from http://www.fao.org/3/a-i4434e.pdf.

FAO (2016a). *The Global Forest Resources Assessment 2015*. Rome, Italy. Retrieved from http://www.fao.org/3/a-i4793e.pdf.

FAO (2016b). *The state of food and agriculture- climate change, agriculture and food security*. Rome, Italy. Retrieved from http://www.fao.org/3/a-i6030e.pdf.

FAO (2017). *The state of food security and nutrition in the world - Building resilience for peace and food security*. Rome, Italy. Retrieved from http://www.fao.org/3/a-I7695e.pdf.

FAO, IFAD, IMF, OECD, UNCTAD, WFP, … UN HLTF (2011). *Price Volatility in Food and Agricultural Markets: Policy Responses*. Retrieved from https://www.oecd.org/tad/agricultural-trade/48152638.pdf.

Foley, J. A., DeFries, R., Asner, G. P., Barford, C., Bonan, G., Carpenter, S. R., … Snyder, P. K. (2005). Global consequences of land use. *Science*, *309*(5734), 570 LP-574. Retrieved from http://science.sciencemag.org/content/309/5734/570.abstract.

GIZ (2015). Climate risk insurance—for strengthening climate resilience of poor people in vulnerable countries. Bonn, Germany. Retrieved from http://www.climate-insurance.org/fileadmin/mcii/pdf/G7_Climate-Risk-Insurance_Background-Paper.pdf.

Godfray, H. C. J., Beddington, J. R., Crute, I. R., Haddad, L., Lawrence, D., Muir, J. F., … Toulmin, C. (2010). Food security: the challenge of feeding 9 billion people. *Science*, *327*(5967), 812–818. doi: 10.1126/science.1185383.

Hansen, J., Ruedy, R., Sato, M., and Lo, K. (2010). Global surface temperature change. *Reviews of Geophysics*, *48*(4). doi: 10.1029/2010RG000345.

He, P., Baiocchi, G., Hubacek, K., Feng, K., and Yu, Y. (2018). The environmental impacts of rapidly changing diets and their nutritional quality in China. *Nature Sustainability*, *1*(3), 122–127. doi: 10.1038/s41893-018-0035-y.

Hertel, T. W., Burke, M. B., and Lobell, D. B. (2010). The poverty implications of climate-induced crop yield changes by 2030. *Global Environmental Change*, *20*(4), 577–585.

HLPE. (2014). *Food losses and waste in the context of sustainable food systems*. Rome: FAO.

Hoffman, T., and Vogel, C. (2008). Climate Change Impacts on African Rangelands. *Rangelands*, *30*(3), 12–17. Retrieved from http://www.jstor.org/stable/25145387.

Huang, H., von Lampe, M., and van Tongeren, F. (2011). Climate change and trade in agriculture. *Food Policy*, *36*, S9–S13. doi: 10.1016/j.foodpol.2010.10.008.

Intergovernmental Panel on Climate Change (ed.) (2014). Food Security and Food Production Systems. In *Climate change 2014—impacts, adaptation and vulnerability: part a: global and sectoral aspects: working group II contribution to the IPCC Fifth Assessment Report: Volume 1: Global and Sectoral Aspects* (Vol. 1), pp. 485–534. Cambridge: Cambridge University Press. doi: 10.1017/CBO9781107415379.012.

IPCC (2007). *Climate Change 2007: Impacts, adaptation and vulnerability*. Cambridge, U.K.

IPCC (2014). *Climate Change 2014: synthesis report. contribution of working groups I, II and III to the Fifth Assessment Report of the Intergovernmental Panel on Climate Change*. Geneva, Switzerland.

Ivanic, M., and Martin, W. (2008). Implications of higher global food prices for poverty in low-income countries. *Agricultural Economics*, *39*, 405–416. doi: 10.1111/j.1574-0862.2008.00347.x.

Kebede, Y. (1992). Risk taking behaviour & new technologies: the case of producers in the Central Highlands of Ethiopia. *Quarterly Journal of International Agriculture, 31*, 269–289.

Khandekar, M. L. (2013). Are extreme weather events on the rise? *Energy & Environment, 24*(3–4), 537–549. doi: 10.1260/0958-305X.24.3-4.537.

Klotz, C., Pee, S. de, Thorne-Lyman, A., Kraemer, K., and Bloem, M. (2008). Nutrition in the perfect storm: why micronutrient malnutrition will be a widespread health consequence of high food prices. *Sight and Life Magazine*, 2, 6-13.

Licker, R., Johnston, M., Foley, J. A., Barford, C., Kucharik, C. J., Monfreda, C., and Ramankutty, N. (2010). Mind the gap: how do climate and agricultural management explain the "yield gap" of croplands around the world? *Global Ecology and Biogeography, 19*(6), 769–782. doi: 10.1111/j.1466-8238.2010.00563.x.

Lipinski, B., and McGray, H. (2010). Summary of Studies Estimating the Cost of Climate Change Adaptation in the Developing World. World Resources Institute. Retrieved from http://pdf.wri.org/cost_of_adaptation_in_the_developing_world.pdf.

Lobell, D. B., Schlenker, W., and Costa-Roberts, J. (2011). Climate trends and global crop production since 1980. *Science, 333*(6042), 616 LP-620. Retrieved from http://science.sciencemag.org/content/333/6042/616.abstract.

Lowder, S. K., Carisma, B., and Skoet, J. (2015). Who invests how much in agriculture in low- and middle-income countries? an empirical review. *The European Journal of Development Research, 27*(3), 371–390. doi: 10.1057/ejdr.2015.39.

Ludi, E., Stevens, C., Peskett, L., and Cabral, L. (2007). *Climate change and agriculture: Agricultural trade, markets and investment.* Overseas Development Institute. Retrieved from https://www.odi.org/sites/odi.org.uk/files/odi-assets/publications-opinion-files/1884.pdf

Mann, M. E., Zhang, Z., Rutherford, S., Bradley, R. S., Hughes, M. K., Shindell, D., … Ni, F. (2009). Global signatures and dynamical origins of the Little Ice Age and Medieval Climate Anomaly. *Science, 326*(5957), 1256 LP-1260. Retrieved from http://science.sciencemag.org/content/326/5957/1256.abstract.

Matson, P. A., Parton, W. J., Power, A. G., and Swift, M. J. (1997). Agricultural intensification and ecosystem properties. *Science, 277*(5325), 504–509. doi: 10.1126/science.277.5325.504.

Mueller, N. D., Gerber, J. S., Johnston, M., Ray, D. K., Ramankutty, N., and Foley, J. A. (2012). Closing yield gaps through nutrient and water management. *Nature, 490*, 254. doi: 10.1038/nature11420.

Myers, S. S., Zanobetti, A., Kloog, I., Huybers, P., Leakey, A. D. B., Bloom, A. J., … Usui, Y. (2014). Increasing CO2 threatens human nutrition. *Nature, 510*(7503), 139–142. doi: 10.1038/nature13179.

Nelson, G. C., Valin, H., Sands, R. D., Havlík, P., Ahammad, H., Deryng, D., … Willenbockel, D. (2014). Climate change effects on agriculture: economic responses to biophysical shocks. *Proceedings of the National Academy of Sciences, 111*(9), 3274–3279. doi: 10.1073/pnas.1222465110.

Niang, I., Ruppel, O. C., Abdrabo, M. A., Essel, A., Lennard, C., Padgham, J., and Urquhart, P. (2014). Africa. In *Regional aspects. contribution of working group II to the Fifth Assessment Report of the Intergovernmental Panel on Climate Change*. Cambridge, U.K.: Cambridge University Press.

O'Brien, K., Leichenko, R., Kelkar, U., Venema, H., Aandahl, G., Tompkins, H., … West, J. (2004). Mapping vulnerability to multiple stressors: climate change and globalization in India. *Global Environmental Change, 14*(4), 303–313. doi: 10.1016/j.gloenvcha.2004.01.001.

Parfitt, J., Barthel, M., and Macnaughton, S. (2010). Food waste within food supply chains: quantification and potential for change to 2050. *Philosophical Transactions of the Royal Society B: Biological Sciences, 365*(1554), 3065 LP-3081. Retrieved from http://rstb.royalsocietypublishing.org/content/365/1554/3065.abstract.

Parry, M., Rosenzweig, C., and Livermore, M. (2005). Climate change, global food supply and risk of hunger. *Philosophical Transactions of the Royal Society B: Biological Sciences, 360*(1463), 2125 LP-2138. Retrieved from http://rstb.royalsocietypublishing.org/content/360/1463/2125.abstract.

Pingali, P. (2007). Agricultural growth and economic development: a view through the globalization lens. *Agricultural Economics, 37*, 1–12. doi: 10.1111/j.1574-0862.2007.00231.x.

Pingali, P. (2010). Chapter 74: agriculture renaissance: making "agriculture for development" work in the 21st century. In Pingali, P., and Evenson, R. (eds.) *Handbook of Agricultural Economics*, Volume 4, pp. 3867–3894. Elsevier. doi: 10.1016/S1574-0072(09)04074-2.

Pingali, P. (2012). Green Revolution: impacts, limits, and the path ahead. *Proceedings of the National Academy of Sciences, 109*(31), 12302–12308. doi: 10.1073/pnas.0912953109.

Pingali, P., and Khwaja, Y. (2004). *Globalisation of Indian diets and the transformation of food supply systems.* ESA Working Paper No. 4–5. Rome, Italy.

Pingali, P., Ricketts, K., and Sahn, D. E. (2015). The fight against hunger and malnutrition: the role of food, agriculture, and targeted policies. In Sahn, D. E. (ed.), *Agriculture for nutrition.* Oxford, UK: Oxford University Press.

Pingali, P., Schneider, K., and Zurek, M. (2014). Poverty, agriculture and the environment: the case of Sub-Saharan Africa. In von Braun, J., and Gatzweiler, F. W. (eds.), *Marginality: addressing the nexus of poverty, exclusion and ecology*, pp. 151–168. New York: Springer.

Pingali, P., Witwer, M., and Abraham, M. (2016). Getting to zero hunger: learning from the MDGs for the SDGs. In Lalaguna, P. D. Y., Barrado, C. M. D., and Liesa, C. R. F. (eds.), *International society and sustainable development goals*, pp. 175–202. Pamplona, Spain: Thomson Reuters Aranzadi.

Porter, J. R., Xie, L., Challinor, A. J., Cochrane, K., Howden, S. M., Iqbal, M. M., ... Travasso, M. I. (2014). *Climate change 2014: impacts, adaptation, and vulnerability. part a: global and sectoral aspects. contribution of working group II to the Fifth Assessment Report of the Intergovernmental Panel on Climate Change.* In C. B. Field, V. R. Barros, D. J. Dokken, K. J. Mach, M. D. Mastrandrea, T. E. Bilir, ... L. L.White (eds.). Cambridge, U.K.: Cambridge University Press, Cambridge.

Poulton, C., Dorward, A., and Kydd, J. (2010). The future of small farms: new directions for services, institutions, and intermediation. *World Development, 38*(10), 1413–1428. doi: 10.1016/j.worlddev.2009.06.009.

Pretty, J., Toulmin, C., and Williams, S. (2011). Sustainable intensification in African agriculture. *International Journal of Agricultural Sustainability, 9*(1), 5–24. doi: 10.3763/ijas.2010.0583.

Ramankutty, N., Evan, A. T., Monfreda, C., and Foley, J. A. (2008). Farming the planet: 1. geographic distribution of global agricultural lands in the year 2000. *Global Biogeochemical Cycles, 22*(1). doi: 10.1029/2007GB002952.

Rapsomanikis, G. (2015). *The economic lives of smallholder farmers: An analysis based on household data from nine countries.* Rome, Italy. FAO.

Reardon, T., Barrett, C. B., Berdegué, J. A., and Swinnen, J. F. M. (2009). Agrifood industry transformation and small farmers in developing countries. *World Development, 37*(11), 1717–1727. doi: 10.1016/j.worlddev.2008.08.023.

Reardon, T., and Minten, B. (2011). Surprised by supermarkets: diffusion of modern food retail in India. *Journal of Agribusiness in Developing and Emerging Economies, 1*(2), 134–161. doi: 10.1108/20440831111167155.

Rojas-Downing, M. M., Nejadhashemi, A. P., Harrigan, T., and Woznicki, S. A. (2017). Climate change and livestock: Impacts, adaptation, and mitigation. *Climate Risk Management, Forthcoming.* doi: 10.1016/j.crm.2017.02.001.

Rosegrant, M. W., K., J., Cenacchi, N., Ringler, C., Robertson, R., Fisher, M., ... Sabbagh, P. (2014). *Food security in a world of natural resource scarcity: the role of agricultural technologies.* Washington, D.C. IFPRI.

Rosenzweig, C., Elliott, J., Deryng, D., Ruane, A. C., Müller, C., Arneth, A., ... Khabarov, N. (2014). Assessing agricultural risks of climate change in the 21st century in a global gridded crop model intercomparison. *Proceedings of the National Academy of Sciences, 111*(9), 3268–3273.

Rosenzweig, C., Iglesias, A., Yang, X. B., Epstein, P. R., and Chivian, E. (2001). Climate change and extreme weather events: implications for food production, plant diseases, and pests. *Glob. Change Human Health, 2*, 90–104.

Ruel, M. T., Deitchler, M., and Arimond, M. (2010). Developing simple measures of women's diet quality in developing countries: overview. *The Journal of Nutrition, 140*(11), 2048S–2050S. doi: 10.3945/jn.110.123695.

Sadoff, C. W., and Muller, M. (2009). *Better water resources management: greater resilience today, more effective adaptation tomorrow.* GWP TEC Perspectives Paper. Stockholm.

Sadoulet, E., and De Janvry, A. (1995). Quantitative development policy analysis. Retrieved from http://www.dipsa.unifi.it/romano/ASP/Sadoulet_de Janvry QDPA.pdf.

Screen, J. A., and Simmonds, I. (2010). The central role of diminishing sea ice in recent Arctic temperature amplification. *Nature, 464,* 1334–1337. doi: 10.1038/nature09051.

Skoufias, E., Rabassa, M., and Olivieri, S. (2011). The poverty impacts of climate change: a review of the evidence. The World Bank.

Skoufias, E., and Vinha, K. (2013). The impacts of climate variability on household welfare in rural Mexico. *Population and Environment, 34*(3), 370–399. doi: 10.1007/s11111-012-0167-3.

Subbarao, G. V, Arango, J., Masahiro, K., Hooper, A. M., Yoshihashi, T., Ando, Y., … Iwanaga, M. (2017). Genetic mitigation strategies to tackle agricultural GHG emissions: The case for biological nitrification inhibition technology. *Plant Science, 262,* 165–168. doi: 10.1016/j.plantsci.2017.05.004.

Swinnen, J., and Squicciarini, P. (2012). Mixed messages on prices and food security. *Science, 335*(6067), 405–406. doi: 10.1126/science.1210806.

Tack, J., Coble, K., and Barnett, B. (2018). Warming temperatures will likely induce higher premium rates and government outlays for the U.S. crop insurance program. *Agricultural Economics, 49*(5), 635–647. doi: 10.1111/agec.12448.

Taub, D. R. (2010). Effects of rising atmospheric concentrations of carbon dioxide on plants. *Nature Education Knowledge, 3*(10), 21.

The World Bank (2005). *Managing agricultural production risk innovations in developing countries.* Washington, D.C.

Thornton, P. K., van de Steeg, J., Notenbaert, A., and Herrero, M. (2009). The impacts of climate change on livestock and livestock systems in developing countries: a review of what we know and what we need to know. *Agricultural Systems, 101*(3), 113–127. doi: 10.1016/j.agsy.2009.05.002.

UNDP (2009). *Climate change in the African drylands: options and opportunities for adaptation and mitigation.* New York.

Webb, P., and Block, S. (2012). Support for agriculture during economic transformation: impacts on poverty and undernutrition. *Proceedings of the National Academy of Sciences, 109*(31), 12309–12314. Retrieved from http://www.pnas.org/content/109/31/12309.abstract.

Welsh, R. (1997). Vertical coordination, producer response, and the locus of control over agricultural production decisions. *Rural Sociology, 62*(4), 491–507. Retrieved from http://search.proquest.com/docview/199357502?accountid=10267.

Wheeler, T., and von Braun, J. (2013). Climate change impacts on global food security. *Science, 341*(6145), 508 LP-513. Retrieved from http://science.sciencemag.org/content/341/6145/508.abstract.

Wollenberg, E., Richards, M., Smith, P., Havlík, P., Obersteiner, M., Tubiello, F. N., … Campbell, B. M. (2016). Reducing emissions from agriculture to meet the 2 °C target. *Global Change Biology, 22*(12), 3859–3864. doi: 10.1111/gcb.13340.

Yohe, G. W., Lasco, R. D., Ahmad, Q. K., Arnell, N. W., Cohen, S. J., Hope, C., … Perez, R. T. (2007). Perspectives on climate change and sustainability. In Parry, M. L., Canziani, O. F., Palutikof, J. P., van der Linden, P. J., and Hanson, C. E. (eds.), *Climate Change 2007: impacts, adaptation and vulnerability. contribution of working group II to the Fourth Assessment Report of the Intergovernmental Panel on Climate Change,* pp. 811–841. Cambridge, U.K.: Cambridge University Press.

Zezza, A., Azzarri, C., Davis, B., Covarrubias, K., Tasciotti, L., and Anriquez, G. (2008). *The impact of rising food prices on the poor.* FAO–ESA Working Paper No. 8–7.

GLOBALIZED AGRICULTURE AND TROPICAL DEFORESTATION

Derek Byerlee
Georgetown University

Introduction

For much of history, tropical forests were regarded as a resource to be exploited through timber extraction and clearing for agriculture, although their value to local people for livelihoods and environmental services was sometimes recognized (Williams, 2003; Byerlee and Rueda, 2015). During the twentieth century, the high values of tropical forests for conserving biodiversity and more recently for mitigating climate change have gradually been acknowledged. Indeed, a letter to *Science* in 1981 signed by twelve leading scientists of the time, including two of the editors of this volume, was influential in drawing attention to the urgency of conserving tropical forests to save their biodiversity (Eisner, 1981). In 1992, this role was enshrined in the international Convention on Biodiversity; later, the role of tropical forests for mitigating climate change was recognized in the UN Framework Convention on Climate Change. In 2015, the global community under the auspices of the United Nations for the first time set a sustainable development goal to achieve zero deforestation by 2020.[1]

Deforestation today largely takes place in the tropics, where an average of 10 million hectares of tree cover was lost each year from 2010 to 2014.[2] Nearly half of this loss occurred in Brazil and Indonesia (Figure 1) and together they accounted for more than half of greenhouse gas emissions

1 United Nations, Sustainable Development Goal 15, Life on Land, https://www.un.org/sustainable development/biodiversity/.

2 World Forest Watch, https://www.globalforestwatch.org/countries/overview. Based on definition of forests as greater than 30 percent tree cover.

resulting from deforestation.[3] Official statistics from the United Nations Food and Agricultural Organization that measure net forest loss after taking into account forest regeneration and tree plantations put the net loss of tropical tree cover at 7 million hectares per year for 2001–2010 (FAO, 2016). During this same period temperate areas gained forested area (FAO, 2016).

Changes in the extent of agricultural areas mirror these trends in deforestation. While temperate areas experienced a steady decline in cropland, tropical areas increased cropland by nearly 250 million hectares, nearly doubling over the period 1961–2015.[4] During the same years, pastureland expansion in tropical Latin America added another 100 million hectares to the total agricultural area in the region. Studies using satellite imagery have estimated that 75 percent of tropical forest loss during 1980–2000 resulted from agricultural expansion (Gibbs et al., 2010).

FIGURE 1. Gross loss in tree cover in the tropics, 2001–2015.

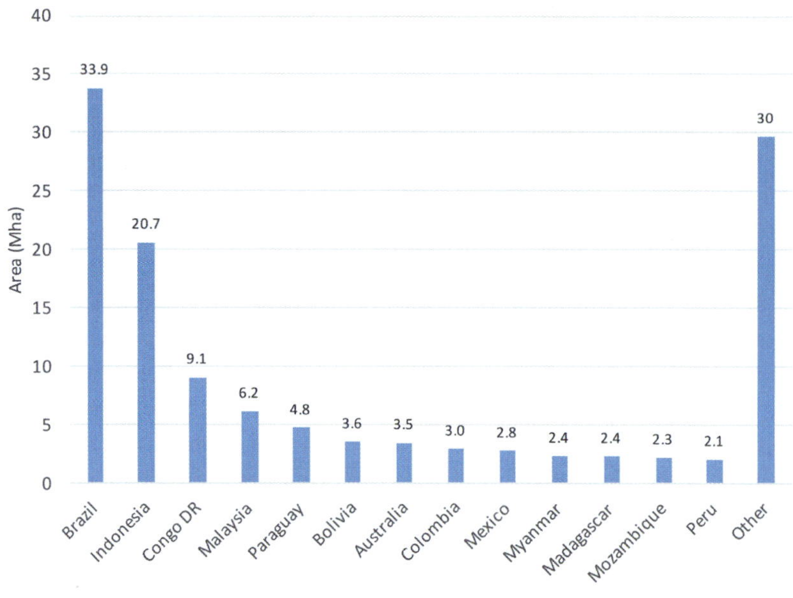

Source: World Forest Watch, https://www.globalforestwatch.org/countries/overview. Based on definition of forests as greater than 30 percent tree cover. Brazil includes only states mostly located in the tropics.

3 World Forest Watch, https://www.globalforestwatch.org/countries/overview.

4 Includes all of the Americas from Mexico to Brazil, all of Southeast Asia, and all of Sub-Saharan Africa except South Africa.

Until the late twentieth century, the expansion of agricultural land in the tropics was primarily driven by the need to supply food to domestic markets for a growing population (Williams, 2003). Starting in the 1980s, however, Nepstad and others identified an important shift in tropical agriculture, when a handful of commodities exported to world markets began to account for the bulk of expansion of cropland and pastureland in the tropics (Nepstad, 2006). The most important of these commodities are beef, soybeans (for animal feed), and palm oil for food and other products (Seymour and Busch, 2016). In the tropics, large-scale commercial farming centered on these commodities now accounts for 40 percent of forest loss versus 33 percent for subsistence farming and 27 percent for other causes (FAO, 2016). For the six countries with available data, beef, soybeans, and oil palm accounted for 50, 13, and 6 percent, respectively, of recorded deforestation (Figure 2).

FIGURE 2. Use of land immediately following clearing of forests in six tropical countries (Argentina, Bolivia, Brazil, Paraguay, Indonesia, and Malaysia), 2001–2013.

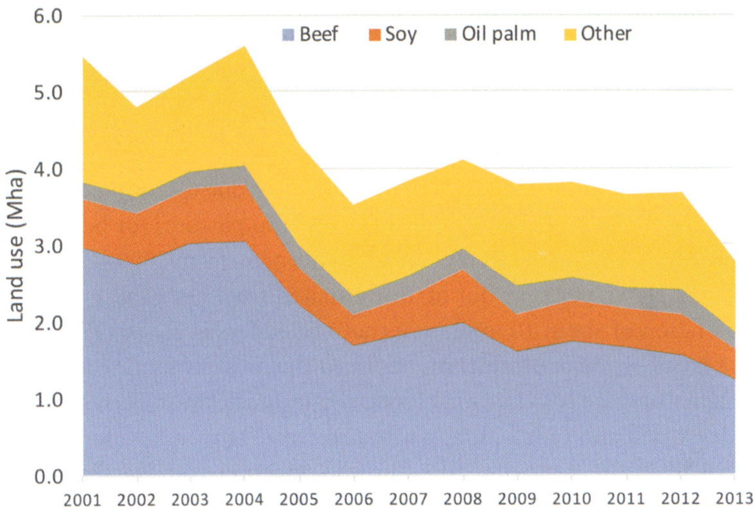

Source: Beckman et al., 2017.

This dramatic expansion of commodity exports from the tropics over the past thirty years has magnified concerns over climate change and biodiversity loss in the tropics. To address these concerns, several environmental groups have launched highly visible media campaigns against tropical deforestation to produce commodities. The rapid pace of tropical agricultural expansion has also negatively impacted the rights of local communities to access and use land and forest resources. Historically, expansion on the frontier was often driven by migrant farmers at the expense of local communities, many of them indigenous. In modern times, the formation of large farms and other commercial enterprises has tended to further limit their ability to sustain their livelihoods (Byerlee and Rueda, 2015).

This chapter reviews the role of producing and exporting global commodities in deforestation in three sections. The first describes the major drivers that have increased the global demand for certain agricultural commodities in the recent past and provides a plausible scenario for demand to 2050. The second part compares recent experiences of commodity expansion on forests in two contrasting cases, Brazil and Indonesia, the two countries experiencing the most extensive amount of tropical deforestation at present. The third part highlights the role of intensification on existing land to supply the global demand for beef, soybeans, and palm oil, versus continued expansion to new areas. We conclude with an outline of a multidimensional approach that will be needed to meet the UN sustainable development goal of zero deforestation.

The Big Three Commodities in Tropical Land Use Change and Their Drivers

The single most important driver of deforestation due to commodity expansion has been conversion of tropical forests into pastures for beef cattle in Latin America where the average rate of conversion from 2000–2011 was 2.7 million hectares per year out of an estimated total global rate of tropical deforestation of about 10 million hectares (Henders et al., 2015). In the savanna areas, soybean production has been the most dynamic force in converting natural areas to farming, especially in Brazil's vast Cerrado savanna and dry woodland biome, that has become the "soy basket" for the world (Figure 3). At the same time, the Cerrado, together with the forests of the Amazon biome, has contributed to the increase in pasture area to produce beef. Some evidence also suggests that soybean expansion in the savanna displaces pastures into the Amazon and indirectly causes deforestation there (Richards et al., 2014). Over the period 2001–2013, for South America as a whole, 17 percent of the new cropland and 57 percent of the new pastureland replaced land previously under forests (Graesser et al., 2015).

In Southeast Asia, palm oil has been an important driver of rapid deforestation in Indonesia and the eastern states of Malaysia. An estimated 33 percent of newly established oil palm in the region during 1990–2010 was on land provided by clearing forests (Gunarso, et al., 2013). Oil palm directly accounted for just 10 to 18 percent of total deforestation in Indonesia in the early 2000s, suggesting there were other important drivers of deforestation as well, mainly unsustainable logging, plantations for pulp, and fire (Busch et al., 2014). However, some of the forest lost to logging may have been motivated by the possibility of planting oil palm after the timber was extracted.

Not surprisingly, beef cattle and oil crops rank high in GHG emissions due to land use change. Out of total emissions from land use change of 3.7 Gt CO2 /y in 2000–2011, beef, soybeans, and oil palm, respectively, accounted for 1.0, 0.1 and 0.2 Gt/y of CO2 emissions due to land use change (Henders et al., 2015). However, these numbers for soybeans and oil palm are certainly underestimated since they do not take account of peatlands (for oil palm) and soil carbon and indirect effects (for soybeans). Overall, it is estimated that efforts to reduce tropical deforestation and reverse degradation could contribute more to reducing GHG emissions to 2030 than the removal of all cars in the world (World Resources Institute, 2017).

FIGURE 3. Expansion of soybean and oil palm area, 1970–2014.

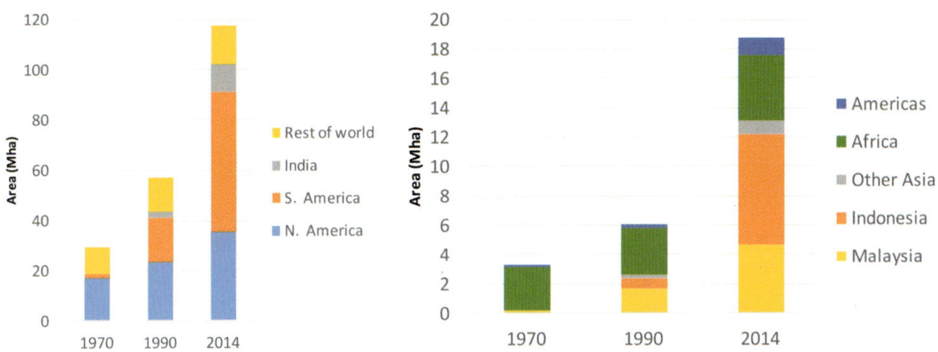

Source: FAOSTAT, http://www.fao.org/faostat/en/#home; note the Y-axis scale for soybeans is six times that for oil palm.

Five factors converged to create a "perfect storm" that drove demand and supply of the "big three" commodities and their associated land use change in the tropics from the early 1990s.

The first was rapid economic growth in middle-income countries at a rate unprecedented in human history. After relatively slow growth through the 1980s, per capita income growth in middle-income countries led by China exploded in the late 1990s to reach a peak of 6.2 percent in 2007 before falling to around 3 percent per year in 2015.[5] Middle-income countries accounting for about three quarters of the world's population are concentrated in Asia, where the per capita consumption of edible oils increased by 35 percent and that of meat by 85 percent over the 1991–2013 period, while per capita food consumption of cereals declined slightly.[6]

Demand for meat, especially poultry and pork, have in turn fueled a rapid increase in demand for feed grains. Poultry was the most dynamic sector based almost entirely on growth of intensive production systems using prepared feed mixes. Thus the production of "oil meal," a major ingredient in feed mixes, has closely tracked the growth of the poultry industry (Figure 4), although a substantial share was also used to feed pigs, especially in China (Gale, 2015). Soybeans provide about 70 percent of global oil meal supply.

FIGURE 4. Indices of oil meal use for feed relative to poultry, pork, and beef consumption at global level, 1990–2013 (1990 = 100).

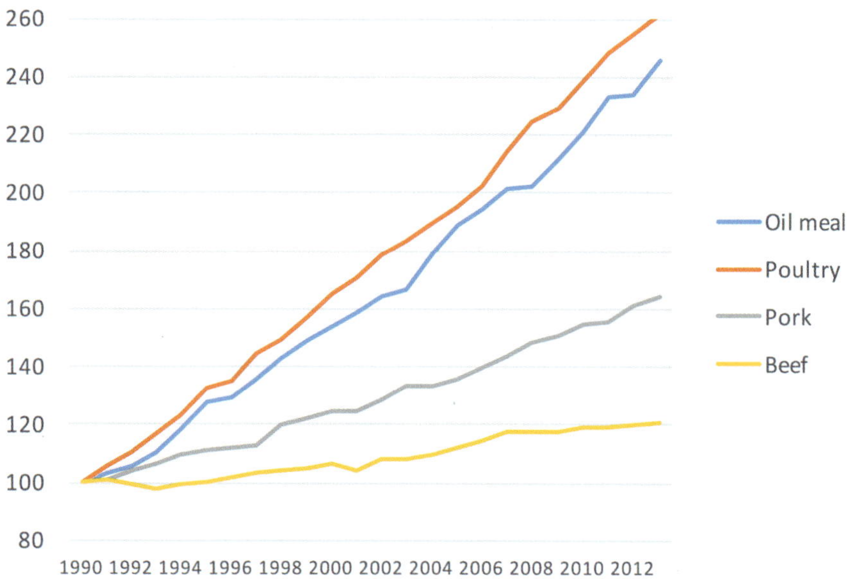

Source: https://apps.fas.usda.gov/psdonline/app/index.html#/app/advQuery.

A second factor on the demand side has been the rise in use of biodiesel fuels from the early 2000s, owing both to a spike in the cost of fossil fuels

5 World Bank, https://data.worldbank.org.

6 FAOSTAT, 2017. www.faostat.fao.org.

combined with the relatively low prices of agricultural commodities, that improved the competitiveness of biofuels and made them politically attractive. Aided by mandates on blending biofuels and fossil fuels for diesel, the use of vegetable oils (including soy oil and palm oil) as a feedstock for biodiesel accounted for over one third of the growth of all vegetable oil consumption during the period 2005–2014 (Byerlee, 2017). Because of the mandates, use of vegetable oils in biodiesel is insensitive to market forces, and their use for biodiesel has continued to rise rapidly, despite falling prices of fossil fuels and rising prices for vegetable oils.

Third, most of the major middle-income countries joined the World Trade Organization around 2000—India in 1995 and, most importantly, China in 2001. As a result, trade in vegetable oils and oilseeds was liberalized more rapidly than for any other agricultural commodity. Reliance on imports of oils and oilseeds to meet rapid demand growth made economic sense, and since they were considered less strategic than the major cereal staples, imports were also politically more palatable (Byerlee, 2017).

The most spectacular growth in trade was in soybeans to China, which imported almost no soybeans in 1995 and now imports nearly 100 Mt. This volume of grain being shipped to one country is without parallel in world history. Brazil has become the major supplier of soybeans to China, with growth in soybean production in the Cerrado closely tracking imports by China (Figure 5) (Byerlee, 2017). During the same period, India became the world's largest consumer of palm oil, imported entirely from Indonesia and Malaysia.

The connection between world trade and tropical deforestation was first recognized with the rapid expansion of cattle pastures in Central America in the 1980s to supply low-grade meat for U.S. hamburgers (Shane, 1986). In the early 2000s, the collapse of the Russian livestock sector stimulated imports of beef from Brazil, much of it from newly cleared areas in the Amazon (Schierhorn et al., 2016). More recently, China has converted from being an exporter of beef to be a major importer, with Brazil its major supplier. However, unlike soybeans and oil palm, about two thirds of the growth in demand for Brazilian beef since 1990 has been domestic, accompanying the Brazilian economic boom to 2008. Consequently, Brazil has become one of the highest consumers of beef per capita in the world (38 kilograms per year).

As a result of trade liberalization, soybeans became the world's most valuable traded agricultural commodity by the late 1990s: The value of the global soybean trade is now almost double that of wheat, which had previously held first place for over a century. During the same period, beef and palm oil have become the third and fourth most valuable traded agricultural commodities after wheat, and they are poised to overtake wheat in the next few years.

These dramatic shifts in world trade can be viewed as trade in "virtual land" (and water) from land-abundant to land-scarce countries. China "imports" about 50 million hectares of land, mostly due to soybean imports from Latin America (Qiang et al., 2013).[7] This change from the production of the major traded commodity, wheat, being centered in temperate regions, to the "new" traded commodities produced exclusively or increasingly in the tropics, underlies the rise of commodity exports in tropical deforestation.

Fourth, on the supply side both technical progress and government subsidies have spurred a decline in the unit costs of production for palm oil and soybeans. For palm oil, the unit cost of production declined by about half from 1966 to 2010 due to increased yield, labor-saving technologies, and improved efficiency in oil extraction (Bruno, 2014). In Brazil, soybean productivity has grown markedly, indicated by yield growth of 2.0 percent annually from 1976 onward (Figure 5). Embrapa, the Brazilian agricultural research corporation, has developed soybean varieties adapted to warmer temperatures at lower latitudes and to the acid soils that are characteristic of much of the Cerrado. Along with the almost universal adoption of soil amendments and conservation tillage in this region, soybeans produced there have become highly competitive in world markets. This is despite their huge disadvantage in high transport costs from distant inland locations to ports (Byerlee et al., 2017).

State subsidies and tax incentives have further reduced the costs of land and capital and reinforced cost reductions due to technical progress in tropical commodities. Indonesia and the states of Sarawak and Sabah in Malaysian Borneo have provided major concessions of state-owned land, often forested, at nominal prices to private investors in oil palm cultivation. In Brazil, the national development bank, *Banco Nacional de Desenvolvimiento Economico e Social* (BNDES), has financed land development and working capital for cattle and soybean expansion at interest rates significantly below market rates. Indeed, they made such development possible in the early years when interest rates were highly negative in real terms until inflation was brought under control. Also in Brazil, tax policies, with taxes on forested land higher than those on agricultural land, further reinforced the incentive to clear forests (Binswanger, 1991).

Fifth and finally, highly competitive prices and trade liberalization have resulted in massive substitutions in food and feed consumption. While 80 percent of the edible oil consumed in Indonesia was coconut oil in the 1970s, today it is nearly 80 percent palm oil. India did not consume palm oil until 1995; today palm oil accounts for nearly half of its oil consumption, and for

7 I have updated the data in Quiang from 2012 to 2017.

FIGURE 5. Intensification of soybean production in the Cerrado, Brazil.

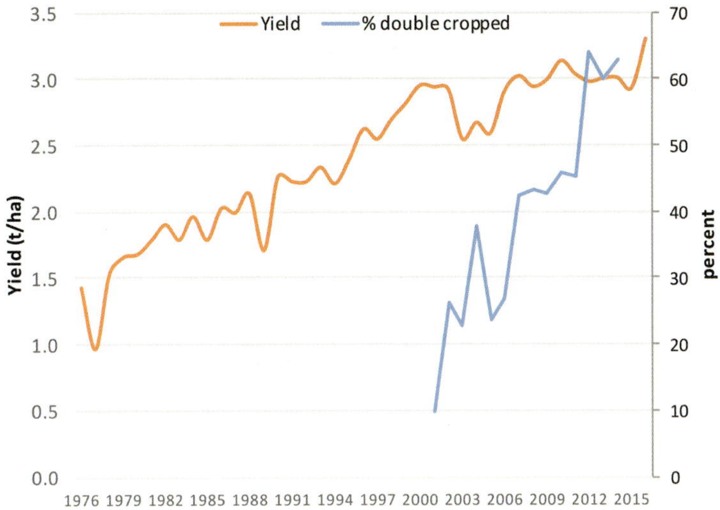

Source: http://www.conab.gov.br/conteudos.php?a=1252&Pagina_objcmsconteudos=3#A_objcms conteudos.

two-thirds of the growth in consumption of all oils since 1995. Similarly, the modern Chinese feed industry has undergone a massive structural change by developing a modern feed industry that substituted soy meal for traditional feeds such as crop by-products (Gale, 2015). Globally, too, poultry consumption overtook that of beef in about 2000; it is now poised to become the most important source of meat, displacing pork. These substitutions have significantly added to the growth in consumption of palm oil and soybeans above the already high rates due to income growth and biofuels.

Nonetheless, these five factors driving unprecedented commodity expansion in world markets are unlikely to be sustained in the future. Projections suggest a sharp slowdown in demand for the two major oil crops, palm oil and soybeans, except in Sub-Saharan Africa where growth may accelerate due to high population growth combined with positive income growth (Figure 6). Slowing soybean demand in turn reflects a decline in the growth rate of poultry consumption related to slower income growth and a falling income elasticity in major middle-income countries, notably China. However, the International Food Policy Research Institute projects a modest increase in demand growth for beef at 1.4 percent annually to 2050 with the largest growth in imports in Sub-Saharan Africa and South Asia (except India) (IFPRI, 2017). The Southern Cone of Latin America (Brazil, Argentina,

Paraguay, and Uruguay) is projected to increase its dominance of world beef exports, rising from 3.8 Mt in 2010 to 11.5 Mt in 2050, and continuing the pressure to expand into forestland (IFPRI, 2017).

Projections to midcentury, 2050, are, of course, fraught with uncertainty. One of the big uncertainties is the growth of meat consumption of the 40 percent of the world's population who still eat less than 20 kilograms per capita, especially India with its cultural constraints on meat consumption, and Sub-Saharan Africa where future economic growth is debated.[8] On the other hand, in rich countries, mounting concerns about the health, ethical, and environmental impacts of meat consumption may accelerate a trend away from meat, especially beef. This trend may also extend to middle-income countries, as evidenced by the recent campaign against meat consumption in China (Milman and Leavenworth, 2016). Furthermore, new industries in the form of cultured or synthetic meat and synthetic proteins to substitute for soy meal could come on stream in the next few years (Shapiro, 2018). Finally, while the IFPRI projections do consider the impacts of climate change, there is inherently much uncertainty about how climate change together with trade policy changes will affect the relative importance of global "baskets" of food production.

FIGURE 6. Projected annual growth in demand in food use of vegetable oils, poultry meat, and beef to 2050 in Sub-Saharan Africa relative to the world.*

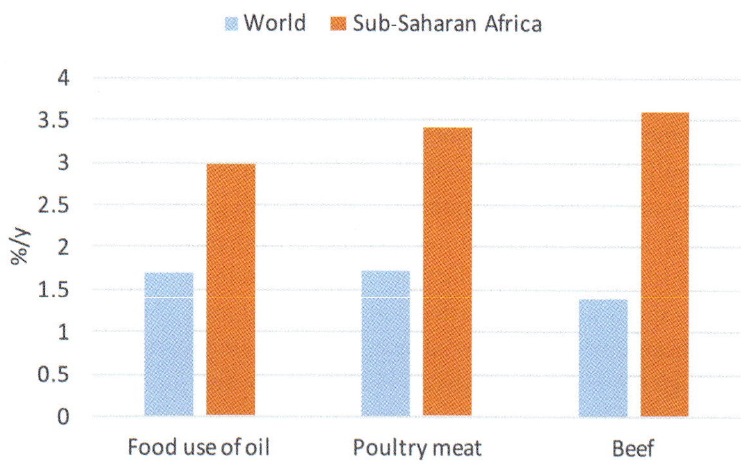

*Projections of oils and soy meal from Byerlee et al., (2017) using 2013 as a base. Projections for poultry and beef from IFPRI (2017) using 2010 as a base.

8 Aquaculture, not considered in this analysis, is one of the world's most dynamic industries and it could become a major consumer of soymeal by 2050.

Contrasting Progress in Halting Deforestation: Brazil and Indonesia

The experiences of Brazil and Indonesia provide sharp contrasts in managing deforestation in the face of global commodity demand. Up to around 2005, Brazil was the largest single source of tropical deforestation. At its peak in 2004, Brazil lost 2.8 million hectares of forest annually in the Amazon and 0.89 million hectares in the Cerrado (Figure 7). Deforestation in the Amazon fell sharply from 2004 and more gradually in the Cerrado. Meanwhile, deforestation in Indonesia rose steadily and now exceeds that in Brazil in many years. A reduction in deforestation (and a halt to the draining of peatlands) in Indonesia comparable to that achieved by Brazil would go a long way to reducing global GHG emissions from land-use change, conserving valuable tropical biodiversity, and providing other environmental benefits.

FIGURE 7. Gross loss of tree cover in tropical Brazil and Indonesia, 2001–2015.*

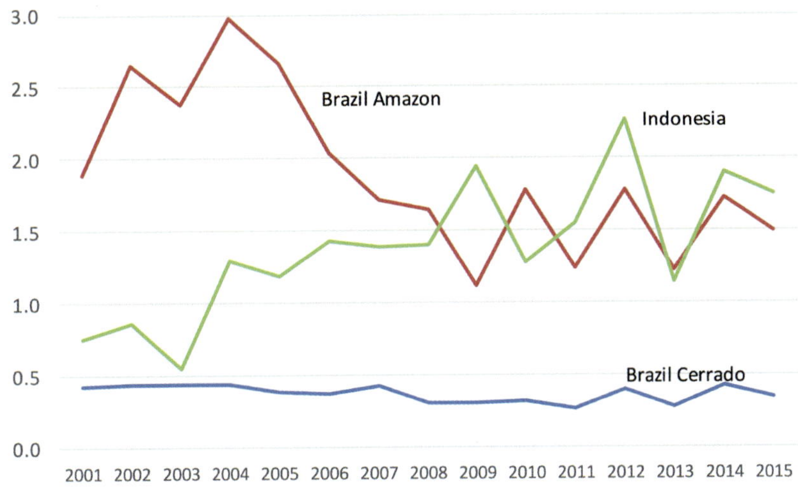

*Same sources as for Figure 1. Loss in the Amazon is an overestimate since it includes loss in the Cerrado as well as in the Amazon biome in the large state of Mato Grosso.

From 1990 to 2012, Brazil added 20 million hectares for crop agriculture and 35 million hectares devoted to pasture and other land uses, while losing 57 million hectares of forest.[9] By far the largest immediate cause of

9 www.faostat.fao.org.

deforestation was the conversion of forests to pasture for cattle, which accounted for over 80 percent of all forest losses (Henders et al., 2015). After 2006, less than 1 percent of new soybean area was sown on newly cleared forest in the Amazon biome (Rudorff et al., 2011). Although soybean area has expanded by another 5 million hectares since 2006, most of this increase came from intensifying production on land that was already cropped and from converting pasture to cropland.

This evidence points to success in slowing deforestation in the Amazon and some progress in stabilizing the clearing of the Cerrado, but in neither case is victory assured. The continued clearing of the Cerrado, although much less controversial than the clearing of the Amazon forests, is a concern due to its high biodiversity; 40 percent of its flora, for example, is endemic (Klink and Machado, 2005). Indeed in 2017, Prince Charles of the U.K. spearheaded a *Cerrado manifesto*, by virtue of which some leading agribusiness and food companies agreed to halt further land clearing.[10]

Long-time observers of deforestation in Brazil have identified at least six major actions that helped to reduce forest losses (Nepstad et al., 2014; Boucher et al., 2013). First, Brazil achieved macro-economic stability in the 1990s that eliminated most of the subsidies on loans to farmers to clear land. Second, the Brazilian government adopted a National Climate Change Plan that specified targets to reduce GHG emissions from the Amazon biome by 80 percent by 2020 and provided the vision and policy framework for reducing deforestation (Nepstad et al., 2014).

Third, civil society campaigned hard against the expansion of cattle and soybean farming into forested areas. A broad coalition of national and international NGOs, stimulated by a 2006 Greenpeace Report, *Eating Up the Amazon*, was instrumental in establishing a moratorium on purchases of soybeans from land deforested in or after 2006. The big five multinational soybean processors and exporters supported this moratorium. A similar moratorium for beef was put into place in 2009. Critically, these moratoriums are enforced by real-time satellite monitoring of forest clearing, a tool pioneered by Brazilian environmental agencies.

Fourth, the new policy framework was backed by greatly increased will and capacity to implement land-use regulations on the ground. The government developed a list of *municipios* (counties) with the highest level of deforestation and withdrew financing and subsidies from agricultural producers in those locations (Nepstad et al., 2014). A new system of property

10 https://corporate.marksandspencer.com/documents/plan-a/statement-of-business-support-for-cerrado-manifesto-2017.pdf.

registration (*Cadastro Ambiental Rural*) and environmental licensing (*Licenciamento Único Ambiental*) linked to satellite monitoring of land use changes was introduced as a prerequisite for receiving financing from banks (Rudorff et al., 2011). These systems enabled the owners of newly cleared areas to be identified immediately and made it possible to evaluate landowners' compliance with requirements for on-farm conservation of natural areas (set at 35 percent of farm area in the Cerrado and 80 percent in the Amazon).

Fifth, payments for environmental services have acted as incentives. At the national level, the Government of Norway under REDD+ committed more than $1 billion to support Brazil's targets for halting deforestation, which was fully paid out in 2015, based on the progress realized. In addition, a government fund for low-carbon agriculture also provided cheap loans for landowners to adopt sustainable practices. This followed the highly successful model in Costa Rica, where a tax on fossil fuel has been used to provide incentives to ranchers to conserve remaining forest areas and to reforest (FAO, 2016).

Finally, indigenous peoples have been granted title to large areas in the Amazon, and land tenure administration has greatly improved. Previously the cheapest route to a secure title to land was to convert forest to pasture (Nepstad et al., 2006). As many as 40 percent of farmers in the Amazon had no land title until a massive titling program was initiated in 2009 (Lapola et al., 2014).

Even so, progress in Brazil has been threatened by a number of factors and deforestation in 2017 doubled from its low in 2014 (although still much less than its high in 2004). The factors behind this rise have still to be fully analyzed but they include fires, weaker political commitment, an economic crisis resulting in budget cuts for enforcement of environmental laws and regulations, and a rise in demand for Brazilian soybeans due to Chinese retaliatory tariffs on U.S. soybeans. The Brazilian elections in November 2018 brought to power a government that has committed to a further relaxation of laws on forest conservation and the rights of indigenous peoples.

In contrast to Brazil (until 2015), the area deforested has continued to increase in Indonesia. Oil palm is undoubtedly a major driver of deforestation in Indonesia but it is one among several drivers such as logging, pulp plantations, and the production of other crops. As in Brazil, the conversion of forested areas to oil palm plantations and other uses has major implications for the richness of biodiversity. Oil palm plantations are not suitable habitats for most forest species (Savilaakso et al., 2014; Fitzherbert et al., 2008); they support half as many vertebrate species as primary forest. The orangutan is the iconic face of the global protest against oil palm, but the problem extends far beyond the plight of a single species. Oil palm plantations, for example, provide suitable habitats for only about 20 percent of the bird species found in tropical forests.

The most visible evidence of the high environmental cost of poor land management is the thick haze that blankets Southeast Asia during many dry seasons. In the El Niño year, 2015, 120,000 fires were reported in Indonesia, causing daily emissions that exceeded the average daily emissions of the entire U.S. economy (The Economist, 2015; Harris et al., 2015). The haze, emanating from uncontrolled burning of forest and peat land in Indonesia, imposes a huge economic and health burden across the entire region (Marlier et al., 2015). In 2015, more than 100,000 deaths from acute respiratory problems were attributed to the fires in Southeast Asia (Koplitz et al., 2016). Economic losses are reported in the tens of billions of dollars and ironically, the haze was so dense in 2015 that it negatively affected yields of oil palm.

While oil palm has undoubtedly been a major boost to the economies of the producing areas in Indonesia, the cost in terms of environmental sustainability has been high. On paper, Indonesia has numerous laws and regulations to promote a sustainable palm oil industry but the implementation of those laws and regulations is highly flawed due to weak capacity, multiple jurisdictions, decentralized governance, and outright corruption (Brockhaus et al., 2012). Many aspects of the governance problem originate with the national land concession program. Two-thirds of all land in Indonesia belongs to the Forest Estate, and 70 percent of the oil palm has been planted on Forest Estate land, mainly through state concessions. However, most land concessions harbor forested areas and in many cases, companies obtain concessions for oil palm but have no intention of planting it, since their major interest is to extract timber (Sheil et al., 2009). Treated as a state secret, the lack of transparency in land concessions further favors non-compliance and rent-seeking. With decentralization of powers of granting concessions to local governments, reports on corruption in local government awards are rife.

Yet recent developments provide some optimism that Indonesia may be turning the corner on deforestation. As in Brazil, the government of Norway under REDD+ negotiated a moratorium by Indonesia on forest clearing in return for payment of up to $1 billion, based on the achievement of agreed performance indicators. However, the agreement has many loopholes, and only about $100 million has been paid out.

At the local level, the governor of Central Kalimantan Province has outlined a specific plan to reduce deforestation by 80 percent. A detailed monitoring system will provide public access to information on plantations to monitor their compliance with agreed sustainability targets. It will also identify and map degraded land that can be used for plantations (Scherr et al., 2015). If successful, these initiatives could provide a model for other provinces and districts for building a sustainable industry.

Major players in the palm oil industry have also committed to voluntary certification under the Roundtable for Sustainable Palm Oil (RSPO), an industry-sponsored scheme. RSPO is one of many voluntary certification schemes that have been set up by industry or by third parties to meet growing consumer demands for environmental sustainability, especially for conservation of tropical forests (such as the Rainforest Alliance Network) and for social justice (such as Fairtrade). Within this growing family of standards, RSPO follows common norms for environmental standards, driven by consumers in rich countries. Today, around 20 percent of traded palm oil is certified although the price premium for certified oil is minimal (less than 3 percent). Since rich countries make up only about 20 percent of world demand for palm oil, the success of RSPO will depend on building demand in middle-income countries. There is some evidence that certification confers modest forest and other environmental benefits (DeFries et al., 2017).

Perhaps the biggest challenge of certification is to extend it to smallholders, who supply over one-third of the market. Certification is costly and there are considerable diseconomies of scale in certification of smallholders as well as in enforcement and traceability. Some regions, notably Sabah State in Malaysia and Central Kalimantan Province in Indonesia, are moving toward "jurisdictional" certification of all palm oil produced in their territory to reduce transaction costs.

Finally, in preparation for the 2015 Paris Climate Conference, Wilmar International and several other leading companies committed to securing their supply chains to comply with "no deforestation, no development on peatlands, and no exploitation of people and local communities" by 2020. Given that Wilmar accounts for a reported 45 percent of traded palm oil, this step is a major advance in enforcing sustainability standards. However, given that Wilmar sources from thousands of suppliers, a much stronger system of certification, monitoring, and traceability will be needed for Wilmar to implement its commitments.

The Emerging Challenge in Africa

The greatest challenge to future land-use expansion in the tropics may lie beyond the big three commodities discussed in this chapter and even beyond global markets. Emerging trends in Africa have been overlooked in the recent global discourse. Already, oil palm companies, many from Asia, are investing in large-scale land acquisition for oil palm in West and Central Africa. Many of these investments threaten tropical forests. One company in Cameroon has had to withdraw after major controversies about deforestation and infringements on land rights, and four major investments in Liberia are stalled (Byerlee et al., 2017). In the savanna areas of Eastern and Southern Africa, such as in Mozambique, there have been a number of investments aimed at large-scale production of soybeans, especially along the new growth corridors.

The greatest threat in Africa may be from production of food crops for rapidly growing domestic markets. The population of Sub-Saharan Africa that reached one billion in 2017 will more than double by 2050. In addition, after decades of negative economic growth per capita, African economies are growing and urbanizing, sometimes quite rapidly, generating demand for meat and other higher value commodities. These domestic market demands, along with investment in infrastructure in the planned growth corridors, will increase commercial pressure on the world's largest remaining tract of intact tropical forests in the Congo basin as well as the world's most extensive areas of uncultivated savannas that are suited to crop and cattle production. The largest threat may be from the emerging class of medium-sized commercial farmers, often urban based, that are quickly responding to the new market opportunities (Jayne et al., 2016). Countries of the region with weak land rights and poor forest governance are ill equipped to withstand these new market pressures in the coming decades. The urgency of economic development and food production in Africa may also require some land expansion. However, with low yields of all food crops and minimal use of fertilizer, the imperative and the potential to intensify are huge.

Prospects for Intensification and Implications for Land Use

At first sight intensification of land use seems an obvious strategy for meeting increased demand, slowing farmland expansion, and saving forests. This is commonly known as "the Borlaug hypothesis" after Norman Borlaug, Nobel laureate and "father of the green revolution," who posited that the green revolution saved about 1 billion hectares of crop land (Borlaug, 2007). More recent analysis using economic models that take account of market responses have placed this figure at about 150 million hectares (Hertel et al., 2014).

However, there is a rich literature that suggests a more nuanced story. Angelsen and Kaimowitz drew attention to "Jevon's paradox," where the availability of more productive technologies for intensifying farming on the frontier provides incentives to farmers to expand agricultural area at the expense of forests and savannahs (Angelsen and Kaimowitz, 2001). One of their most convincing cases is that of soybeans, where new varieties suited to low latitudes, combined with soil improvement technologies, led to rapid yields gains in the Brazilian Cerrado and together with strong global markets for soybeans, promoted the takeoff of soybeans in the Cerrado (Kaimowitz and Smith, 2001; Barretto et al., 2013).

Economists have reconciled these conflicting views in terms of the nature of the market facing producers. If farmers on the frontier are selling into a global market and do not affect world prices, as in the early years of soy expansion in Brazil, Jevon's paradox is likely to prevail. However, for major players in world markets, intensification on the forest frontier reduces world prices and slows land expansion. Also, land saving at a global level may still involve land expansion on the tropical frontier. For example, Villoria et al. modeled the impacts of a closing of the oil palm yield gap in Southeast Asia. This did lead to a small expansion of oil palm in Indonesia at the expense of forests but due to lower palm oil prices, area under oil crops in other parts of the world declined even more (Villoria et al., 2013). Although Villoria et al. (2013) showed a net reduction of greenhouse gas emissions by raising oil palm yields they were not able to evaluate net impacts on biodiversity, which could be negative due to the high biodiversity of tropical forests.

Finally, intensification must consider reverse causality—that is, the availability of cheap land to convert to agricultural uses acts as an incentive to "extensify" rather than intensify. In Brazil, there is evidence that as the transaction costs of clearing new land rises due to better monitoring and enforcement of conservation requirements and moratoriums on buying harvests from recently deforested land, farmers have rapidly intensified beef cattle through improved pastures and herd improvement and management, and intensified soybeans through adoption of double cropping (mostly soybeans with maize) (Figure 5 and Figure 8). There is also evidence that these innovations have contributed to substantial land saving (Lapola et al., 2014; Cohn et al., 2014; Vale 2014). This intensification was partly the result of better governance of forest resources, which reduced the effective supply of land and provided the incentive to intensify (Merry and Soares-Filho, 2017).

The potential for further intensification varies substantially among the three commodities and by region. In the big producers and exporters of soybeans, yields are already around 3t/ha and the yield gap is the lowest of the world's major crops (Fischer et al., 2014). Modest genetic gains in yield potential should contribute to an overall yield growth of 0.5 percent per year (Fischer et al., 2014).

FIGURE 8. Intensification of beef production in Brazil, especially since 1990.

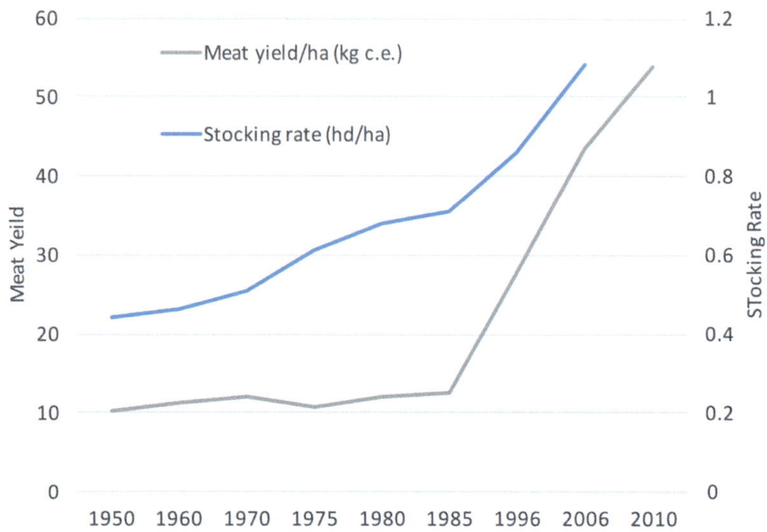

*Stocking rates updated from FAOSTAT, www.faostat.fao.org.
Source: Martha et al., 2012.*

Yield gaps in oil palm grown on large plantations are also fairly modest and estimated at about 30 percent relative to the top-yielding plantations (Fischer et al., 2014; Woittiez et al., 2017). Part of this yield gap reflects labor scarcity and rising wages that have reduced the frequency of harvesting. Progress in mechanical or robotic harvesting could significantly reduce this part of the yield gap. In Indonesia, the yield gap is largest for independent smallholders, averaging around 50 percent relative to attainable yields, with suboptimal use of fertilizer as the major contributor (Euler et al., 2016). A major program of replanting with new varieties and extension advice to improve management will be needed to reduce this gap. Genetic gains in yield potential of oil palm have averaged around 1.0 percent annually and this should continue or even accelerate with new genomic breeding and tissue culture methods. However, oil palm trees are replaced only once every twenty-five years, and translation of genetic advances to the field is slow. Still, overall, growth in palm oil yields of 1.0 percent annually should be quite feasible to 2050.

Although substantial gains have been made in raising productivity of beef cattle production in Brazil, observers agree that there is much potential for further intensification given that productivity is still only about one third of potential (Latawiec et al., 2014). Another measure of this potential is the fact that the average stocking rate per hectare is only a bit over half of that achieved by the top 5 percent of producers (Dias et al., 2016). Most ranchers now sow improved pastures (mostly with grasses of the genus *Brachiaria*), but they are soon degraded by prevailing management practices. Further gains in beef productivity per head through herd management and genetic upgrading are also possible. Policies such as enforcing regulations on conservation, incentives tied to credit to intensify, payments for environmental services, and improved access to technical advice are needed to continue intensification in the beef industry.

TABLE 1. Summary of projections on intensification and area as components of future growth in major commodities.*

	Area expansion, 1991-2015 (%/y)	Yield or stocking rate gap, 2015	Projections to 2050	
			Genetic gain per ha or head	Needed area expansion (%/y)
Soybeans (tropical)	5.0	Low	Med (0.7%/y)	1%
Oil palm	4.7	Med for smallholders	Med (1.0%/y)	1.5%
Beef in Brazil	0.2	High	Med to high (2.5%?)	- ve (Cerrado) + ve (Amazon)

*Soybeans and oil palm expanded and updated from Byerlee et al. (2017). Beef projections constructed by the author. The decline in pasture area to 2050 in the Cerrado is due to intensification of beef and conversion of pastures to soybeans.

Given projections of demand and likely gains in intensification, Table 1 provides a summary of the needed expansion in land area to meet demand. For soybeans, the needed area expansion is very modest compared to the recent past and with large areas of degraded pasture land, this rate of expansion can be achieved through conversion of degraded pastures to soybeans and intensifying beef cattle on remaining pasturelands. This still leaves two major concerns in the pasture-soybean nexus. First, is the continued conversion of bio-diverse savannas on the eastern edge of the Cerrado which is now coming under the spotlight of Brazilian and international environmental groups. Second, is the need to intensify beef production broadly in all countries in the Amazon biome to prevent "leakages" to other countries. This is especially critical since South America is projected to become the major source of beef in world markets in the coming decades (IFPRI, 2017).

For oil palm, a projected growth rate of 1.5 percent annually in area is a sharp slowdown from the recent past but this still poses a challenge in terms of reaching zero deforestation targets in Southeast Asia. Observers agree that there is more than sufficient degraded land available to meet this challenge but spatial planning of land use is difficult under the complex and often overlapping land laws in Indonesia that results in poorly defined land tenure. Recent initiatives in the provinces of Central Kalimantan in Indonesia and the state of Sabah in Malaysia to introduce territorial jurisdictions on certification with strong state support may show the way.

The Way Forward—Multi-Dimensional Approaches

Apart from Africa, the slowdown in demand for agricultural commodities will make it easier to manage future agricultural expansion more sustainably, especially if recent trends in intensification are maintained. Yet even this more benign market environment will not guarantee success in reaching the SDG of zero deforestation without a multi-dimensional approach that involves public, private, and local actors (McCarthy, 2012; Lambin et al., 2014). In Figure 10, the essential elements of such an approach are delineated between market-based approaches and those initiated by states and communities.

Given inherent weaknesses of the state in much of the tropics, international organizations and civil society have placed much faith in market-based approaches. On the supply side, these are largely centered around payments for environmental services (PES) and various types of offset programs such as cap and trade carbon programs and biodiversity offsets. Costa Rica, the only tropical country that has reversed deforestation, has pioneered PES to provide incentives to landowners to conserve remaining forests and to reforest or convert to sustainable agro-sylvi-culture systems. There is growing evidence that these programs are increasing biodiversity at the landscape level, while benefiting poor farmers (Ibrahim et al., 2010). Several sources of financing have been employed, but the use of a tax on fossil fuels to finance PES has been a bold step toward establishing sustainable markets.

The Reduced Degradation and Deforestation program (REDD+) was conceived by the United Nations Framework Convention on Climate Change with much fanfare in 2007 as a global mechanism to provide compensation for environmental services, especially carbon sequestration and conservation of biodiversity.[11] Over time, the emphasis has shifted from the

11 The "+" was added to REDD to include activities aimed at conservation of forest carbon stocks, sustainable management of forests, and enhancement of forest carbon stocks.

development of private markets for carbon toward performance-based payments by governments to countries that commit to reduce deforestation. The large anticipated flow of international funding for REDD+ has not materialized, but the resources available appear to have heightened national commitments and helped build capacity to improve the governance of forest resources (Seymour et al., 2015).

FIGURE 9. Multidimensional approach to sustainable agricultural expansion on the frontier

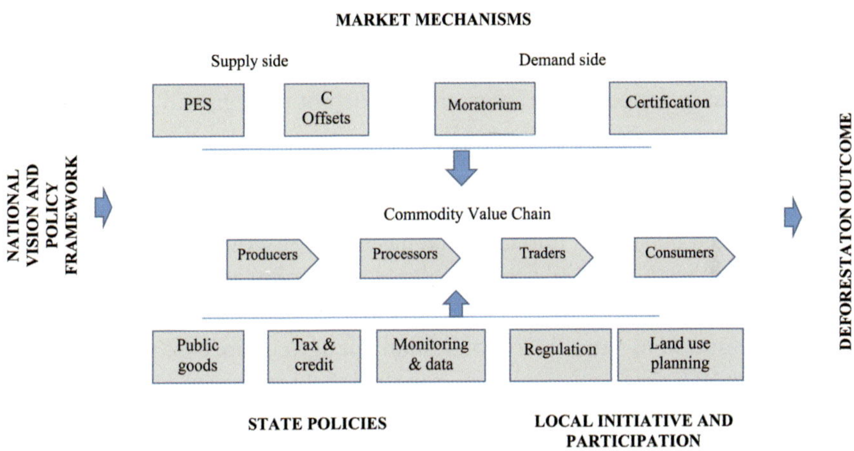

On the demand side, voluntary certification using industry or third-party mechanisms has been increasing but still has limited coverage. Some 40 percent of coffee is now certified and about 20 percent of palm oil, but very little of other commodities associated with deforestation, including rubber, beef, and soybeans. The major challenge for scaling up these programs is the limited demand for certified products and hence a very modest price premium. Already, a significant share of certified products are sold into the market for uncertified products, thus eliminating the premium. It seems that the role of certification will only expand gradually as consumers, especially those in emerging economies, are willing to pay premium prices for sustainable harvests. However, the imposition of moratoriums by large traders on buying produce from recently deforested areas may give a new energy to certification programs as a way for traders to most effectively implement their commitments.

Nonetheless, it is now clear that market-based approaches alone cannot work without better governance of forest and land resources. A clear policy framework that cuts across competing interests in agriculture, land use, food security, and forests is a starting point in sustainable agricultural expansion. Within this framework there is no substitute for well-formulated laws and transparent mechanisms to clarify and implement tenure security for land and forest resources for local communities (FAO, 2016). Likewise the state has important roles in investing in public goods, especially in research to stimulate intensification and build information systems on natural resources and their tenures. Modern information technologies based on satellite imagery of land use changes and mapping of the ownership and use rights of forest and farm lands on the ground with geographic referencing systems offer cheap, powerful, and transparent ways to improve the implementation of forest and land laws on the ground. It is not a coincidence that Costa Rica, a country that has successfully "regreened" much of its degraded areas, has the best land and forest governance in the tropics.

Given that agricultural expansion in the tropics is inevitable and indeed desirable for economic development and poverty reduction, an obvious priority is to avoid expansion into land that has high conservation value, high carbon stocks, or that provides other valuable environmental services. The so-called Bonn Challenge of 2011 calls for 150 million hectares of degraded land to be restored to natural areas or sustainable agriculture by 2020. Landscape approaches are increasingly being applied that look beyond individual land parcels to consider the whole landscape in ways that enhance environmental services such as biodiversity and hydrological functions. By integrating sustainability into economic opportunities, participatory landscape approaches ostensibly put people at center stage (Koh et al., 2009; Sayer et al., 2014).

Although the global community has raised awareness of the threats to sustainability and offers incentives to reward more sustainable action, it will be "the people who live in the tropics who will forge the future fate of tropical forests" (Putz and Romero, 2014). Success on the ground therefore critically depends on involvement and commitment of local people. Ultimately, success will depend on making forest conservation politically attractive by recognizing the benefits of conservation (Nepstad, 2018).

Zero deforestation is a worthy, achievable target that should not be an impediment to expanding agriculture. The environmental and social sustainability stakes are high, and incentive systems will have to be coupled with long-term efforts to resolve the deep, underlying problems of weak land and forest governance. Here, the state is the major player and even if governance improves in one major producing country, production may well shift to another, unless governments and multinational companies commit to coordinated international action. The reassuring news is that more progress has been made in the past five years than in the previous twenty-five in managing agricultural expansion for world markets in a more sustainable way.

References

Angelsen, A., and Kaimowitz, D. (2001) *Agricultural technologies and tropical deforestation*. Wallingford: CABI.

Barretto, A. G. O. P., Berndes, G., Sparovek, G., and Wirsenius, S. (2013). Agricultural intensification in Brazil and its effects on land-use patterns: an analysis of the 1975–2006 period. *Global Change Biology, 19*(6), 1804–1815.

Beckman, J., Sands, R. D., Riddle, A. A., Lee, T., and Walloga, J. M. (2017). International trade and deforestation: potential policy effects via a global economic model. United States Department of Agriculture, Economic Research Service.

Binswanger, H. P. (1991). Brazilian policies that encourage deforestation in the Amazon. *World Development, 197*, 821–829.

Boucher, D., Roquemore, S., and Fitzhugh, E. (2013). Brazil's success in reducing deforestation. *Tropical Conservation Science*, 6(3), 426–445.

Brockhaus, M., Obidzinski, K., Dermawan, A., Laumonier, Y., and Luttrell, C. (2012). An overview of forest and land allocation policies in Indonesia: Is the current framework sufficient to meet the needs of REDD+? *Forest Policy and Economics, 18*, 30–37.

Bruno, L. C. (2014). *Essays on technological progress in East Asia*. Dissertation. Oslo: Norwegian School of Economics.

Busch, J., Ferretti-Gallon, K., Engelmann, J., Wright, M., Austin, K. G., Stolle, F., Turubanova, S., Potapov, P. V., Margono, B., Hansen, M. C., and Baccini, A. (2015). Reductions in emissions from deforestation from Indonesia's moratorium on new oil palm, timber, and logging concessions. *Proceedings of the National Academy of Sciences of the United States of America, 112*(5), 1328–1333.

Byerlee, D., and Rueda, X. (2015). From public to private standards for tropical commodities: a century of global discourse on land governance on the forest frontier. *Forests*, 6(4), 1301–1324.

Byerlee, D., Falcon, W. P., and Naylor, R. L. (2017). *The tropical oil crop revolution: food, feed, fuel, and forests*. Oxford University Press.

Cohn, A. S., Mosnier, A., Havlík, P., Valin, H., Herrero, M., Schmid, E., O'Haref, M., and Obersteiner, M. (2014). Cattle ranching intensification in Brazil can reduce global greenhouse gas emissions by sparing land from deforestation. *Proceedings of the National Academy of Sciences of the United States of America, 111*(20), 7236–7241.

DeFries, R. S., Fanzo, J., Mondal, P., Remans, R., and Wood, S. A. (2017). Is voluntary certification of tropical agricultural commodities achieving sustainability goals for small-scale producers? A review of the evidence. *Environmental Research Letters, 12*(3), art. no. 033001.

Dias, L. C., Pimenta, F. M., Santos, A. B., Costa, M. H., and Ladle, R. J. (2016). Patterns of land use, extensification, and intensification of Brazilian agriculture. *Global Change Biology, 22*(8), 2887–2903.

Eisner, T., Eisner, H., Meinwald, J., Sagan, C., Walcott, C., Mayr, E., Wilson, E. O., Raven, P. H., Ehrlich, A., Ehrlich, P. R., Carr, A., Odum, E. P., and Gans, C. (1981). Conservation of tropical forests. *Science, 213*(4514), 1314.

Euler, M., Hoffmann, M. P., Fathoni, Z., and Schwarze, S. (2016). Exploring yield gaps in smallholder oil palm production systems in eastern Sumatra, Indonesia. *Agricultural Systems, 146*, 111–119.

Fischer, R. A., Byerlee, D., and Edmeades, G. (2014). *Crop yields and global food security: will yield increase continue to feed the world?* Canberra: Australian Centre for International Agricultural Research.

Fitzherbert, E. B., Struebig, M. J., Morel, A., Danielsen, F., Brühl, C. A., Donald, P. F., and Phalan, B. (2008). How will oil palm expansion affect biodiversity? *Trends in Ecology and Evolution, 23*(10), 538–545.

Food and Agriculture Organization of the UN (2016). *State of the World's Forests, 2016*. Rome: FAO.

Gale, F. (2015). Development of China's feed industry and demand for imported commodities. US Dep. Agric. FDS-15K-01. Retrieved from https://www.ers.usda.gov/webdocs/publications/36929/55490_fds-15k-01.pdf.

Gibbs, H. K., Ruesch, A. S., Achard, F., Clayton, M. K., Holmgren, P., Ramankutty, N., and Foley, J. A. (2010). Tropical forests were the primary sources of new agricultural land in the 1980s and 1990s. *Proceedings of the National Academy of Sciences, 107*(38), 16732–16737.

Graesser, J., Aide, T. M., Grau, H. R., and Ramankutty, N. (2015). Cropland/pastureland dynamics and the slowdown of deforestation in Latin America. *Environmental Research Letters, 10*(3).

Gunarso, P., Hartoyo, M. E., Agus, F., and Killeen, T. J. (2013). Oil palm and land-use change in Indonesia, Malaysia and Papua New Guinea. *Reports from the Technical Panels of the 2nd Greenhouse Gas Working Group Roundtable on Sustainable Palm Oil, Kuala Lumpur, Malaysia, 2013*. Retrieved from http://www.tropenbos.org/resources/publications/oil+palm+and+land+use+change+in+indonesia,+malaysia+and+papua+new+guinea.

Harris, N., Minnemeyer, S., Stolle, F., and Payne, O. A. (2015). Indonesia's fire outbreaks produce more emissions than entire U.S. economy. World Resources Institute. Retrieved from http://www.wri.org/blog/2015/10/indonesia%E2%80%99s-fire-outbreaks-producing-more-daily-emissions-entire-us-economy.

Henders, S., Persson, U. M., and Kastner, T. (2015). Trading forests: land-use change and carbon emissions embodied in production and exports of forest-risk commodities. *Environmental Research Letters, 10*(12),

Hertel, T. W., Ramankutty, N., and Baldos, U. L. C. (2014). Global market integration increases likelihood that a future African Green Revolution could increase crop land use and CO2 emissions. *Proceedings of the National Academy of Sciences, 111*(38), 13799–13804.

Ibrahim, M., Casasola, F., Villanueva, C., Murgueitio, E., Ramírez, E., Sáenz, J., and Sepúlveda, C. (2011). Payment for environmental services as a tool to encourage the adoption of silvo-pastoral systems and restoration of agricultural landscapes dominated by cattle in Latin America. In Montagnini, F., and Finney, C. (eds.), *Restoring degraded landscapes with native species in Latin America*, pp. 197–219. New York: Nova Science Publishers.

Indonesia's forest fire haze. (2015, November 6). *The Economist*. Retrieved from http://www.economist.com/blogs/graphicdetail/2015/11/daily-chart-3

International Food Policy Research Institute (IFPRI) (2017). IMPACT projections of food production, consumption, and net trade to 2050, with and without climate change: extended country-level results for 2017 GFPR annex table 7. Harvard Dataverse, V3. doi: 10.7910/DVN/8GYEHI.

Jayne, T. S., Chamberlin J., Traub, L., Sitko, N., Muyanga, M., Yeboah, F. K., Anseeuw, W., Chapoto, A., Wineman, A., Nkonde, C., and Kachule, R. (2016). Africa's changing farm size distribution patterns: the rise of medium-scale farms. *Agricultural Economics, 47*(S1), 197–214.

Kaimowitz, D., and Smith, J. (2001). Soybean technology and the loss of natural vegetation in Brazil and Bolivia. In Angelsen, A., and Kaimowitz, D. (eds.), *Agricultural technologies and tropical deforestation*, pp. 195–211. Wallingford: CABI.

Klink, C. A., and Machado, R. B. (2005). Conservation of the Brazilian Cerrado. *Conservation Biology, 19*(3), 707–713.

Koh, L. P., Levang, P., and Ghazoul, J. (2009). Designer landscapes for sustainable biofuels. *Trends in Ecology & Evolution, 24*(8), 431–438.

Koplitz, S. N., Mickley, L. J., Marlier, M. E., Buonocore, J. J., Kim, P. S., Liu, T., Sulprizio, M. P., DeFries, R. S., Jacob, D. J., Schwartz, J., Pongsiri, M., and Myers, S. S. (2016). Public health impacts of the severe haze in Equatorial Asia in September–October 2015: demonstration of a new framework for informing fire management strategies to reduce downwind smoke exposure. *Environmental Research Letters, 11*(9), art. no. 094023.

Lapola, D. M., Martinelli, L. A., Peres, C. A., Ometto, J. P. H. B., Ferreira, M. E., Nobre, C. A., Aguiar, A. P. D., Bustamante, M. M. C., Cardoso, M. F., Costa, M. H., Joly, C. A., Leite, C. C., Moutinho, P., Sampaio, G., Strassburg, B. B. N., and Vieira, I. C. G. (2014). Pervasive transition of the Brazilian land-use system. *Nature Climate Change*, *4*(1), 27–35.

Latawiec, A. E., Strassburg, B. B. N., Valentim, J. F., Ramos, F., and Alves-Pinto, H. N. (2014). Intensification of cattle ranching production systems: socioeconomic and environmental synergies and risks in Brazil. *Animal*, *8*(8), 1255–1263.

Marlier, M. E., DeFries, R. S., Kim, P. S., Koplitz, S. N., Jacob, D. J., Mickley, L. J., and Myers, S. S. (2015). Fire emissions and regional air quality impacts from fires in oil palm, timber, and logging concessions in Indonesia. *Environmental Research Letters*, *10*(8), art. no. 085005.

Martha Jr., G. B., Alves, E., and Contini, E. (2012). Land-saving approaches and beef production growth in Brazil. *Agricultural Systems*, *110*, 173–177.

Merry, F., and Soares-Filho B. (2017). Will intensification of beef production deliver conservation outcomes in the Brazilian Amazon? *Elementa Science of the Anthropocene 5*, 24.

Milman, O., and Leavenworth, S. (2016, June 20). China's plan to cut meat consumption by 50% cheered climate campaigners. *The Guardian*. Retrieved from https://www.theguardian.com/world/2016/jun/20/chinas-meat-consumption-climate-change.

Nepstad, D. C., Stickler, C. M., and Almeida, O. T. (2006). Globalization of the Amazon soy and beef industries: opportunities for conservation. *Conservation Biology*, *20*(6), 1595–1603.

Nepstad, D., McGrath, D., Stickler, C., Alencar, A., Azevedo, A., Swette, B., Bezerra, T., DiGiano, M., Shimada, J., Da Motta, R. S., Armijo, E., Castello, L., Brando, P., Hansen, M. C., McGrath-Horn, M., Carvalho, O., and Hess, L. (2014). Slowing Amazon deforestation through public policy and interventions in beef and soy supply chains. *Science*, *344*(6188), 1118–1123.

Nepstad, D. (2018, March 8). Tropical deforestation: the need for a strategy adjustment. Mongabay.com. Retrieved from https://news.mongabay.com/2018/03/tropical-deforestation-the-need-for-a-strategy-adjustment-commentary/.

Putz, F. E., and Romero, C. (2014). Futures of tropical forests (Sensu Lato). *Biotropica*, *46*(4), 495–505.

Qiang, W., Liu, A., Cheng, S., Kastner, T., and Xie, G. (2013). Agricultural trade and virtual land use: The case of China's crop trade. *Land Use Policy*, *33*, 41–15. I have updated the data in Quiang from 2012 to 2017.

Richards, P. D., Walker, R. T., and Arima, E. Y. (2014). Spatially complex land change: the indirect effect of Brazil's agricultural sector on land use in Amazonia. *Global Environmental Change*, *29*, 1–9.

Rudorff, B. F. T., Adami, M., Aguiar, D. A., Moreira, M. A., Mello, M. P., Fabiani, L., Amaral, D. F., and Pires, B. M. (2011). The soy moratorium in the Amazon biome monitored by remote sensing images. *Remote Sensing*, *3*(1), 185–202.

Savilaakso, S., Garcia, C., Garcia-Ulloa, J., Ghazoul, J., Groom, M., Guariguata, M. R., Laumonier, Y., Nasi, R., Petrokofsky, G., Snaddon, J., and Zrust, M. (2014). Systematic review of effects on biodiversity from oil palm production. *Environmental Evidence*, *3*(1), art. no. 4.

Sayer, J. A., Margules, C., Boedhihartono, A. K., Dale, A., Sunderland, T., Supriatna, J., Saryanthi, R. (2014). Landscape approaches: What are the pre-conditions for success? *Sustainability Science*, *10*(2), 345–355.

Scherr, S. J., Mankad, K., Jaffee, S. M., and Negra, C. (2015). *Steps toward green: policy responses to the environmental footprint of commodity agriculture in East and Southeast Asia*. Washington, D.C.: EcoAgriculture Partners and the World Bank.

Schierhorn, F., Meyfroidt, P., Kastner, T., Kuemmerle, T., Prishchepov, A. V., and Müller, D. (2016). The dynamics of beef trade between Brazil and Russia and their environmental implications. *Global Food Security*, *11*, 84–92.

Seymour, F., Birdsall, N., and Savedoff, W. (2015). *The Indonesia-Norway REDD+ Agreement: a glass half-full*. Policy Paper 56. Washington, D.C.: Center for Global Development.

Seymour, F., and Busch, J. (2016). *Why forests? Why now? The science, economics, and politics of tropical forests and climate change*. Washington, D.C.: Brookings Institution Press.

Shane, D. R. (1986). *Hoofprints in the forest: cattle ranching and the destruction of Latin America's tropical forests.* Philadelphia, PA: Institute for the Study of Human Issues.

Shapiro, P. (2018). *Clean meat: how growing meat without animals will revolutionize dinner and the world.* New York: Gallery Books.

Sheil, D., Casson, A., Meijaard, E., van Noordwijk, M., Gaskell, J., Sunderland-Groves, J., Wertz, K., and Kanninen, M. (2009). *The impacts and opportunities of oil palm in Southeast Asia: what do we know and what do we need to know?* Bogor: Center for International Forestry Research.

Vale, P. M. (2014). The conservation versus production trade-off: does livestock intensification increase deforestation? Evidence from the Brazilian Amazon [Working paper]. London: Gratham Research Institute on Climate Change and the Environment. Retrieved from: http://www.lse.ac.uk/GranthamInstitute/publication/the-conservation-versus-production-trade-off-does-livestock-intensification-increase-deforestation-evidence-from-the-brazilian-amazon/.

Villoria, N. B., Golub, A., Byerlee, D., and Stevenson, J. (2013). Will yield improvements on the forest frontier reduce greenhouse gas emissions? A global analysis of oil palm. *American Journal of Agricultural Economics, 95*(5), 1301–1308.

Williams, M. (2003). *Deforesting the earth: from prehistory to global crisis.* Chicago: University of Chicago Press.

Woittiez, L. S., van Wijk, M. T., Slingerland, M., van Noordwijk, M., and Giller, K. E. (2017). Yield gaps in oil palm: A quantitative review of contributing factors. *European Journal of Agronomy, 83*, 57–77.

World Resources Institute (2017). Conserving forests could cut carbon emissions as much as getting rid of every car on earth. Retrieved from http://www.wri.org/blog/2017/11/conserving-forests-could-cut-carbon-emissions-much-getting-rid-every-car-earth.

ADVANCES IN THE POTENTIAL YIELD OF GRAIN CROPS

Tony Fischer
CSIRO Agriculture and Food, Canberra, Australia

Introduction

Progress in food crop yield (herein called farm yield, FY) since 1960 has been central to improved global food security, lower real grain prices, and to sparing new land from the plow. Continuing, or preferably accelerating, yield progress remains a vital goal for agricultural science for the same reasons, as has been urged by many (e.g., Godfray et al., 2010; Fischer et al., 2014; Fischer and Connor, 2018). In the 1960s yield of the main grain crops grew at 3 percent per annum[1] but since then it has gradually slowed, as seen in values for the period 1986 to 2001 in Table 1. However, in the following fifteen-year period, up to 2016 being the latest date for statistics, there has been a highly significant ($P<0.001$) upturn in growth of world crop production (to around 2.3 percent per annum). This is dominated, unlike in the previous forty years or so, by a significant increase in crop area growth, but also accompanied by an upturn in yield growth in wheat and some key crops not shown separately in Table 1 (e.g., pulses and rapeseed). This may be explained by the low level of grain stocks around 2000–2008 and the jump in world real grain prices in 2008 and again 2010–2011, prices which have now eased to around 50 percent above the record lows of 1995–2005 (World Bank, 2017). Whatever the explanation, prices are still high for poor consumers and crop area increase is undesirable for environmental reasons to the extent that it is achieved from plowing new land especially in Sub-Saharan Africa and South America. Therefore, crop yield increase continues to be a primary goal in the food security equation.

1 Most measures of change in the food security debate should be expressed in relative terms, because that is how demand and supply are usually expressed.

TABLE 1. Rate of increase in world production, area and yield averaged for two recent 15 year periods (1986–2001 and 2002–2016) for four staple grain crops and all staple food crops, and average yields 2014 to 2016.

Crop	Average rates of growth (% p.a.) over 15 year periods						Mean yield 2014-16 t/ha[b]
	1986-2001			2002-2016			
	Production	Area	Yield	Production	Area	Yield	
Wheat	0.90	-0.30	1.15	1.63	0.26	1.42	3.35
Rice	1.57	0.45	1.19	1.76	0.65	1.20	4.60
Maize	2.08	0.45	1.72	3.05	2.00	1.40	5.60
Soybean	3.26	2.13	1.54	3.12	2.35	1.12	2.68
All crops	1.53	0.25	1.31	2.35	1.12	1.43	3.46

[a]All slopes shown are significantly positive at $P < 0.01$.

[b]Yields throughout this paper are metric tonnes of grain at standard field moisture content per hectare.

[c]This includes the four main staples plus barley, sorghum, all pulses, rapeseed, oil palm, sugar cane and all roots and tubers from FAOSTAT; production and yields converted to wheat energy equivalents, total harvested area in 2014–16 was 1005 Mha.

Source: from FAOSTAT Crops, accessed 7 April 2018 (FAOSTSAT 2018); linear regression against year, slope expressed relative to average values in 1999–2001 or 2014–2016, respectively.

The other side to the food challenge is to note what is happening to global food demand. It was estimated to be growing 1.6 percent per annum in 2012, comprising 1.20 percent population growth, with the remainder being demand per capita growth (Fischer and Connor, 2018). This is somewhat more than crop yield growth, with the area growth making up the supply difference as mentioned. The important point is that world population growth is expected to decline to 0.74 percent by 2035 (UN, 2017), a number about which there can be some confidence. In addition, although with less certainty, per capita consumption growth rate is unlikely to increase further.[2] Thus food demand growth rate should be at least be one quarter less by 2035 and further declining thereafter. Meeting food demand is today's imperative, and this chapter will place emphasis on what is happening to yield now and likely to happen in the near future (twenty years) rather than what may

2 Another demand driver over the last fifteen years also expected to decline is biofuel from grain, vegetable oil, and sugar, which had risen rapidly to use about 7 percent of total food production by 2012 (Fischer and Connor, 2018).

happen in the more distant future. As argued recently Fischer and Connor (2018) the world's food supply problem must be solved in the next twenty years or the consequences are likely to be dire.[3]

FY reflects two components: crop yield with latest available technologies (cultivars and agronomic management, called potential yield, PY), and the gap between this and the FY based on technologies that farmers have actually adopted. FY progress therefore depends on two steps, namely progress in PY plus closing the PY to FY gap, the latter requiring farmers to adopt new technologies faster than they are invented (see Figure 1). Because of lagging FY growth, yield gap determination and gap-closing has recently received much attention (e.g., van Ittersum et al., 2013; Fischer and Connor, 2018; Fischer, 2019). Yield gaps will not, however, be discussed further here, where the focus is on potential yield prospects. It suffices to point out that gap-closing has a huge potential for lifting FY in high-yield gap regions such as Sub-Saharan Africa, West Asia, North Africa, and parts of South Asia (gaps >100 percent or often > 200 percent as a percent of FY), where further PY increase is thus not the highest priority. Where yield gaps are smaller (< 100 percent of FY), which is much of the rest of the world and at least three-quarters of the world's food production (Fischer and Connor, 2018), PY increase has a more important role, and in fact an essential one where yield gaps approach 30 percent, a generally accepted economic limit for farmers facing world prices and exhibiting normal risk aversion, as they do in much of the New World and Europe. This chapter will explore PY progress, focusing first on carefully defining and understanding PY growth measurement, with some examples, before passing to recent estimates of current rates of PY progress across crops and regions of the world, based wherever possible on progress only over the last twenty to thirty years for greatest relevance to the near future. Finally, prospects for lifting, or at least maintaining PY progress, through employing conventional and the most modern tools of crop research will be explored. This is of special interest not only for the reasons given above, but also because some researchers are claiming that PY progress is ceasing as biological limits are being approached. It is concluded that although limits to progress are not in sight, current rates of PY progress, ranging from 0.4 to 1.0 percent per annum, are unlikely to increase.

3 While climate change will be happening over the next twenty years, rates of change are low and although negative for yield in many situations, there are also positive effects, especially when the benefit of CO2 increase is considered for C3 crops. This paper doesn't explicitly consider climate change. I argue that research conducted every season in the field to advance yield is automatically facing this challenge, and that the scope for pre-emptive research on crop adaptation is limited.

Some Definitions and Clarifications

We have come some distance since Evans (1993) established a foundation for understanding progress in crop yields, but his definitions and approach have stood the test of time. Potential yield (PY) is defined as the yield obtained with the best available cultivar, grown under the best management (free of manageable abiotic and biotic stresses) and experiencing the natural resources (climate, soil) representative of the region of interest (Evans and Fischer, 1999; van Ittersum et al., 2013; Fischer, 2015).[4] It is also useful to define a water-limited potential yield (PYw) applied to rainfed crops and clearly only makes sense when water supply is inadequate to meet the unstressed crop demand (< say 80 percent of this). PYw needs always to be related to the actual amount of water, usually the crop evapotranspiration (which is limited to available soil water at sowing plus rainfall to maturity, but may be less if all the available soil water is not used). PY can only be measured in field experiments adequately sampling the region of interest over several years from which results from unusual weather damage (e.g., lodging events or hail, etc.) are excluded. Such experiments are the only way to calibrate crop simulation models, which in turn can be used to extend PY estimations in time and space but with caution (Grassini et al., 2015a), and to explore the sensitivity of PY to some genetic traits and agronomic management factors. The very high yields in various crop competitions and theoretical yields, calculated using unrealized optimistic physiological relationships, cannot claim to be reliable estimations of PY, which they always exceed, although they contain useful lessons.

Field trials, therefore, are essential for estimating PY progress, and this deserves some attention here because many published progress estimates are inadequate for my purposes. Several approaches are common (Fischer, 2015):

1. Vintage trials. In these trials, cultivars representing different dates of release are grown side-by-side under the best agronomy soon after the most recent cultivar is released; PY is measured and regressed against year of release. The cultivars chosen should have been commercially successful in past years.
2. Breeders trials or better, independently managed, trials designed to test potential new cultivars along with several recurrent check

4 This is not to be confused with potential yield as used by some maize scientists when referring to yield per plant under low planting density. The tillering and branching plasticity of most crops makes this a measure of little relevance to plants in crops at normal densities.

cultivars. The yields of new cultivars are expressed relative to the check cultivar yield(s), and these relative yields are plotted against year of release. A variant of this expresses the mean yield of the top four or so entries of such trials (which may have thirty or more new entries each year) as a percentage of the recurrent check yield(s). This is only satisfactory if the numbers of entries in trials and the error levels don't change over time.

3. Regional and national variety trials in which there is considerable turnover of entries and even checks, but many entries must be present for at least two years. Mixed model linear regression can handle such data in which as little as 10 percent of the cultivar x year x location matrix might actually exist (e.g., Mackay et al., 2011; Piepho et al., 2014; Laidig et al., 2017). Cultivar yields can be estimated and then regressed against year of release, or a vintage coefficient can be estimated from the full model. A coefficient for the year of growing can also be estimated to reflect effect of time trends in agronomy and in climate on yield.

For unbiased estimates of PY progress several rules must be obeyed. Firstly, disease levels must either be minimal (e.g., due to the environment or the use of fungicides) or yields corrected for disease scores. In earlier variety releases breeders often paid less attention to disease resistance and in many crops recently released cultivars, almost always with good disease resistance at the outset, quickly succumb to new races of diseases and lose yield accordingly (e.g., Mackay et al., 2011; Piepho et al., 2014). If either of these circumstances applies in any of the above methods, progress in PY, as defined here, will be overestimated (see later). Resistance to soilborne diseases, which cannot readily be observed, introduces a greater problem by requiring testing in fields free of soil diseases (e.g., via suitable crop rotation). Secondly, it is essential that reasonably large field plots, preferably permitting edge trimming, be used. There is growing evidence that the yield of just the edges of plots not only confounds the true plot yield, but can change with breeding for yield in the opposite direction to true yield. Sadras and Lawson (2011) and Sadras et al. (2016) showed in vintage trials with wheat (*Triticum* ssp.) and chickpea (*Cicer arietinum*), respectively, that edge row yield was negatively correlated with true inner plot yield. Edge trimming is therefore essential, and a change in plot design in the course of any time series trials can obviously upset measurements of the trend in true yield. Third, calculation of relative progress as the linear slope relative to the mean, or worse still, the initial yield level in the time series is misleading. To be relevant to the future, progress should be calculated relative to the predicted yield level for the latest year of release, remembering that PY progress is rarely exponential, notwithstanding the models of economists. Also for future relevance to PY progress, vintage series should be restricted to cultivars released in the last thirty years or so.

It turns out that PY progress is easier to estimate than PY itself because a vintage series of cultivars rarely shows relative progress changing much with changes in the environment or in agronomic management. This is also the reason why the relative impact of new cultivars on FY is the same as that measured in PY trials (e.g., Evans and Fischer, 1999). Silvey (1981) had earlier pioneered this approach to estimate the effect of breeding on FY progress by calculating the expected effect of cultivar change on FY in the UK, where information on the cultivars grown by farmers each year is readily available. Combining this with the relative PY of each cultivar and weighting for cultivar area, permits estimation of how much of any relative change in FY is due to cultivar.

Finally, we come to agronomic management and weather effects on trials used to estimate PY progress. This has been discussed at length in Fischer (2015) using the greater response of modern cultivars to nitrogen (many crops) and to higher planting density (maize; *Zea mays*) as examples. The former example is explored more fully in the next section. In general, however, it is obvious from the definition of PY that trials receive the best agronomic management known at the time, and this will always be an improvement on the management that earlier vintage cultivars received when first released. Two things usually arise from this: (1) the management change lifts the yield of the oldest cultivar in the vintage trial and presumably all other cultivars by a given amount (akin to the management "main effect"), and (2) newer cultivars may respond more to the management improvements (a cultivar × management interaction). Method one does not measure the main effect of agronomic change and lumps the main effect of genetic change with the interaction term. However, these can be separated in methods two and three above, because trials are run over the whole period of interest, although Cv × M interaction terms have not to date been disaggregated from the mixed linear modeling used in method three. The following section gives an example of cultivar × management improvement.

In modern agriculture the positive interactions between new cultivars and new management affecting PY appear to be less than those during earlier phases of modernization of cropping (e.g., increased fertilizer, water, or plant density), with agronomy nowadays concentrating more on improving the efficiency of input use, with often small consequent changes in PY (e.g., herbicide resistant cultivars for no-till). Nevertheless, there are some agronomic changes underway which are lifting PY. An example is the extra available soil water arising from conservation agriculture in drier areas, especially when greater seed-zone moisture permits earlier sowing. Some other new developments, possibly lifting PY, are discussed in section 5.

Another issue with measurement of PY progress arises when unmanageable factors (the natural resource base, the climate and the soil) show time trends during the release period of the cultivars studied; this can arise from trends in rainfall, temperature, solar radiation or levels of atmospheric pollutants (e.g., ozone). Significant positive and negative time trends have been recorded in all these factors, as well as the direct benefits of the steady increase in CO2, currently running at about 2 ppm/yr. Soil can be subject to hidden toxin buildup and/or minor element depletion. It is possible that these changes affect all cultivars equally or that breeding has adapted the more recent cultivars to such changes (a G × natural resource interaction). In either case PY progress is measured correctly in the methods outlined, for even maintaining yield in the face of a deteriorating environment is progress. However, a full understanding of yield change demands that studies of yield trends are always alert to such possibilities.

Irrigated Wheat in the Yaqui Valley: A Valuable Example

The 180,000 hectares of irrigated wheat in the Yaqui Valley in northwest Mexico illustrates many issues relating to PY progress, and these have fortunately been well documented over the years. This was the wheat breeding target of Dr. Normal Borlaug, Nobel Peace Prize winner in 1970 for the impact of his efforts on the green revolution in Asia. This intense breeding effort has been maintained by the Global Wheat Program of the International Maize and Wheat Improvement Center (CIMMYT), Mexico, in collaboration with the Mexican Government agricultural research body, as a core part of CIMMYT's global breeding effort. The increase in wheat PY and FY in the Valley since 1950 (Figure 1) reflect what happened during the green revolution in South Asia and elsewhere, as does the increase in N fertilizer use, the major agronomic change in these irrigated regions (there were many other important but smaller agronomic changes for PY). Figure 1 shows FY change since 1951 in the Yaqui Valley, and it should be noted that after 1995 production changed, for economic reasons, from bread wheat (BW) to more than 50 percent durum wheat (DW) (*T. durum*), reaching over 85 percent in the last decade. The PY change results in Figure 1 came from vintage trials in the early-mid 1970s (PY1, method 1 above, Fischer and Wall, 1976) and in the early 1990s (PY2, Sayre et al, 1997), both with bread wheat (*T. aestivum*) cultivars. Since around 2006 to 2016, there were a mixture of vintage and recurrent check trials, but this time with durum wheat cultivars are shown (PY3, K. D. Sayre and R. P. Singh, unpublished data).

All PY trials had top management including disease control for the periods of evaluation. The rate of PY progress was 0.96 percent per annum for PY1 and 0.79 percent for PY2, slowing further in the last study interval (PY3, 0.41 percent per annum), but it has not ceased. This was for DW; the value for BW for this interval was 0.47 percent (see also Table 3). The study period encompasses one of significant warming in the 1970s and 1980s (Bell and Fischer, 1994) followed by one of nighttime cooling (1983–2013) which significantly affected the climate component of PY and FY trends (first negatively, then positively). It does not, however, influence the PY progress shown (unless they contain cultivar × temperature interactions, which was considered unlikely). Many other aspects of these yield changes are discussed in Fischer et al. (2014).

FIGURE 1. The progress (1950 to 2017) in potential yield (PY), farm yield (FY) and change in N fertilizer use for irrigated spring wheat in the Yaqui Valley of northwest Mexico. See text for explanation of the three PY relationships.

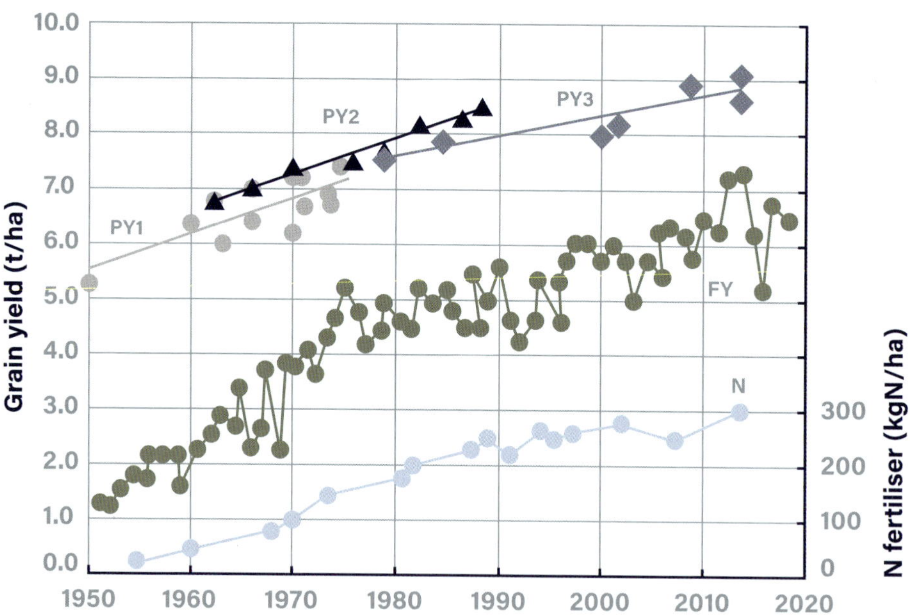

The Yaqui Valley also provides an illustration of a typical cultivar by agronomy interaction, in this case the cultivar × N (Ortiz-Monasterio et al., 1997). Figure 2 shows the response of two typical cultivars of markedly different vintage (Yaqui50, a tall variety, released in 1950, and Genaro81, a second generation semidwarf cultivar released in 1981): Delineated are the main effects of cultivar and N and the positive interaction between cultivar and N. The true PY progress from 1950 to 1981 in Figure 2 is given by the yield increase from point A to point C, whereas a vintage trial conducted at 1980s N levels would measure B to C, missing the PY progress due to the effect of N (A to B). It is generally a small error, but readers need to be aware that agronomy (main effect and interaction effect) has contributed much to PY advance, and researchers need to be alert to this and other complications already mentioned that can arise in practically every yield progress study.[5]

FIGURE 2. Yield of two irrigated wheat cultivars of 1950 and 1981 vintage, respectively, in response to N fertilizer levels. Dotted vertical lines show the yields at the N fertilizer rates typical of their release times (35 kg N/ha in 1950s and 200 kg N/ha in 1980s), and horizontal dashed lines showing the yield steps separating the cultivars (see text).

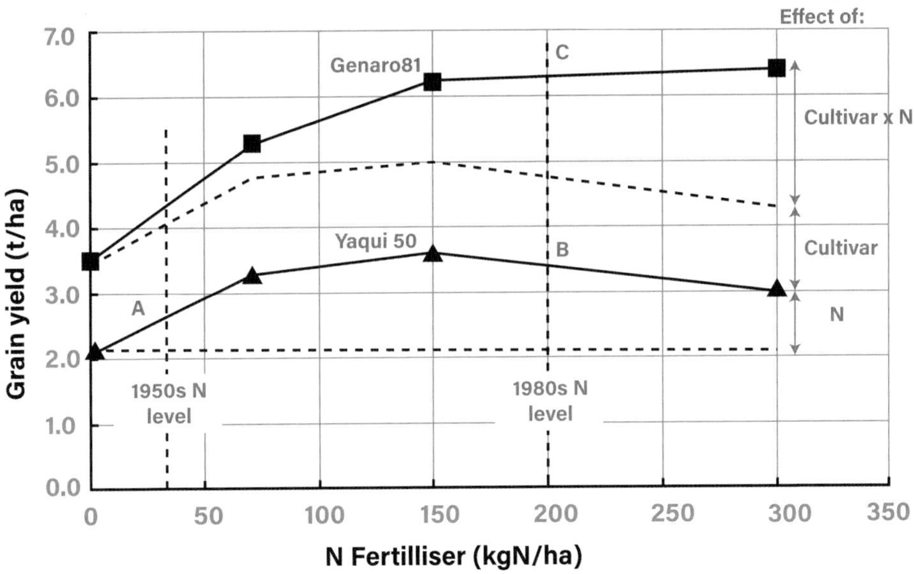

Source: Ortiz-Monasterio et al., 1997.

5 It is impossible to split any G × management interaction into G and M components; they are too intimately linked in the interaction.

Recent estimations of PY progress

Introduction

Fischer et al. (2014), attempting to avoid the above-mentioned sources of bias, collated world results on PY progress up to 2010, and included wheat, rice (*Oryza sativa*), and maize, for which there were many reports and other important crops for which thorough reports were less common (Table 2). Mean values were generally between 0.5 and 1.0 percent p.a., but there were some higher ones (maize, canola (*Brassica napus*), sugar beet (*Beta vulgaris*), cassava (*Manihot esculenta*)). It was noted that progress in PYw was no less than that in PY. Data was also presented on morphological and physiological changes accompanying the yield progress.

TABLE 2. Summary of relative rates of potential yield (PY, PYw) progress for important grain crops from "bread basket" case studies from around the world (from Fischer et al., 2014, Table 9.1).

Crop	PY		PYw		Source In Fischer et al. (2014)
	Number of case studies	Mean rate progress (%, ±SEM)	Number of case studies	Mean rate progress (%, ±SEM)	
Wheat	6	0.55±0.10	5	0.54±0.11	Table 3.5
Rice	10	0.80±0.09	1	0.70	Table 4.5
Maize	6	1.10±0.13	7[a]	1.13±0.15	Table 5.7 and Section 5.2A
Barley	NA	NA	2	0.70±0.14	Section 7.2
Soybeans	3	0.50±0.10	NA	NA	Table 6.2
Canola	3	1.37±0.03	1	1.40	Section 7.4
Sunflower	NA	NA	4	0.95±0.42	Section 7.5

[a]These came from the Era hybrids under several managed droughts in Chile and California.

Smith et al. (2014a) collated all yield progress data for thirteen U.S. grain crops in a CSSA Special Publication, although PY progress was not adequately quantified in several cases. Sound PY progress estimations from that book are reported below (Table 3), along with all more recent publications (post 2010), grouped under their respective commodities. Where the cultivars studied substantially overlap with studies reported in Table 2, the results are excluded; thus Table 3 represents the most recent estimate possible for progress. Preference was given to vintage ranges no longer than the last thirty years but longer vintage sets are included if there is clearly no change in slope with vintage (such cases are reported separately). Generally linear rates of progress were calculated by the authors; as outlined above, here they are converted to relative slopes by dividing by the predicted PY in the last year of each vintage series. Space doesn't permit a full discussion of all studies but some are highlighted.

Wheat (*T. aestivum, T. durum*) and Triticale (*X Triticosecale*)

For wheat (Table 3) the average rates of progress across PY plus PYw were 0.60 ± 0.062 percent and 0.52± 0.050 percent per annum for spring and winter wheat, respectively, clearly not significantly different, and means were similar to those in Table 2. Again PY progress was in general no different to PYw progress. The U.S. studies mentioned above (Graybosch et al., 2014) took advantage of the USDA-ARS coordinated regional wheat performance nurseries comprising multilocational trials of different wheat types; progress appears to be slower in the Pacific Northwest. Irrigated spring wheat (first 4 rows of Table) showed remarkably similar rates (0.42 to 0.49 percent p.a.). There is no explanation of the highest rates (three at around 0.90 percent), but the Canadian results deserve comment. Thomas and Graf (2014) reported a doubling of the rate of progress around 1992, coinciding with more resources in the main wheat breeding program, including a doubling in the population size and the introduction of some modern techniques such as doubled haploidy and molecular markers for difficult traits such as percent grain protein (de Pauw et al., 2011). It is also noteworthy that Cormier et al. (2014) in northwest France had a parallel measure of yield progress with less N fertilizer which gave identical progress (0.38 percent p.a.) to that in Table 3, at approximately 25 percent lower yield, a common observation in wheat when progress is tested at lower soil N levels (Fischer et al., 2014). Finally, the one study with triticale (Losert et al., 2017) thoroughly describes this unique manmade cereal, which after starting about a century ago with very low PY due especially to sterility and grain shriveling, is now showing PY and PY progress not unlike wheat. Table 3 excludes three wheat studies that claim to have reached yield plateaus around 1990–2000, because they had too few recent cultivars and none showed statistically significant breaks in the uptrend.

• • •

TABLE 3. Recent studies of PY progress in spring wheat and winter wheat.

PY type	Location	Vintage range	Most recent PY (t/ha)	Slope (% p.a.)a	Method (M)b and comment	Reference
			Spring wheat (bread wheat unless noted otherwise)			
PY	Yaqui Valley, Mex	1966-2009	7.7c	0.42	M1, bed plots 2009-11	Aisawi et al., 2015
PY	Yaqui Valley, Mex	1982-2014	8.6	0.47	M1, bed+flat plots 2006-16	K. D. Sayre, R. P. Singh, pers comm.
PY	Yaqui Valley, Mex	1979-2013	8.7	0.41d	M1, spring durum wheat, bed + flat plots 2006-2016	K. D. Sayre, R. P. Singh, K. Ammar, pers comm.
PY	Chillan, Chile	1960-2008	9.0	0.49	M1, plots, 2008-2011	Del Pozo et al., 2014
PYw	Southern NSW, Aus	1942-2014	4.7	0.53	M1, plots, 2014-16	Hunt et al., 2017
PYw	South Australia	1958-2007	4.6	0.40	M1, plots, 2010 and 2011	Sadras and Lawson 2013
PYw	Western Australia	1997-2016	3.1	0.65	M1, NVT plots, 2012-2017	R. A. Fischer, unpubl.
PYw	Parana, Brazil	1987-2009	3.8	0.93	M1, plots, 2010-2011	Beche et al., 2014
PYw	Northern Great Plains, USA	1980-2010	3.96	0.64	M2, USDA-ARS HRS trials, check PY + 23%E	Graybosch et al., 2014
PYw	Man., Sask., Canada	1991-2011	na	0.90	M2, Seed guides, 1995-2011	Thomas and Graf, 2014
			Winter wheat (all bread wheat)			
PY	China, Henan	1950-2012	8.40	0.68	M1, plots, 2013-15	Gao et al., 2017
PY	France	1982-2010	10.2	0.90	M2, INRAT, 1982-2010	Oury et al., 2012
PY	NW France	1986-2010	8.72	0.40	M1, 2007-2009	Cormier et al., 2013
PY	Germany	1983-2014	10.6	0.53	M3, official VCU, 1983-2014	Laidig et al., 2017
PY	UK	2004-2018	11.4	0.44	M2, official AHDB trials, 2015-2017	R. A. Fischer, unpubl.
PY	Eastern USA	1980-2010	5.29	0.51	M2, USDA-ARS SRW Trials when	Graybosch et al., 2014

PY type	Location	Vintage range	Most recent PY (t/ha)	Slope (% p.a.)a	Method (M)b and comment	Reference
Winter wheat (all bread wheat) (*cont.*)						
PY	Southern USA	1980-2010	4.57	0.85	M2, USDA-ARS, SRW trials	Graybosch et al., 2014
PY	Pacific Northwest, USA	1980-2010	7.50	0.23	M2, USDA-ARS SWW trials, check + 38%[e]	Graybosch et al., 2014
PY	Pacific Northwest USA	1980-2010	5.85	0.37	M2, USDA-ARS HWW trials, check + 48%[e]	Graybosch et al., 2014
PYw	Southern Great Plains USA	1980-2010	3.75	0.57	M2, USDA-ARS HRW trials, check cv + 20%[e]	Graybosch et al., 2014
PYw	Nebraska, USA	1980-2010	3.71	0.41	M2, Uni Neb. HRW trials	Graybosch et al.,2014
PYw	Northern Great Plains, USA	1980-2010	4.29	0.69	M2, USDA-ARS HRW trials	Graybosch et al., 2014
Winter triticale						
PY	Southern Germany	1983-2014	9.01	0.59	M1, plot trials, 2014, 15	Losert et al., 2017

[a]Using as denominator PY predicted for latest vintage year from PY vs release year linear regression.

[b]See Section 2 for a description of the 3 methods (M1, M2, M3) used to determine PY progress.

[c]PY yields low probably due to uncorrected bed system used, but physiology sound.

[d]See Figure 1 where these data are plotted as PY3.

[e]The amount by which the recurrent check cultivar yield increased over the trial period = improved agronomy plus any climate trend.

Several of the wheat reports measured other changes, both in grain quality or in morphological and physiological traits, associated with progress. Thus Laidig et al. (2017) measured an 8 percent fall in protein over the thirty-two years of yield improvement in Germany, although baking quality was maintained or improved slightly. Aisawi et al. (2015) is important because it targeted physiological changes in CIMMYT breeding to reveal a definite shift in the early 1990s from progress in harvest index (HI) and grain number (GN=grains/m^2) seen with earlier progress, to that from greater biomass and grain weight (GW). Progress was also associated with increased stomatal conductance, already reported, and longer stay-green. Curiously most of the later cultivars, although carrying one or another major dwarfing gene, were quite tall (up to 110 centimeters), which may reflect unwitting selection pressure for performance on the bed planting system adopted at CIMMYT since the 1990s. Del Pozo et al. (2014), Beche et al. (2014), and Gao et al. (2017) found progress more akin to that reported by others, namely associations with reduced stature, higher HI, and greater GN, but in addition all found increases in stomatal conductance and or greater chlorophyll content, and the last two papers also reported greater biomass. Gao et al. (2017) also found greater stem water-soluble carbohydrate contents with newer cultivars, something that had been reported before for winter wheats in the UK, but not so far for spring wheats in Mexico (e.g., Aisawi et al., 2015 and earlier studies).

Rice

All reports for irrigated rice show similarly high PY values, but progress varies greatly (Table 4). The lower Yangtse in China has two reports of high progress (average 1.2 percent per annum). Chen et al. (2014) in Hangzhou compared super rices to checks under full season planting, a range of N levels and two planting methods: Only at high N (225 kgN/ha) and with transplanting did the progress with super rice become evident. At a nearby site in Yangzhou, Jiangsu, Gu et al. (2015) report remarkable (and accelerating) PY progress with full season rice *O. japonica* inbreds. In Louisiana, U.S., McKenzie et al. (2014) showed reasonable PY progress (0.73 percent) that probably does not yet reflect the small boost in PY to be expected from the introduction of hybrids there.

TABLE 4. Recent studies of PY progress in irrigated rice.

PY	Location	Vintage Range	Most recent PY (t/ha)a	Slope (% p.a.)b	Method (M) and comment	Reference
PY	Hangzhou, China	1984–2008	10.2	1.00	M1, plots 2011+12, single season, transplant, N225	Chen et al., 2014
PY	Yangzhou, China	1955–2005	11.1	1.38	M1, plots, 2012 and 2013	Gu et al., 2015
PY	Louisiana	1977–2011	9.80	0.73	M3, Uniform Regional Rice Nursery	McKenzie et al., 2014
PY	Northern California	1986–2005	10.2	0.23	M1 (but only 3 cvs), multilocation trials	McKenzie et al., 2014
PY	Biggs, California	1981–2016	11.3	0.35	M3, plot trials on-station, 1984-2016	Espe et al., 2018

[a]Yields refer to paddy or rough rice.

[b]Using as denominator PY predicted for latest vintage year from PY vs release year linear regression.

[c]See Section 2 for a description of the three methods used to determine PY progress.

The final entries in Table 4 are from California. McKenzie et al. (2014) found significant but slow progress in regional trials, a rate similar to that calculated here from separate on-station data of Espe et al. (2018). The results of Espe et al. (2018) are however somewhat puzzling, with the authors declaring that the PY of the cultivars released since 1981 declined at 29.5 kg/ha for every year since release (even though in any year there was still progress between them). Yield declines after releases have also been reported at the International Rice Research Institute (IRRI) in the Philippines in a tropical paddy rice environment (Peng et al., 2010).[6] Neither occurrence has a satisfactory explanation, although Fischer et al. (2014) argue that the IRRI rice environment (not the cultivars) may have deteriorated perhaps due to the rise in Tmin or in atmospheric pollution. McKenzie et al. (2014) did not record regional yield declines in commercial yields of three key cultivars of the Espe et al. (2018) study over 1996 to 2010. Australia and Egypt produce rice under similar conditions to California: There are no recent PY studies but FY in Australia over the last fifteen years increased at 1.6 percent per annum, to reach 10.3 t/ha, suggesting but not proving there is still breeding progress; there has been no recent FY progress in Egypt (FY in 2016 was about 9.4 t/ha).

Maize

The mean PY progress for maize (Table 5) was 0.66 ± 0.048 percent per annum, after excluding the lowest value (0.11 percent per annum), which may be the only evidence available of a PY plateau (since 1988) but has very low accuracy (r squared = 0.02). Either way the rates are significantly lower than the 1.1 percent presented previously (Table 2). Smith et al. (2014b) discuss the U.S. results in depth, noting how improved insect resistance arising from GMO hybrids with Bt resistance may have contributed a small part of the recent progress (because insecticide control in earlier non-Bt hybrids was never 100 percent). They also reported greater relative rates of progress (around 1.0 percent per annum) in competition-winning yields, at levels twice those of FY.[7] In Table 5, only progress in the tropical OPVs approach 1 percent; progress elsewhere in African breeding programs is likely to still be at such levels or better (Fischer et al., 2014).

6 For example, with the first famous semidwarf rice cultivar, IR8, released in 1966; fortunately the arrival of tropical rice hybrids starting about 2000 have offered a boost to PY of 10 percent or better in such environments.

7 Verified winning yields in USA are now over 25 t/ha (2017) and remain a totally unexplored source for greater understanding of yield determination, including that of PY.

TABLE 5. Recent studies of PY progress in maize.

PY or PYw	Location	Vintage range	Most recent PY (t/haa	Slope (% p.a)b	Method (M) and comment	Reference
PY	Indiana	1967–2005	11.3	0.66	M1, plots 2013, 2014 DeKalb hybrids	Chen et al., 2016
PY	Champaign, Illinois	1967–2006	11.7	0.74	M1, plots 2009, 2010 DeKalb hybrids	Haegele et al., 2013
PY	Iowa	1963–2011	12.4	0.74	M1, plots Pioneer Era hybrids, date not given by reference	Smith et al., 2014b
PY	Woodland, California	1963–2011	15.3	0.53	M1, plots Pioneer Era hybrids, irrigated	Smith et al., 2014b
PYw	Woodland, California	1963–2011	9.0	0.65	M1, plots Era hybrid, partially irrigated	Smith et al., 2014b
PY	Viluco, Chile	1963–2011	17.3	0.55	M1, plots Pioneer Era hybrids, irrigated	Smith et al., 2014b
PY	Viluco, Chile	1988–2011	16.9	0.11ns	MI, plots Pioneer Era Hybrids, irrigated	Smith et al., 2014b
PY	Iowa, Illinois	1995–2013	14.6	0.52	M1, plots 2013, 2014 Era hybrids Pioneer	DeBruin et al., 2016
PY	Nigeria, Benin, Ghana	1988–2010	4.0	0.98	M1, plots 2009–2011 IITA OPVs	Badu-Apraku et al., 2014
PY	Balcarce, Argentina	1993–2010	15.1	0.61	M1, plots, 2009–2013, DEKALB hybrids	Di Matteo et al., 2016

Again, some studies report changes in crop traits. Smith et al. (2014b), with the Pioneer hybrid series, found that the long-recognized steady improvement in resistance to lodging and barrenness were plateauing at full resistance, while leaf angle and stay-green were also plateauing (the former close to vertical). Anthesis to silking interval and tassel size continued on their long recognized decreasing trajectory. DeBruin et al. (2016) working on the same material additionally reported that recently GW increase was more important than GN increase. Haegele et al. (2013) and Chen et al.

(2016) studied a DEKALB hybrid vintage series in the U.S., and Di Matteo et al. (2016) another in Argentina. Biomass increases were common and dominant in all studies, and especially post-anthesis in the very detailed Chen et al. study, in which it was associated with longer stay green. While GN increase occurred in Argentina, and in Haegele et al. (2013), GW increase tended to dominate in the United States, especially in the most recent hybrids (Chen et al., 2106). The latter authors emphasized that increases in grain sink strength were driving greater partitioning of ear DM increase to grains. Haegele et al. (2013) and others have found that the most modern hybrids show greater post-anthesis uptake of soil N.

Soybeans, Dry beans (*Phaseolus vulgaris*), Canola

For soybean the average PY progress was 0.69 percent ± 0.075 percent per annum (Table 6), slightly more than that summarized earlier (Table 2). Results from the huge collaborative vintage trial (2010 and 2011) of Specht et al. (2014), are reported in more detail by Rincker et al. (2014). They showed remarkably similar progress (0.71 to 0.78 percent) for each major Maturity Group, and again without sign of slowing progress. These authors also found greater linear progress with earlier sowing and higher density, but it is unlikely that relative progress increased in either case. The earlier study of Fox et al. (2013) showed equal relative progress in a maize-soybean rotation, compared to soybean grown after ten years of continuous maize, a sequence history well known to benefit soybean yield, in this case by 12 percent.

Rincker et al. (2014) reported a small decrease in lodging score with year of release (–0.12/ year on a 1–5 scale) but while important, it is probably not enough to invalidate the study as one of PY progress.[8] The observed significant decrease in foliar disease incidence with vintage (reported in a separate study in Illinois, Hartman et al., 2015) does, however, suggest that the progress in Figure 3 is not without some bias due to more disease with older cultivars. Nonetheless the effect is expected to be small in cultivars released in the last forty years when disease resistance traits were more strongly sought by breeders (B.W. Diers, pers. comm.), and from the Fox et al. (2013) results, probably doesn't involve a significant change in resistance to soil borne diseases. That absolute progress in FY was no more than PY progress also supports the notion that there is no disease bias in the PY results (Figure 3), and shows how closely linked are PY and FY progress in an intense modern cropping system, with a yield gap in 2008 of barely 20 percent (of FY).

8 Harman et al. (2015) found the mean lodging score for post-1970 cultivars to be consistently < 2.

TABLE 6. Recent studies of PY progress in soybeans, dry beans, and canola.

PY or PYw	Location	Vintage range	Most recent PY (t/ha)	Slope (% p.a)	Method (M) and comment	Reference
			Soybeans			
PY	Urbana, Illinois USA	1958–2007	4.4	0.80	M1 plots, 2012 and 2013	Koester et al., 2014
PY	Arkansas, USA	1950–2000	3.5	0.40	M1, plots 2007, 2008 MG IV-VI	Rogers et al., 2015
PY	Midwest USA	1968–2009	4.0	0.77	M1, plots MGII, 2010 and 2011	Specht et al., 2014
PY	Midwest USA	1964–2009	4.2	0.71	M1, plots MGIII, 2010 and 2011	Specht et al., 2014
PY	Midwest USA	1971–2009	3.4	0.78	M1, MGIV, plots 2010 and 2011	Specht et al., 2014
			Dry Beans			
PY	USA	1981–2012	3.4	0.36	M unclear, pinto bean	Vandemark et al., 2014
PY	USA	1981–2012	2.8	0.94	M unclear, navy bean	Vandemark et al., 2014
PY	USA	1987–2012	3.2	0.75	M unclear, black bean	Vandemark et al., 2014
PY	USA	1990–2012	2.9	1.45	M unclear, kidney bean	Vandemark et al., 2014
			Canola			
PYw	Bool Lagoon, South Australia	1978–2007	2.2	0.63	M1, plots2008, corrected for black leg	Potter et al., 2016
PYw	Western Canada	2003–2009	3.3	2.16	M2, early years of F1 hybrids in Prairie Canola Perf. Trials	Morrison et al., 2016

Rincker et al. (2014) reported a small decrease in lodging score with year of release (–0.12/ year on a 1–5 scale); while the small reduction in lodging is important, it is probably not enough to invalidate the study as one of PY progress.[9] The observed significant decrease in foliar disease incidence with vintage, reported in a separate study in Illinois, Hartman et al. (2015), does however suggest that the progress in Figure 3 is not without some bias due to more disease with older cultivars. Nonetheless the effect is expected to be small in cultivars released in the last forty years when disease resistance traits

9 Harman et al. (2015) found the mean lodging score for post-1970 cultivars to be consistently < 2.

were more strongly sought by breeders (B. W. Diers, pers. comm.), and probably doesn't involve a significant change in resistance to soilborne diseases (Fox et al., 2013). Similar progress in FY and PY progress also supports the absence of disease bias in the PY results (Figure 3), and shows the close linkage of PY and FY progress in an intense modern cropping system, with a yield gap in 2008 of barely 20 percent (of FY).

FIGURE 3. Soybean potential yield (PY), determined in 2010 and 2011 trials and plotted against year of release, and farm yield (FY) plotted against actual year for Iowa, USA.

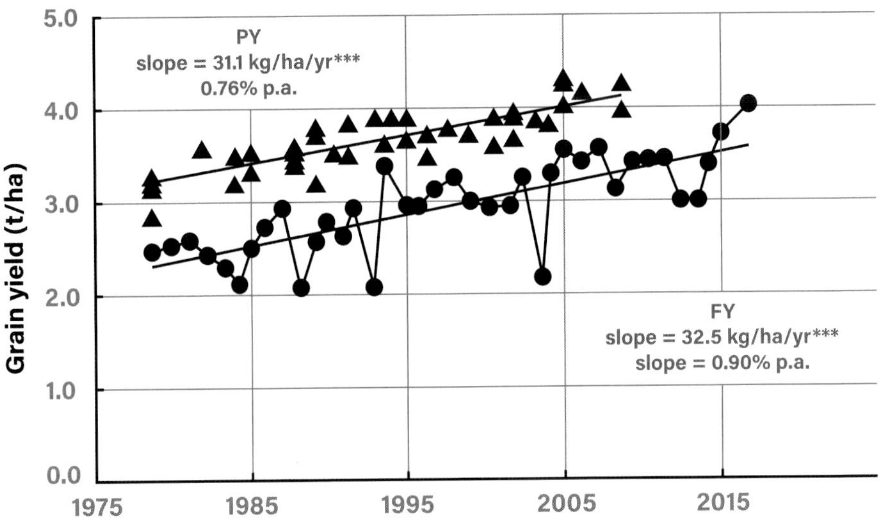

Source: PY Rincker et al. (2014) Group II cvs; FY USDA NASS accessed 24 April 2018.

In the U.S. soybean studies of Table 6, Rincker et al. (2014) noted that associated with year of release were also small but significant increases in crop duration (+0.102 d/year), and decreases in height (–0.133 cm/year). Grain weight (GW) did not change although seed oil content tended to increase while seed protein content declined slightly. Such seed composition changes were more evident in the MGIV to VI material (Rogers et al., 2015). Koester et al. (2014) measured the well-recognized components of light harvesting, namely PAR interception, efficiency of PAR conversion (RUE), and partitioning of biomass to seed (HI). Their analysis revealed that PY progress was related to improvement in all three components when examined over their whole vintage series (1923 to 2007), but post 1958 (as shown for PY in Table 6) the relationships were less clear. Nevertheless, the results support the many studies showing the longer cycle of newer cultivars (= greater interception), their greater photosynthetic rate (= higher RUE), and frequent increases in HI (several studies).[10]

A largely irrigated crop of minor importance in the United States, dry or common bean (*Phaseolus vulgaris*), averaged 0.88 percent per annum PY progress across the four market types of bean. This crop is, however, of much greater importance in Latin America and Sub-Saharan Africa, where it frequently encounters dry spells during production. It seems that CIAT and its collaborators are making PYw progress (Beebe et al., 2014), although it is not thoroughly quantified.

In Australia canola showed reasonable PYw progress (Table 6), in a data set with minimal influence of hybrids that were just becoming available when the trial was conducted. Progress included oil and protein concentrations (up 0.09 percent and 0.05 percent per annum, respectively) but resistance to black leg remains the major target for breeders (Potter et al., 2016). In contrast, in Canada hybrids were having a large effect on PYw progress, at least in their initial years, although industry sensitivity limits access to objective data. This agrees with the high rates of progress reported in Europe earlier (Table 2).

Prospects for Accelerating PY Progress

Raising PY (and PYw) Through Breeding

The review of PY and PYw progress in Section 4 suggests that the annual relative rates of yield progress are definitely not stagnant, rather they are declining slowly. This is not only because PY (the denominator in calculating relative progress) is rising but primarily because, as almost all breeders will admit, progress is becoming more difficult, especially with the staple crops. There are a few other lessons from the review. Wheat and soybeans still appear to be making steady if not spectacular PY progress. Rice, at least irrigated rice, is seeing low progress except possibly where F1 hybrids are being deployed. In temperate maize, after almost a century of progress, and despite the absence of grain quality requirements that in other crops slow yield progress and the current heavy investments of the private sector, progress has

10 HI is difficult to measure in soybean because of leaf fall before full physiological maturity of the seed.

slowed notably. In tropical maize, however, progress appears better. Hybrid canola, a relatively new crop, is still showing good progress. In only one case was there recently a notable increase in rate of progress: As mentioned this was for Canadian Western Red Spring wheat around 1990, attributed to more resources, modern techniques as well as doubling population sizes in that breeding program, confirming the importance of numbers in breeding. All the other new breeding technologies of the last thirty years (biometrics, automation, molecular markers for difficult traits) have at best helped maintain progress (at least in terms of the linear slope, kg/ha/yr).

Fischer and Connor (2018) and earlier Fischer et al. (2014) attempted to take a broad look at PY (and PYw) prospects including new cultivars, new agronomy, and their common positive interaction. A broad perspective is useful because many papers on prospects push the authors' areas of skill and lack perspective and balance. For breeding, various positives are frequently mentioned:

- A host of new breeding technologies including better biometrics, molecular markers, and culminating in genomic selection (Bernardo, 2016)
- Steadily more efficient and accurate phenotyping under the general banner of high-throughput phenotyping (Araus and Cairns, 2014)
- Crop simulation models and weather and soil databases that permit quantification of environmental patterns to support breeding (Chenu, 2015), and unprecedented, if poorly validated, exploration of gene to trait to yield relations (e.g., Chenu et al., 2009)
- Considerable unexploited genetic material safely stored and increasingly well documented in the world's gene banks, with some evidence of yield advances arising from exploiting such materials (e.g., the new synthetic wheats)
- The possibility of a modest one-off yield boost through commercialization of F1 hybrid seed with some crops (e.g., wheat, tropical rice)

Not included in this list is genetic engineering (GE) for greater PY, because to date no such cultivars have been released with higher intrinsic PY—and only one case for higher PYw, in maize (Nemali et al., 2015). It is hard not to be cynical about GE prospects, for there have been so many claims over the last twenty years or more, none of which have delivered a direct impact on yield. For several reasons this target is probably too difficult for impact in the next twenty years or so, notwithstanding continual claims of imminent success, most of which failed in taking results for spaced plants in controlled environments to plots in the field. This view is supported in the thorough review of Hall and Richards (2013). Much effort focuses on genetically engineering photosynthesis, but research again reveals great complexity and lack of understanding (Furbank et al., 2015); the much heralded C4 rice project, despite steady progress, does not expect to have cultivars in farmers' fields before 2040 (J. Langdon, pers. comm., 2018). It should be noted, however, that GE has indirectly and modestly increased FY via insect and herbicide resistance in some crops (such as maize), but possibly not others (such as soybean) (Klümper and Qaim, 2014; Xu et al., 2013), and by earlier planting with glyphosate-resistant cultivars. This is mostly one-off yield-gap closure, nevertheless with enduring environmental and financial benefits.

A major part of future of breeding for greater PY relates to those yield-enhancing plant and crop traits still amenable to change through breeding. This has been widely discussed by crop physiologists (e.g., Foulkes et al., 2009; Reynolds and Langridge, 2016) and in Fischer et al. (2014), and traits associated with yield increase reported in several of the studies mentioned in section 4. There appears to be a shift in PY progress initially driven by more grains/m^2 to some examples of recent progress arising from heavier grains, something usually associated with greater stay-green. Not surprisingly there is also a shift away from examples of progress driven by HI increase (now that crops are shorter and HI approaches 0.5) to ones associated with greater biomass; several of the latter examples are supported by increases in stay-green, radiation use efficiency (RUE), leaf greenness or chlorophyll content, stomatal conductance, and/or leaf photosynthesis. See Fischer and Rebetzke (2018) for small grain cereals, and Roche (2015) for C3 crops in general. Thus there are sound precedents for continuing to pursue even greater biomass, combined as always with maintaining high HI and of lodging resistance.[11]

Biomass can also be increased by extending crop duration and with it light capture, but in most situations this is already close to the seasonal limits set by low temperature or water supply or set by the cropping system.

11 In some crops like spring wheat HI has not consistently reached 0.50 under potential conditions, so there is scope for further small gains, but the bigger gains in all crops are probably to be found in HI as it relates to PYw, when water stress around and after flowering can have major negative effects on HI.

However, a few possibilities do still exist, as illustrated by attempts to breed sugar beet adapted to late autumn planting. The crop is typically planted in spring at latitudes of 40–45°N, whereas for autumn planting it must survive winter freezing and not be triggered to flower by vernalization during that time. If this could be achieved, sugar beet should develop more leaf area sooner in the spring, deeper roots to exploit water and nitrate, and produce more biomass and yield by harvest in the following autumn (Jaggard et al., 2010). The other pathway for greater biomass is of course greater RUE. Modeling suggests that depending on prevailing sun angles and diffuse to total solar radiation ratios, small erect leaves should, other things equal, give greater canopy photosynthesis. However, most modern cultivars seem to have achieved this canopy structure already, and it remains for increases in leaf photosynthetic activity under full light (Pmax) to lift RUE.[12] There are many GE proposals for increasing Pmax in the pipeline (Furbank et al., 2015), but none have to date delivered in the field, and better prospects may be found in the existing natural genetic variation, which has already been unwittingly exploited by breeders seeking higher yield. This is especially so when it is considered, for example, that wheat wild relatives have higher Pmax (probably because of higher specific leaf N), and that this trait can be brought into modern cultivars through introgression of novel *T. tauschii* alleles (Tang et al., 2017). It should be pointed out that none of these proposed physiological improvements will be easily or quickly achieved, and that hidden trade-offs could render them ineffective.

Turning to water-limited conditions, PYw has definitely improved even recently relative to crop evapotranspiration (ET) (e.g., wheat in Australia, Sadras and Angus, 2006). This could partly be a spillover of PY progress that can be expected in most situations other than under extreme water shortage. At the same time, traits specifically targeted to improve performance under water shortage have received much attention from physiologists over a long period. Some have probably been exhausted by breeding (e.g., earliness) while others remain to be thoroughly validated in the field (e.g., deeper roots, tolerance of grain setting to reproductive stage drought) and then converted into usable selection criteria. Others are unlikely to ever leave the greenhouse.

There does, however, seem to have been some significant progress in seeking higher transpiration efficiency in wheat (Richards et al., 2002), deeper rooting in rainfed rice (Venuprasad et al., 2008), and especially with targeted

12 Based on some early experience, in particular with perennial ryegrass, it has been proposed that selection for low respiration could contribute to RUE increase without, of course, negative side effects: The International Wheat Yield Partnership is investing in this notion.

selection for drought tolerance in maize (Edmeades, 2013). The last-mentioned efforts, maintained for over 40 years by CIMMYT and lately its African collaborators, have delivered improved cultivars for farmers under dry and low soil N situations in eastern Africa. Finally, a new trait gathering interest in some rainfed crop species, and of possible value to others, is stomatal sensitivity to vapor pressure deficit (vpd), for which genetic variation has been found in many crop species (Vadez et al., 2014). It appears that some genotypes restrict the expected linear rise in transpiration with increasing vpd at some moderate threshold, with the effect of conserving soil water ahead of critical reproductive events to the benefit of yield, especially if rains falter and then return. This mechanism appears to be the basis for success for a new suite of Aquamax® maize hybrids that exhibit modest yield gains (6 percent) under water shortage across the U.S. corn belt (Gaffney et al., 2015).

Concluding this brief discussion of breeding and PY prospects, one cannot be very optimistic about a boost in the current rate of breeding progress. The transformative changes so frequently heralded, such as a breakthrough for example in photosynthesis, are likely to take a long time to impact PY and FY. At best we can hope that breeding and new agronomy (see below) will maintain current rates in the range of 0.4 to 1.0 percent per annum.

New Agronomy and Breeding × Agronomy Interactions for Greater PY (PYw)

New agronomy is a part of overall PY increase, commonly in the past interacting very positively with breeding, such that most studies conclude both agronomy and breeding have contributed more or less equally to past PY (and FY) increase. It is difficult to say whether breeding has led to this or if agronomy has been the driver, but close communication between the two has been essential for its realization. The interaction is often best summed up by Leibscher's Law, which states that to achieve the most for an advance such as in genetic PY, all other yield inputs (weather, water, nutrients, other management factors) need to be optimized (de Wit, 1992).

In maize, however, the highly positive G x density interaction may be coming to an end as densities become high enough to maximize biomass production in the available crop cycle, and new hybrids tolerate these densities without becoming barren. Maize at today's 80,000 plants/ha in Iowa is not so far away from maximum densities of 100,000 plants/ha as used, for example, in New Zealand. There may, however, remain some scope for further PY gains using more regular planting arrangements (e.g., triangular or honeycomb designs) that minimize early interplant competition. Such precision planting is possible via seed singulation (precisely spacing individual seeds along the row), but has yet to be widely explored in maize or other field crops and will likely require new plant types from the breeders for greatest gain. Curiously, rice production in Asia (apart from that in the System of Rice Intensification) is proceeding in the opposite direction as increased labor costs mean that direct seeding in rows replaces transplanting on a typical 20 x 20 centimeter grid. In this case it is unclear if current cultivars are the best adapted to the new planting system, but the work of Chen et al. (2014) mentioned earlier suggest not.

Generally, it seems that new opportunities for yield increase and/or for positive interactions with new genotypes keep emerging in agronomy, often unanticipated. Conservation agriculture (CA), in particular in subhumid regions with fallow periods, has generally provided significantly more stored soil water at seeding and hence higher PYw in any given region. In the Great Plains of North America this has actually led to shorter fallow periods that allow additional crops in the rotation. In this case PYw of individual crops may suffer but cropping intensity and overall average annual system production increases as has happened in for example Saskatchewan with a general shift away from one crop every second year (after "summer fallow") to a crop every year under CA. Further south and east, in the U.S. corn belt, earlier sowing has been aided by reduced tillage, and herbicide- and cold-resistant cultivars (plus some climate warming), creating opportunities for longer-cycle, higher-yielding maize and soybean cultivars (Grassini et al., 2015b). In southern Australia CA has increased fallow soil-water storage creating opportunities for earlier sowing of wheat in the autumn. This in turn demands new sowing drills (for reaching deeper moisture), cultivars with longer coleoptiles (for better emergence for deep sowing in warm soils), and new cultivars with appropriate phenological development controls (sensitivity to vernalization rather than to photoperiod), and bringing prospects of greater grain yield (Flohr et al., 2018), and of extra winter forage in grazing-grain enterprises. Such crop research is noteworthy because it was reinforced by recent climate trends and mapped out preemptively with crop simulation modeling (Kirkegaard and Hunt, 2010). CA is at the early adoption stage in Asia, bringing many new challenges for small-scale farmers. In Asia primarily it not only offers water and fuel savings, but also significant increases in PY of wheat from earlier sowing, again requiring new cultivars, and also facilitating increased intensity of cropping (Hobbs et al., 2017).

Other agronomic innovations that may lift PY concern soil management. Understanding of soil microbiology has been boosted by the advent of molecular tools, and this new knowledge may eventually lead to positive impacts on yield additional to those already exploited through *Rhizobium* biology for legumes, mycorrhiza management for some crops, and crop sequencing. Certainly the benefit of crop rotation is well-known if not fully understood, and more complex sequences or longer intervals between given commodities would probably be desirable. As mentioned in section 4.5, soybean after ten maize crops yielded 12 percent more than in a two year rotation with maize: That advance exceeds twenty years of PY breeding progress!

It may, however, turn out that soil chemistry and physics have more to offer than soil biology. Already PY of the Brazilian Cerrado soils has been lifted more than two-fold with massive doses of lime and phosphorus. Other such one-off boosts to productivity are seen with tile drainage, and in rainfed environments such as Australia, amelioration of coarse topsoil with addition of clay ("claying"), or heavy topsoil with gypsum, and of heavy subsoils with ripping and deep placement of lime and organic materials. It is a moot point whether such major investments in the natural resource base of cropping represents an increase in PY rather than yield-gap closure. To the extent that the prior condition was not recognized as lacking a manageable (and profitable) input (e.g., lime and P in the case of the Cerrado), its initial recognition is clearly PY increase. Once the improvement is recognized and adopted by some, it becomes gap closure for the non-adopters. In this context, there is much evidence that soil compaction can restrict root penetration and PY and, in addition, the occasional outstanding performance of crops in "loose" soils (Fischer et al., 2014) which deserves follow-up. Precision guidance and controlled traffic can restrict compaction to a small fraction of any field and lift yields.

Developments for crop agronomy from outside of agriculture could include improved short-term and seasonal weather forecasts. Partly this is closing the yield gap between FY and PYw by reducing farmers' risk of losses when using higher inputs in rainfed subhumid environments, but agronomy, which targets the forecast seasonal weather rather than the long-term mean weather, is actually lifting PYw toward that possibility with perfect knowledge of forthcoming seasonal weather (Rodriguez et al., 2018). Another potential "left field" development is the use of ultra-thin biodegradable plastic films, sprayed between crop rows to increase soil warming in spring at higher latitudes or to reduce soil evaporation in low rainfall locations. Already ordinary plastic film is widely used for this purpose in rainfed cropping in northern China, with large benefits to yield, but its low biodegradability renders the practice environmentally unacceptable.

Finally, it should be noted that agronomy extends beyond individual crops to cropping and farming systems, involving crop sequences and the inclusion of grazing and stall-fed animals. There may be benefits to system productivity with crop rearrangements (Guilpart et al., 2017), or as mentioned for individual crop PY, in particular from diversifying wheat, maize, and soybean cropping systems with different crops such as the other oilseeds and pulses, less common cereals, or pastures. Often such diversification awaits a greater breeding and marketing development effort with the newer crops.

Concluding Remarks

Potential yield increase both drives and limits farm yield increase in modern cropping. This up-to-date review of PY progress indicates that for most situations and most crops recent progress is running at between 0.4 to 1.0 percent per annum relative to the latest PY values, with staple crops subject to the longest history of intense breeding showing the lowest rates of gain (e.g., wheat, rice, and maize). Rates of gains have clearly slowed and are proving harder (more expensive) to realize, but they have not ceased anywhere, although some uncertainty surrounds progress in irrigated rice. Despite the steady stream of advanced breeding technologies, the only apparent jump in progress appears to be associated with a large increase in size and technology of public breeding programs (for Canadian hard red spring wheat)—in other words more investment, or the arrival of F1 hybrids (canola, possibly *indica* rice), again involving more investment in breeding, in this case from the private sector.

PY gains derive from breeding, agronomy, and their positive interaction, and while the analyses presented relate especially to breeding gains, they stand in stark contrast to the overly optimistic predictions over the last decades regarding the impact of the new breeding technologies. Indeed experience now suggests that genetic engineering for yield will not be likely to deliver before the twenty-year time horizon suggested here as being the most critical for world food security. Prospects for new technologies in agronomy via improved soil management (especially soil microbiology and physics as they relate to crop sequences and soil compaction, respectively) might contribute more but receive relatively little research investment. The review by Smith et al. (2014a) sought comment from active breeders on yield prospects for each U.S. crop, but there was little attention to this difficult subject especially to agronomy; at least there were few of the unrealistically optimistic claims of molecular biologists, many opting for greater attention to utilization of existing, often-untested genetic diversity. One balanced exercise in twenty-year foresight has recently appeared for rainfed wheat in Australia (Robertson et al., 2016), predicting yield gains of 0.8 to 1.0 percent per annum arising from an array of new cultivars and agronomy (PY advance plus some yield gap closure). It is instructive that the biggest feasible potential genetic gain could come from greater resistance to late spring frost damage at flowering (as found for example in barley); agronomic changes would follow to take greatest advantage of such a breakthrough. Another breakthrough was predicted to come from more reliable and remunerative break crops.

Acceleration of PY progress is possible but will require more investment in conventional as well as "new" breeding technologies, and in novel agronomic technologies. This will be best achieved by collaborative effort between breeders, agronomists, and farmers. It is not clear, however, if the privatization of crop breeding, which has increased investment in crop R&D, is compatible with the open collaboration that is needed.

References

Aisawi, K. A. B., Reynolds, M. P., Singh, R. P., and Foulkes, M. J. (2015). The physiological basis of the genetic progress in yield potential of CIMMYT spring wheat cultivars from 1966 to 2009. *Crop Science, 55*, 1749–1764.

Araus, J. L., and Cairns, J. E. (2014). Field high-throughput phenotyping: the new crop breeding frontier. *Trends in Plant Science, 19*, 52–61.

Badu-Apraku, B., Fakorede, M. A. B., and Oyekunle M. (2014). Agronomic traits associated with genetic gains in maize yield during three breeding eras in West Africa. *Maydica, 59*, 49–57.

Beebe, S. E., Rao, I. M., Devi, M. J., and Polania, J. (2014). Common beans, biodiversity, and multiple stresses: challenges of drought resistance in tropical soils. *Crop & Pasture Science, 65*, 667–675.

Beche, E., Benin, G., Lemes da Silva, C., Munaro, L. B., and Marchese, J. A. (2014). Genetic gain in yield and changes associated with physiological traits in Brazilian wheat during the 20th century. *European Journal of Agronomy, 61*, 49–59.

Bell, M. A., and Fischer, R. A. (1994). Using yield prediction models to assess yield gains: a case study for wheat. *Field Crops Research, 36*, 161–166.

Bernardo, R. (2016). Bandwagons I, too, have known. *Theor. Appl. Genet., 129*, 2323–2332.

Chen. S., Wang, D., , Xu, C., Ji. C., Zhang, X., Zhao, X., Zhang, X., Singh Chauhan, B. (2014). Responses of Super Rice (*Oryza sativa* L.) to different planting methods for grain yield and nitrogen-use efficiency in the single cropping season. *PLoS ONE.* doi: 10.1371/journal.pone.0104950

Chen, K., Camberato, J. J., Tuinstra, M. R., Kumundini, S. V., Tollenaar, M., and Vyn, T. J. (2016). Genetic improvement in density and nitrogen stress tolerance traits over 38 years of commercial maize hybrid release. *Field Crops Res., 196*, 438–451.

Chenu, K. (2015). Characterizing the crop environment—nature, significance and applications. In Sadras, V. O., and Calderini, D. F. (eds.), *Crop physiology: applications for genetic improvement and agronomy*, pp. 321–348. San Diego: Academic Press.

Chenu, K., Chapman, S. C., Tardieu, F., McLean, G., Welcker, C., and Hammer, G. L. (2009). Simulating the yield impacts of organ-level quantitative trait loci associated with drought response in maize: A "gene to phenotype" modeling approach. *Genetics, 138*, 1507–1523.

Cormier, F., Faure, S., Dubreuil, P., Heumez, E., Beauchêne, B., Lafarge, S., Praud, S., and Le Gouis, J. (2013). A multi-environmental study of recent breeding progress on nitrogen use efficiency in wheat (Triticum aestivum L.). *Theor. Appl. Genetics.,126*, 3035–3048.

DeBruin, J. L., Schussler, J. R., Mo, H., and Cooper, M.(2017). Grain yield and nitrogen accumulation in maize hybrids released during 1934 and 2013 in the U.S. Midwest. *Crop Sci., 57*, 1431–1446.

Del Pozo, A., Matus, I., Serret, M. D., and Araus J. L. (2014). Agronomic and physiological traits associated with breeding advances of wheat under high-productive Mediterranean conditions. The case of Chile. *Environmental and Experimental Botany, 103*, 180–189.

De Pauw, R. M., Knox, R. E., Humphreys, D. G., Thomas, J. B., Fox, S. L., Brown, P. D., Singh, A. K., Pozniak, C., Randhawa, H. S., Fowler, D. B., Graf, R. J., and Hucl, P. (2011). New breeding tools impact Canadian commercial farmers. *Czech. J. Plant Breed., 47*, S28–S34.

Di Matteo, J. A., Ferreyra, J. M., Cerrudo, A. A., Echarte L., and Andrade, F. H. (2016). Yield potential and yield stability of Argentine maize hybrids over 45 years of breeding. *Field Crops Research, 197*, 107–116.

de Wit, C. T. (1992). Resource use efficiency in agriculture. *Agricultural Systems, 40*, 125–151.

Edmeades, G. O. (2013). Progress in achieving and delivering drought tolerance in maize—an update. In *Global status of commercialized biotech/GM crops: 2012*. ISAAA Brief No. 44, pp. 239–272. Ithaca, NY: International Service for the Acquisition of Agri-biotech Applications.

Espe, M. B., Hill, J. E., Leinfelder-Miles, M., Espino, L. A., Mutters, R., Mackill, D., van Kessel, C., and Linquist, B. A. (2018). Rice yield improvements through plant breeding are offset by inherent yield declines over time. *Field Crops Research, 222*, 59–65.

Evans L. T. (1993). *Crop evolution, adaptation and yield*. Cambridge: Cambridge University Press.

Evans L. T., and Fischer R. A. (1999). Yield potential: its definition, measurement, and significance. *Crop Sci., 39*, 1544–1551.

FAOSTAT. (2018). Food and Agriculture Organization, Data. Retrieved from http://www.fao.org/faostat/en/#data.

Fischer, R. A. (2015). Definitions and determinations of crop yield, yield gaps, and of rates of change. *Field Crops Res., 182*, 9–18.

Fischer, R. A. (2019). Closing the yield gaps. In Zeigler, R. (ed.), *Modern plant and agricultural sciences and public policies: creating a sustainable global food supply*. Melbourne: CSIRIO Publishing (in press).

Fischer, R. A., and Wall, P. C. (1976). Wheat breeding in Mexico and yield increases. *J. Aust. Inst. Agric. Sci., 42*, 139–148.

Fischer, R. A., and Connor D. J. (2018). Issues for cropping and agricultural science in the next 20 years. *Field Crops Research, 222*, 121–142.

Fischer, R. A., and Rebetzke, G.J. (2018). Indirect selection for potential yield in early generation, spaced plantings of wheat and other small grain cereals: a review. *Crop & Pasture Sci., 69*, 439–459.

Fischer, R. A., Byerlee, D., and Edmeades, G. O. (2014). Crop yield and global food security: will yield increase continue to feed the world? *ACIAR Monograph 158*. Canberra: Australian Centre for International Research. Retrieved from http://aciar.gov.au/publication/mn158.

Flohr, B. M., Hunt, J. R., Kirkegaard, J. A., Evans, J. R., Trevaskis, B., Zwart, A., Swan, A., Fletcher, A. L., and Rheinheimer, B. (2018). Fast winter phenology in wheat can stabilise flowering date over wide sowing dates and maximize yield in semi-arid environments. *Field Crops Res., 223*, 12–25.

Foulkes, J. M., Reynolds, M. P., and Sylvester-Bradley, R. (2009). Genetic improvement of grain crops: Yield potential. In Sadras V., and Calderini, D. (eds.) *Crop physiology: application for genetic improvement and agronomy*, pp. 306–325. Sydney: Academic Press.

Fox, C. M., Cary, T. R., Colgrove A. L., Nafziger E. D., Haudenshield J. S., Hartman G. L., Specht J. E., and Diers, B. W. (2013). Estimating soybean genetic gain for yield in the northern United States—Influence of cropping history. *Crop Sci., 53*, 2473–2482.

Furbank R. T., Quick, W. P., and Sirault, X. R. R. (2015). Improving photosynthesis and yield potential in cereal crops by targeted genetic manipulation: prospects, progress and challenges. *Field Crops Research, 182*, 19–29.

Gaffney, J., Schussler, J., Loffler, C., Cai, W., Paszkiewicz, S., Messina, C., Groeteke, J., Keaschall, J., and Cooper, M. (2015). Scale evaluation of maize hybrids selected for increased yield in drought-stress conditions of the U.S. corn belt. *Crop Sci., 55*, 1608–1618.

Gao, F., Ma, D., Yin, G., Awais, R., Dong, Y., Xiao, Y., Xia, X., Wu, X., He, Z. (2017). Traits in Chinese wheat cultivars of Southern Yellow and Huai Valley since 1950. *Crop Sci., 57*, 760–773.

Godfray, M. C. J., Beddington, J. R., Crute, I. R., Haddad, L., et al. (2010). Food security: The challenge of feeding 9 billion people. *Science, 327*, 812–818.

Grassini, P., van Bussel, L. G. J., van Wart, J., Claessens, L., Yang, H., Boogaard, H., de Groot, H., van Ittersum, M. K., and Cassman, K. G. (2015a). How good is good enough? Data requirements for reliable crop yield simulations and yield-gap analyses. *Field Crops Res., 177*, 49–63.

Grassini, P., Specht, J. E., Tollenaar, M., Ciampitti, I., and Cassman K. G. (2015b). High yield maize-soybean cropping systems in the U.S. Corn Belt. In Sadras, V.O., and Calderini, D. (eds.), *Crop physiology: applications for genetic improvement and agronomy*. 2nd Edition, pp. 17–41. Sydney: Academic Press.

Graybosch, R., Bockelman, H. E., Garland-Campbell, K. A., Garvin, D. F., and Regassa, T. (2014). Wheat. In Smith, S., Diers, B., Specht, J., and Carver, B. (eds.), *Yield gains in major U.S. field crops*, pp. 459–485. CSSA Special Publication 33, Madison, Wisconsin.

Gu, J., Chen, J., Chen, L., Wang, Z., Zhang, H., and Yang, J. (2015). Grain quality changes and responses to nitrogen fertilizer of japonica rice cultivars released in the Yangtze River Basin from the 1950s to 2000s. *The Crop Journal, 3*, 285–297.

Guilpart, N., Grassini, P., Sadras, V., Timsina, J., and Cassman, K. G. (2017). Estimating yield gaps at the cropping system level. *Field Crops Research, 206*, 21–32.

Haegele, J. W., Cook, K. A., Nichols, D. M., and Below, F. E. (2013). Changes in nitrogen use traits associated with genetic improvement for grain yield of maize hybrids released in different decades. *Crop Science, 53*, 1256–1268.

Hall, A. J., and Richards, R. A. (2013). Prognosis for genetic improvement of yield potential and water-limited yield of major grain crops. *Field Crops Research, 143*, 18–33.

Hartman, G. L., Bowen, C. R., Haudenshield, C. M., Fox, C. M., Cary, T. R., and Diers, B. W. (2015). Evaluation of disease and pest damage on soybean cultivars released from 1923 through 2008 under field conditions in central Illinois. *Agronomy J., 107*, 2373–2380.

Hobbs, P., Gupta, R., Jat, R. K., Malik, R. K. (2017). Conservation agriculture in the IndoGangetic Plains of India: Past, present and future. *Expl. Agric.*, 1–17. doi: 10.1017/S0014479717000424.

Hunt, J., Flohr, B., Fettell, N., Stear, M., Swan, T., Rheinheimer, B., Haskins, B., and Kirkegaard, J. (2017). Genetic yield progress from 1901–2014 in popular SNSW bread wheat cultivars. *Proc. 2017 Australian Agronomy Conference, 24–28 September 2017*, Ballarat, Australia. Retrieved from www.agronomyconference.com

Jaggard, K. W., Qi, A., and Ober, E. S. (2010). Possible changes to arable crop yields by 2050. *Philosophical Transactions of the Royal Society B, 365*, 2835–2851.

Kirkegaard, J. A., and Hunt, J. R. (2010). Increasing productivity by matching farming system management and genotype in water-limited environments. *Journal of Experimental Botany, 61*, 4129–4143.

Koester, R. P., Skoneczka, J. A., Cary, T. R., Diers, B. W., and Ainsworth, E. A. (2014). Historical gains in soybean (Glycine max Merr.) seed yield are driven by linear increases in light interception, energy conversion, and partitioning efficiencies. *Journal of Experimental Botany, 65*, 3311–3321.

Klümper, W., and Qaim, M. (2014). A meta-analysis of the impacts of genetically modified crops. *PLoS ONE, 9*, e111629.

Laidig, F., Piepho, H-P., Rentel, D., Drobek, T., Meyer, U., and Huesken, A. (2017). Breeding progress, environmental variation and correlation of winter wheat yield and quality traits in German official variety trials and on-farm during 1983–2014. *Theor. & Applied Genetics, 130*, 223–245.

Losert, D., Maurer, H. P., Marulanda, J. J., and Wurschum, T. (2017). Phenotypic and genotypic analyses and breeding progress in European triticale (X *Tritosecale* Wittmack). *Plant Breed., 136*, 18–27.

Mackay, I., Horwell, A., Garner, J., White, J., McKee, J., Philpott, H. (2011). Reanalyses of the historical series of UK variety trials to quantify the contributions of genetic and environmental factors to trends and variability in yield over time. *Theor. & Applied Genetics, 122*, 225–238.

Mackenzie, K. S., Sha, X., Moldenbauer, K. A. K., Lindscombe, S. D., Lynham, N. B., and Nalley, L. L. (2014). In Smith, S., Diers, B., Specht, J., and Carver, B. (eds.), *Yield gains in major U.S. field crops*, pp. 267–292. CSSA Special Publication 33, Madison, Wisconsin.

Morrison, M. J., Harker, K. N., Blackshaw, R. E., Holzaphel, C. J., and O'Donovan, J. T. (2016). Canola yield improvement on the Canadian Prairies. *Crop & Pasture Science, 67*, 245–252.

Nemali, K. S., Bonin, C., Dohleman, F. G., Stephens, M., Reeves, W. R., Nelson, D. E., Castiglioni, P., et al. (2015). Physiological responses related to increased grain yield under drought in the first biotechnologically derived drought-tolerant maize. *Plant, Cell and Environment, 38*, 1866–1880.

Ortiz-Monasterio, J. I., Sayre, K. D., Rajaram. S., and McMahon, M. (1997). Genetic progress in wheat yield and nitrogen use efficiency under four nitrogen rates. *Crop Science, 37*, 898–904.

Oury, F-X., Godin, C., Maillard, A., Chassin, A., Gardet, O., Giraudd, A., Heumez, E., Morlais, J-V., Rolland, B., Rousset, M., Trottet, M., and Charmetaet, G. (2012). A study of genetic progress due to selection reveals a negative effect of climate change on bread wheat yield in France. *Europ. J. Agronomy, 40*, 28–38.

Peng, S., Huang, J., Cassman, K. G., Laza, R. C., Visperas, R. M., and Khush, G.S. (2010). The importance of maintenance breeding—a case study of the first miracle rice variety—IR8. *Field Crops Research, 119*, 342–347.

Phiepo, H-P., Laidig, F., Drobek, T., and Meyert, U. (2014). Dissecting genetic and non-genetic sources of long-term yield trend in German official variety trials. *Theor & Appl Genetics, 127*, 1009–1018.

Potter T., Burton, W., Edwards, J., Mailer, R., Salisbury, P., and Pearce, A. (2016). Assessing progress in breeding to improve yield, quality, and black leg (*Leptosphaeria maculans*) in selected Australian canola cultivars (1978–2012). *Crop & Pasture Science, 67*, 308–316.

Reynolds, M. P., and Langridge, P. (2016). Physiological breeding. *Current Opinion in Plant Biology, 31*, 162–171.

Richards, R. A., Rebetzke, G. J., Condon, A. G., and van Herwaarden, A. F. (2002). Breeding opportunities for increasing the efficiency of water use and crop yield in temperate cereals. *Crop Science, 42*, 111–121.

Rincker, K., Nelson, R., Specht, J., Sleper, D., Cary, T., Cianzio, S. R., et al. (2014). Genetic improvement of U.S. soybean in Maturity Groups II, III, and IV. *Crop Science, 54*, 1–14.

Robertson, M., Kirkegaard, J., Rebetzke, G., Llewellyn, R., and Wark, T. (2016). Prospects for yield improvement in the Australian wheat industry: a perspective. *Food and Energy Security 5*, 107–122.

Roche, D. (2015). Stomatal conductance is essential for higher yield potential of C3 crops. *Critical Reviews in Plant Science, 34*, 429–453.

Rodriguez, D., de Voil, P., Hudson, D., Brown, J. N., Hayman, P., Marrou, H., and Meinke, H. (2018). Predicting optimum crop designs using crop models and seasonal climate forecasts. *Nature Scientific Reports, 8*, 22–31.

Rogers, J., Chen, P., Shi, A., Zhang, B., Scaboo, A., Faye Smith, S., and Zeng, A. (2015). Agronomic performance and genetic progress of selected historical soybean varieties in the southern USA. *Plant Breeding, 134*, 85–93.

Sadras, V. O., and Angus, J. F. (2006). Benchmarking water-use efficiency of rainfed wheat in dry environments. *Australian Journal of Agricultural Research, 57*, 847–856.

Sadras, V. O., and Lawson, C. (2011). Genetic gain in yield and associated changes in phenotype, trait plasticity and competitive ability of South Australian wheat varieties released between 1958 and 2007. *Crop & Pasture Science, 62*, 1–17.

Sadras, V. O., and Lawson, C. (2013). Nitrogen and water-use efficiency of Australian wheat varieties released between 1958 and 2007. *European J. Agronomy, 46*, 34–41.

Sadras, V.O., Lake, L., Li, Y., Farquarson, E.A., Sutton, T. (2016). Phenotypic plasticity and its genetic regulation for yield, nitrogen fixation and δ13C in chickpea crops under varying water regimes. *J. Exp. Botany, 67*, 4331–4351.

Sayre. K. D., Rajaram, S., and Fischer, R.A. (1997). Yield potential progress in short bread wheats in Northwest Mexico. *Crop Science, 37*, 36–42.

Silvey, V. (1981) The contribution of new wheat, oat and barley varieties to increasing yield in England and Wales 1947–1978. *J. Natl. Inst. Agric. Bot., 15*, 399–412.

Smith, S., Diers, B., Specht, J., and Carver, B. (2014a). *Yield gains in major U.S. field crops*. CSSA Special Publication 33, Madison, Wisconsin.

Smith, S., Cooper, M., Gogerty, J., Löffler, C., Borcherding, D., and Wright, K. (2014b). Maize. In Smith, S., Diers, B., Specht, J., and Carver, B. (eds.), *Yield gains in major U.S. field crops*, pp. 125–171. CSSA Special Publication 33, Madison, Wisconsin .

Specht, J. E., Diers, B. J., Nelson, R. L., de Toledo, J. F. F., Torrion J. A., and Grassini, P. (2014). Soybean. In Smith, S., Diers, B., Specht, J., and Carver, B. (eds.), *Yield gains in major U.S. field crops*, pp. 311–355. CSSA Special Publication 33, Madison, Wisconsin.

Tang, Y., Wu, X., Li, C., Yang, W., Huang, M., Ma, X., and Li, S. (2017). Yield, growth, canopy traits and photosynthesis in high-yielding, synthetic hexaploid-derived wheats cultivars compared with non-synthetic wheats. *Crop & Pasture Sci.*, *68*, 115–125.

Thomas, J. B., and Graf, R. J. (2014). Rates of yield gain of hard red spring wheat in western Canada. *Can. J. Plant Sci.*, *94*, 1–13.

UN (2017). United Nations Department of Economics and Social Affairs, Population Division. Retrieved from http://www.un.org/en/development/desa/population/.

Vadez, V., Kholova, J., Medina, S., Kakkera, A., and Anderberg, H.(2014). Transpiration efficiency: new insights into an old story. *J. Exp. Botany*, *65*, 6141–6153.

van Ittersum, M. K., Cassman, K. G., Grassini, P., Wolf, J., Tittonell, P., and Hochman, Z. (2013). Yield gap analysis with local to global relevance. *Field Crops Research*, *143*, 4–17.

Vandermark, G. J., Brick, M. A., Osorno, J. M., Kelly, J. D., and Urrea, C. A. (2014). Edible grain legumes. In Smith, S., Diers, B., Specht, J., and Carver, B. (eds.), *Yield gains in major U.S. field crops*, pp. 87–123. CSSA Special Publication 33, Madison, Wisconsin.

Venuprasad, R., Sta Cruz, M. T., Amante, M., Magbanua, R., Kumar, A., and Atlin, G. N. (2008). Response to two cycles of divergent selection for grain yield under drought stress in four rice breeding populations. *Field Crops Research*, *107*, 232–244.

World Bank (2017) Global Economic Monitor (commodities). Retrieved from http://databank.worldbank.org/data/reports.aspx?source=global-economic-monitor-commodities.

Xu, Z., Hennessy, D.A., Sardana, K., Moschini, G-C. (2013). The realized yield effect of genetically engineered crops: U.S. maize and soybean. *Crop Sci.*, *53*, 735–745.

PHYSIOLOGICAL BREEDING FOR CLIMATE CHANGE

Matthew Reynolds, Sivakumar Sukumaran,
Francisco Pinto, and Gemma Molero
Global Wheat Program, International Maize and
Wheat Improvement Center (CIMMYT), Mexico

Introduction

Given current trends in crop improvement (Ray et al., 2013), global demand for cereals is projected to exceed supply by 2050. Cereals currently achieve genetic yield gains of 0.5 to 1 percent annually (van Ittersum, 2016), but climate change could reduce this due to increased temperatures and less predictable rainfall patterns (Challinor et al., 2014; Asseng et al., 2015, 2017). Achieving the yields needed to match demand in increasingly unfavorable environments requires a better understanding of crops' adaptive characteristics, so that genetic variation for these value-added traits can be identified and incorporated into breeding programs. Physiological traits of varying complexity must be considered in crop improvement strategies (Chapman et al., 2012; Mir et al., 2012; Long et al., 2015). Genetic gains in yield *per se* are currently achieved via unspecified recombinations of minor-effect genes within elite germplasm pools, and the introgression of novel genetic diversity typically associated with disease resistance and grain quality (Braun et al., 2010; Eagles et al., 2014). Physiological interventions in conventional breeding add two main elements: the design of crossing strategies involving complementary traits achieved by thorough characterization of candidate parents, and secondly enrichment of favorable alleles in progeny by high-throughput phenotyping, genotyping, or their combination (Rutkoski et al., 2016; Tattaris et al., 2016). The physiological trait (PT)-based approach increases the probability of accumulating favorable alleles for higher yield under adverse growing conditions, compared to crossing physiologically uncharacterized lines. Unlike conventional breeding approaches that tend to consider few, easy to evaluate agronomic traits, physiological breeding considers a much larger range of traits, including traits that are genetically complex (Reynolds et al., 2009; Richards et al., 2010; Cossani and Reynolds, 2012; Reynolds et al., 2012). The physiological approach also differs from molecular breeding by encompassing both phenomic and genomic information. This is important as the genetic basis of many complex value-added traits remains poorly studied and usable markers are generally lacking. As shown in Figure 1, the key steps of a physiological breeding or pre-breeding pipeline are:

1. **Design crops** with combinations of traits that theoretically improve adaptation to specific target environments.
2. **Identify genetic resources** with new and/or complementary allelic variation.
3. **Develop phenotyping protocols** under appropriate experimental conditions, which will allow for desired physiological traits to be readily identified among panels of genetic resources.
4. **Genetically dissect** complex traits to estimate how different genetic germplasm pools may be complementary, and develop gene-based approaches for progeny selection.
5. **Make strategic crosses** considering complementary physiological traits to achieve cumulative gene action for progeny yield, using high-throughput phenotyping and genomic selection to identify the best performing sister lines.
6. **Multi-location trials** to identify trait/allele combinations that achieve environmentally robust genetic gains.

FIGURE 1. Key steps in a physiological breeding pipeline: 1) crop design, 2) genetic resource exploration, 3) phenotyping (mainly in field environments), 4) genetic analysis to enable marker-assisted breeding, 5) hybridization and progeny selection, 6) evaluation of genetic gains via multi-location testing systems, like the International Wheat Improvement Network (Braun et al., 2010), and 7) informatics services underpinning all activities. The latter underlines the interdependence of each step where, for example, results from multi-location trials combined with genetic analysis permit dissection of genotype × environment interaction, better-targeted phenotyping of genetic resources, and refined crossing strategies (Reynolds and Langridge, 2016).

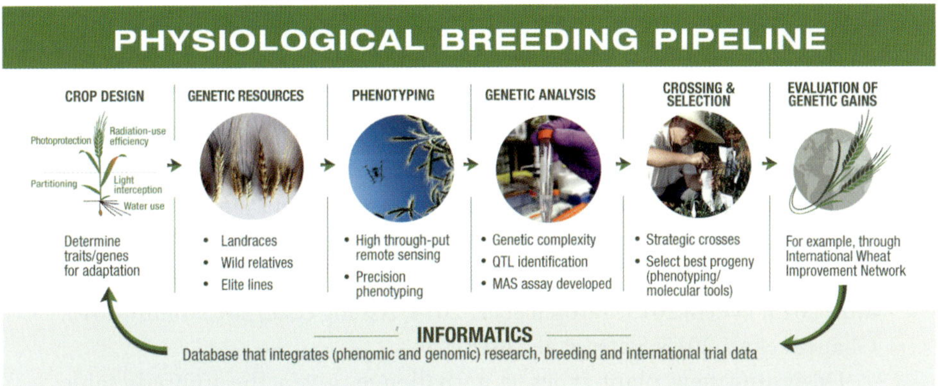

Strategic crossing involving complex traits will rely heavily on phenomics until there is a more comprehensive genetic basis selection tool explaining cultivar level differences in yield. Nonetheless, it currently utilizes markers associated with genes of major effects, such as *Ppd, Vrn,* and *Rht* in wheat (*Triticum* spp.) (Eagles et al., 2014). In addition, progeny selection can be facilitated by genomic selection (Cooper et al., 2014; Rutkoski et al., 2016), which is highly complementary to the physiological approach, considering the difficulties involved in measuring most physiological traits at the scale needed for breeding (which can involve thousands of plants from a single cross). Marker-assisted selection associated with minor effect genes is also expected to become increasingly useful as the knowledge base increases (Griffiths et al., 2015; Langridge and Reynolds, 2015). Genetic dissection to demonstrate successful trait combinations and the trade-offs among them (Sukumaran et al., 2017a) can also help design crosses, for which quantitative trait loci (QTL) analysis adds considerable weight (Pinto et al., 2010; Rebetzke et al., 2013; Cooper et al., 2014; Griffiths et al., 2015). This chapter presents the primary steps involved in testing new value-added trait gene combinations and achieving genetic gains for yield and adaptation via physiological breeding.

Designing Improved Crops

Physiological breeding begins with designing conceptual models of improved plant types, based on scientific literature, knowledge of useful traits used in mainstream breeding programs, outputs from simulation models (Chenu et al., 2013), and other experimental data from genetic resource evaluations. One of the key purposes of physiological breeding is to rigorously test the yield impacts of novel trait combinations, across a range of target environments. Except for yield itself, plant breeding has traditionally been dominated by selection for genetically simple traits such as disease resistance. However, developments in genomics and phenomics are helping to unravel more complex traits (Zhao et al., 2011; Li et al., 2012; Araus and Cairns, 2014; Ghanem et al., 2015), generating renewed interest in designing improved plant types for yield potential and stress adaptation under current and future climates (Richards et al., 2010; Chapman et al., 2012; Cossani and Reynolds, 2012; Reynolds et al., 2012; Wilkinson et al., 2012; Hall and Richards, 2013; Lynch, 2013; Tardieu et al., 2014; Asseng et al., 2015; Blum, 2015; Ghanem et al., 2015; Slafer et al., 2015).

Designing new plant types is a challenge, given the unpredictable genotype-environment-management (G×E×M) interactions and gaps in knowledge of the physiological and genetic bases of adaptation (G), further exacerbated by interactions with E (Chenu et al., 2013). Clearly, E has many dimensions and must be defined in terms of the variables that most strongly affect growth. These are temperature, radiation, and rainfall, which are in flux, predictable only within certain limits and subject to climate change. Other environmental factors (e.g., photoperiod and soil properties) are more predictable and constant, though this information is often lacking for many target environments, hardly surprising considering that crops are grown on around 1 billion hectares of land worldwide. Furthermore, determining which variables most strongly affect growth, and their interactions, is an active research subject (Hodson and White, 2010).

These factors have discouraged breeders from explicitly selecting for complex traits that express a high degree of G×E, and led to the notion that "big data" type approaches would more readily lead to genetic gains than integrated approaches based on knowledge of genetics (Langridge and Fleury, 2011). Similarly, research using controlled environments to reduce G×E and employing rapid life-cycle model species (favoring those with small genomes), were expected to extrapolate to crops (Gehan et al., 2015). However, these approaches have not yet initiated a paradigm shift in mainstream plant breeding (Langridge and Reynolds, 2015). In fact, controlled environments generate their own confounding factors, such that the response of complex traits is not comparable with field results (Poorter et al., 2016). An obvious example is root growth in pots, which cannot be expected to resemble that in the field, where temperature and depth profiles are completely different (Füller et al., 2012). Furthermore, a reliance on stochastic approaches tends to be retrospective, in that the phenotypic and genotypic data is typically generated using extant germplasm (Daetwyler et al., 2013). On the other hand, hypothesis-driven and mechanistic research does at least suggest novel levels of trait expression that can be subsequently explored among genetic resources in appropriate screening environments.

Nonetheless, the dissection of complex traits in terms of their physiology and genetic markers—aided in some cases by high-throughput field phenotyping—can help in designing new ideotypes; for example, by indicating the genetic bases for trait synergies or tradeoffs, which facilitate more refined crossing strategies (Reynolds et al., 2009; Pinto et al., 2010; Richards et al., 2010; Zhao et al., 2011; Bennett et al., 2012a; Li et al., 2012; Bonneau et al., 2013; Bustos et al., 2013; Rebetzke et al., 2013; Vadez et al., 2013; Cossani and Reynolds, 2015; Griffiths et al., 2015; Lopes et al., 2015; Pinto and Reynolds, 2015; Sukumaran et al., 2015a; Tattaris et al., 2016; van Ginkel and Ortiz, 2018). New lines developed using strategic crossing can be retrospectively analyzed to determine the most successful constellations of traits/alleles for crop design (Sukumaran et al., 2017a; Reynolds et al., 2018). In combination with empirical data, crop simulation models can test hypotheses about the potential genetic gains associated with new levels of complex trait expression (Chenu et al., 2013; Vadez et al., 2013; Tardieu et al., 2014). Thus, improved genotypes can be designed and validated either by recombining traits/genes present in existing gene pools, or by introgressing alleles from related gene pools in genetic resource collections worldwide (Feuillet et al., 2008; Ortiz et al., 2008).

Utilizing Genetic Resources

Genetic resources from primary, secondary, and tertiary gene pools have been used for crop improvement. The primary gene pool refers to lines with the same genome as the crop, typically encompassing modern and obsolete cultivars (including landraces). The secondary gene pool encompasses non-domesticated relatives with the same genome (or one or more of the same genomes, in the case of polyploids). The tertiary gene pool encompasses wild relatives with distinct but related genomes; in wheat, for example, this would include any species in the *Triticeae* tribe.

While interspecific hybridization is extremely challenging and typically results in sterility (in the absence of extreme measures), it has achieved significant impacts (Ortiz et al., 2008) and has been accepted as a risk-free way to move genes between related species (Harper et al., 2011). Polyploidy facilitates the use of genetic resources from the secondary and tertiary gene pools, most likely by providing genetic buffering. In the case of breadwheat (*Triticum aestivum*), wide crossing with wild relatives has been a highly successful strategy, perhaps in part because of the presence of homoeologous genomes (Trethowan and Mujeeb-Kazi, 2008). So-called synthetic wheats—which combine the durum (*T. durum*) (AB) genome with wild sources of the D genome, both from the secondary gene pool of hexaloid wheat—can yield as much as modern cultivars and express good radiation use efficiency (RUE), though agronomic type is poor. However, selected progeny from one or more backcrosses with elite hexaploids have resulted in significant improvements in yield and heat and drought adaptation, as well as disease resistance, though their genetic bases are not yet established (Reynolds et al., 2007; Trethowan and Mujeeb-Kazi, 2008; Lopes and Reynolds, 2011; Cossani and Reynolds, 2015; Reynolds et al., 2018).

Landraces provide novel sources of allelic diversity typically from the primary crop gene pool, and most countries have extensive and overlapping collections (e.g., Wingen et al., 2014). Focused Germplasm Identification Strategy (FIGS; http://www.figs.icarda.net/) is a tool combining "passport information" (i.e., any data collected at the time of collection) with geographic information systems data to identify landraces that have been collected in environments relevant to breeding targets. Landraces are quite common in the pedigrees of modern wheat and, while previously used primarily as sources for disease resistance (Smale et al., 2002; Ortiz et al., 2008), have recently shown value in increasing yield potential and heat tolerance across a range of key wheat growing environments (Reynolds et al., 2018).

Gene pool exploitation has been limited by a lack of appropriate technologies for evaluating large numbers of lines under abiotic stresses. Wheat, for example, with its large polyploidy genome, is biologically amenable to the broadening of its relatively narrow genetic base. Yet, to date only a small fraction of the nearly 0.5 million wheat gene bank accessions available worldwide (encompassing traditional landrace varieties, non-domesticated species, and advanced and obsolete cultivars) have been used in strategic crosses (Ortiz et al., 2008). The recent and accelerated development of high-throughput phenotyping (allowing thousands of lines to be screened) is expected to spur the selection and generation of new genotypes using existing genetic resources.

Phenotyping

High-throughput field phenotyping is already making significant contributions to physiological breeding (Araus and Cairns, 2014; Chapman et al., 2014; Tattaris et al., 2016). Rapid development of new sensors, platforms, and data processing algorithms has increased our capacity to measure environmental variables and various structural and physiological traits at unprecedented scales and frequencies. These complex datasets can potentially improve models predicting G×E, but they also present a challenge in data analysis and management.

High-throughput phenotyping generally necessitates the use of non-invasive or remote sensing of plant spectral reflectance. From the spectral profiles reflected by green tissue, indices are derived that relate to two main classes of traits: thermal/hydration properties measured in the infrared part of the electromagnetic spectrum, and pigment related indices assessed in visible bands (Araus and Cairns, 2014). Spectroscopic analysis covers traits like water relations, photosynthesis, plant temperature, nutrient status, and agronomic traits. Among the spectral indices, the Normalized Difference Vegetation Index (NDVI)—relating to the amount of green photosynthetic active material—has been widely used for assessing plants' performance, nitrogen content, and yield predictions (Haboudane et al., 2002; Hansen and Schjoerring, 2003; Fava et al., 2009; Tattaris et al., 2016). Meanwhile water indices predict differences in leaf and soil water potential, as well as yield and biomass (Gutierrez et al., 2010). Red-green-blue imagery is being used for growth analysis and 3D reconstruction of the canopy architecture, providing relevant information to model biomass accumulation and light interception and usage (Walter et al., 2018). Combinations of indices may ultimately substitute for gas exchange measurements (Serbin et al., 2012), perhaps in combination with sun-induced chlorophyll fluorescence (Zarco-Tejada et al., 2013), or using the whole spectra at leaf level (Silva-Perez et al., 2018). More complex empirical approaches integrate spectral information from several wavelengths to better predict complex traits, such as photosynthesis. These approaches involve complex statistics and modeling, including: multivariate linear regression (Hoerl and Kennard, 1970; Hernandez et al., 2015), partial least square regression (Feilhauer et al., 2010), principal component analysis (Bateson and Curtiss, 1996), and simplex volume maximization (Kersting et al., 2012; Römer et al., 2012).

Remotely sensed traits are ideal for large-scale screening and some may have adequate precision for detailed genetic or physiological dissection, necessary for identifying QTL or conducting genome-wide association studies (GWAS). Analyzing canopy temperature, for example, has helped identify QTL associated with drought and heat stress tolerance (Pinto et al., 2010; Rebetzke et al., 2012), while QTL common to both of these stresses were also shown to be associated with roots' adaptive responses (Pinto and Reynolds, 2015). Sensors dedicated to specific sets of wavelengths, such as the GreenSeeker that measures NDVI, can be applied to simple growth analysis (Araus and Cairns, 2014; Fahlgren et al., 2015).

Manned or unmanned low-flying platforms (Chapman et al., 2014; Tattaris et al., 2016) provide two main benefits over ground-based systems for increasing throughput and precision. Firstly, including hundreds of plots in a single image avoids the effects of environmental changes that may confound data acquired during lengthy, plot-to-plot ground measurements. Second, digital image analysis allows data to be curated via simple outlier analysis of pixel distributions within plots, which may be caused by damaged plants or heterogeneities associated with the soil, for example. Compared to ground-based data, these factors improve the correlations typically observed between remote-sensed traits (Tattaris et al., 2016). As technologies improve, the precision and scope of remote sensing will increase. Nonetheless, there are phenotypic traits that cannot easily be measured using remote sensing, such as detailed growth analysis, where destructive harvests are required for precision and to quantify partitioning among organs and other non-visible traits, such as root architecture.

Traits that have yet to be incorporated into reliable high-throughput protocols include aboveground biomass, fruiting efficiency, harvest index (HI), and source-sink balance. These are among the "lowest hanging fruit" in terms of increasing genetic gains, since they are frequently associated with yield and show considerable genetic variation within elite cultivars (Reynolds and Langridge, 2016). Identifying material with optimal expression of these traits will facilitate the adoption of strategic physiological approaches into mainstream breeding.

Biomass is one clear example of integrative trait value. While the biomass of elite cultivars has increased modestly in recent years (Shearman et al., 2005; Aisawi et al., 2015; Reynolds et al., 2018), genetic resource screenings have identified biomass values well over those expressed in the best modern cultivars (Reynolds et al., 2015). Table 1 shows data from a set of primary wheat synthetic lines evaluated under yield potential conditions in Mexico during two consecutive seasons, with the best line showing 31 percent more biomass than highest biomass check line (Sokoll) and 37 percent more biomass than the highest-yielding check (Borlaug 100).

TABLE 1. Ten primary wheat synthetics showing higher biomass than the best elite check (Sokoll) in yield trials of 160 pre-selected lines, evaluated across two seasons (2015–2016 and 2016–2017) under favorable irrigated conditions in northwest Mexico, unpublished data.

Primary Synthetic Cross	Yield	Maturity	Biomass*	% of Sokoll Biom*
	g/m2	Days	g/m2	(%)
68.111/RGB-U//WARD/3/FGO/4/RABI/5/AE.SQUARROSA (778)	500	113	2236	131
ALTAR 84/AE.SQUARROSA (895)	510	129	2069	121
CETA/AE.SQUARROSA (796)	520	127	2054	120
CETA/AE.SQUARROSA (273)	418	125	2046	120
68.111/RGB-U//WARD/3/FGO/4/RABI/5/AE.SQUARROSA (778)	406	122	2025	118
CROC_1/AE.SQUARROSA (466)	494	128	2021	118
GARZA/BOY//AE.SQUARROSA (350)	525	124	2019	118
68.111/RGB-U//WARD/3/FGO/4/RABI/5/AE.SQUARROSA (788)	457	116	2007	117
68.111/RGB-U//WARD/3/AE.SQUARROSA (325)	387	125	1992	116
DOY1/AE.SQUARROSA (517)	552	123	1974	115
Check (SOKOLL)	514	117	1710	-
Check (BORLAUG100 F2014)	581	117	1636	-
Heritability	0.604	0.848	0.656	
Grand Mean	482	120	1663	
LSD	166	7	301	
CV	11.5	1.2	9.5	
Genotype significance	0.000	0.000	0.000	
*adjusted with DTA as COV				

There is clear evidence for expression of superior biomass and biomass-related traits (such as light interception and RUE) in genetic resources, and a strong case for significant underutilized photosynthetic capacity in existing wheat germplasm (Molero et al., 2016). The overarching goal of exploring genetic diversity for biomass is to introduce allelic sources that can contribute to the expression of a high final biomass (and other photosynthetic-related traits) in a range of wheat elite genetic backgrounds. It is not trivial to perform growth analysis on large numbers of lines, as conventional processes are not high-throughput, though good associations have been demonstrated between NDVI and Water Index with in-season biomass (Babar et al., 2006). Furthermore, biomass can now be estimated at high-throughput under a range of environmental conditions using airborne platforms (Chapman et al., 2014; Tattaris et al., 2016). The development of new methodologies based on remote sensing techniques (3D reconstructions, LiDAR, etc.) will be crucial in selecting lines with higher biomass to accelerate genetic gains.

HI is another "low hanging fruit" of great importance. HI was greatly increased during the green revolution, thanks to the introduction of semi-dwarf wheat, but it still expresses significant genetic variation in most modern cultivars (e.g., Aisawi et al., 2015). This variation in the ability to partition resources to grain represents significant untapped yield potential, especially given the generally negative association between HI and biomass in most modern cultivars (Aisawi et al., 2015; Rivera-Amado et al., 2016). Table 2 shows potential yield gains associated with more optimal expression of HI in elite CIMMYT material, from the currently achievable level of 0.5 up to 0.6, which is close to the theoretical limit (Foulkes et al., 2011). These data indicate that step changes in yield are attainable at current levels of biomass, if HI can be better exploited.

TABLE 2. Wheat germplasm lines from the pre-breeding pipeline showing the highest biomass in replicated yield trials, northwest Mexico, 2013, unpublished data.

Cross	Yield	Maturity	Biomass	HI	Norm Biom**	% of check Biom**	Yield (g/m2) with theoretical HI		
	g/m2	Days	g/m2		g/m2	(%)	HI=0.5	HI=0.55	HI=0.6
WORRAKATTA/2*PASTOR	684	127	1766	0.39	1801	116	883	971	1059
WBLL1// PUB94.15.1.12/WBLL	718	133	1803	0.40	1753	113	901	992	1082
HE1/2*CNO79// BAV92/3/ROLF07	732	128	1728	0.42	1749	113	864	950	1037
SOKOLL// PUB94.15.1.12/WBLL1	685	126	1703	0.40	1746	113	851	936	1022
PASTOR//HXL7573/ 2*BAU/3/WBLL1	749	128	1720	0.44	1737	112	860	946	1032
WBLL1// YANGLING SHAANXI/ ESDA/3/ROLF07	731	129	1719	0.43	1732	112	860	946	1031
D67.2/PARANA 66.270// AE.SQUARROSA (320)/ 3/CUNNINGHAM	734	134	1791	0.41	1730	112	896	985	1075
BCN/WBLL1// PUB94.15.1.12/ WBLL1	757	133	1766	0.43	1719	111	883	971	1060
WBLL1// PUB94.15.1.12/WBLL	741	134	1773	0.42	1710	111	887	975	1064
CHEN/AE.SQ// WEAVER/3/SSERI1	730	129	1689	0.43	1698	110	844	929	1013
SOKOLL (Check in synthetics)	677	124	1609	0.42	1678	108	804	885	965
Check	710	131	1560	0.46	1548	100	780	858	936
Mean	721	129.7	1719	0.41	1717	111	859	945	1031
CV	4.6	1.2	5.9	5.0	-	-	-	-	-
LSD (5%)	64.5	3.2	188	0.04	-	-	-	-	-

*Best elite line from bread wheat breeding program (three-year data)
**Biomass normalized to average cycle length

In the future, to ensure that HI is expressed at >0.5 across the major wheat agro-ecosystems, assimilates must be optimally partitioned to reproductive structures (without sacrificing functional integrity of the roots and aboveground canopy). The physiological and genetic basis for assimilate partitioning among different plant organs is, relatively, poorly understood, though recent studies of wheat yield potential have reported large genetic ranges for dry matter partitioning among plant organs and some promising leads have been identified (Foulkes et al., 2015). Examples of the former include reducing structural dry matter investment to stem internodes 2 and 3 to enhance spike growth during stem elongation (Rivera-Amado, 2016), and increasing fruiting efficiency, essentially reducing chaff dry weight cost per unit of grain (Slafer et al., 2015). Genetic variation for these traits has recently been identified in elite populations and exotic germplasm (Molero et al., 2017).

It is well established that there is a dynamic source-sink interaction in plants and previous studies have highlighted genetic variation in source:sink balance and its importance in boosting yield and RUE (Reynolds et al., 2001; Sukumaran et al., 2015b). Experiments in cereals have demonstrated that a high demand for assimilates determined by sink strength of the grains—can stimulate the supply of photo-assimilates (Reynolds et al., 2005) and vice-versa (Calderini et al., 2001). These studies have demonstrated that wheat plants growing under yield potential conditions carry excess photosynthetic capacity, a probable conservative response to the natural risk of losing leaf area or photosynthetic function. As a result, increasing photosynthetic potential does not necessarily optimize grain number (Sadras, 2007), as shown by the negative association between HI and biomass (Aisawi et al., 2015; Bustos et al., 2013). To achieve full expression of yield potential, the source-sink dynamic must be optimized by ensuring that grain set expression matches the photosynthetic potential of current and future genotypes.

Recent studies using the most popular durum wheat variety in the Yaqui Valley (northwest Mexico), CIRNO C2008, found that its high yield potential was explained by a well-balanced source:sink, as previously observed in lines with the *Lr19* translocation from *Agropyron elongatum* (Reynolds et al., 2001). CIRNO establishes a strong grain-sink (large grain number, large grain weight potential, and no infertile spikelets per spike) that results in drastically increased RUE during grain-filling, high grain-filling rate, and therefore high yield (Molero, pers comm). These results highlight the importance of studying source- and sink-related traits at the same platform to boost yield potential.

Genetic Bases of Abiotic Stresses

The G×E×M interaction is mutually inclusive, and without understanding G or E or M, a comprehensive understanding of the genetic basis of traits under abiotic stresses is not feasible. Research on G is usually conducted in controlled growth chambers and greenhouse experiments, where E can be fixed, but their relevance to agronomic environments is questionable. This has paved the way for multi-locational trials to better understand trait G×E and heritability (H^2) estimates. Genomic regions determining the traits (in relation to E) have been studied mostly through multi-location trials (Sukumaran et al., 2018), but another level of complexity is added through recent epigenetic studies, which may be important for phenotypic plasticity and adaptation (Gardiner et al., 2015; Gallusci et al., 2017).

Genetic studies on economically important traits aim to understand: 1) amenability of traits to phenome- or genome-based breeding (H^2 being a key factor); 2) traits' relationship with environmental fluxes; and 3) marker-trait associations (MTA) or QTL-trait associations in elite lines and cultivars (Eagles et al., 2014). These studies also attempt to identify alternate sources of superior alleles and their combinations for marker-assisted breeding to break the existing wheat yield barrier (Eagles et al., 2014). The traits can be divided into different categories based on H^2 estimates (high/low) and ease of phenotyping (hard/easy).

While methods differ in statistical complexity, both crossed and natural populations are used to dissect traits' genetic bases (Sukumaran and Yu, 2014). Crossed populations include bi-parental populations (e.g., Seri/Babax), multi-parent advanced generation inter-crosses (e.g., MAGIC) (Kover et al., 2009), nested association mapping (NAM) (Yu et al., 2008), and panels that are generally an intelligent assembly of association-mapping populations for GWAS (Zhu et al., 2008). The bi-parental populations normally detect QTL-trait associations with higher mapping power but lower resolution (10–20 cM), whereas association mapping populations detect MTA—or even genes, e.g., Eps genes in WAMI (Sukumaran et al., 2016) with lower power but higher resolution (2–3 cM). Resolution and power also varies depending on population size and allele frequency. Minor alleles are not detected in GWAS due to statistical issues, whereas MAGIC and NAM populations can be effective in discovering minor alleles. NAM provides both genetic diversity and QTL detecting power, which is a combination of association mapping and bi-parental mapping approaches. MAGIC populations have a complex pedigree structure and can be used for both genetic dissection of traits and simultaneous genetic improvement of lines, though it is still too early to judge them on their effectiveness in genetic dissection and crop improvement (Huang et al., 2015).

The most popular approach in recent years has been GWAS, which uses statistical models to account for confounding effects of population structure and spurious associations (Yu et al., 2006; Zhang et al., 2010). Another confounding effect is associated with the variation in phenology in field measurements, where physiological traits depend on crop development, such as flowering time. This can be avoided using phenology-controlled populations, which have effectively detected QTL and MTA (Pinto et al., 2010; Sukumaran et al., 2015a). One such population, the Wheat Association Mapping Initiative (WAMI), has been used in several studies related to physiological traits (Lopes et al., 2015; Sukumaran et al., 2015a, 2015b, 2016, 2017a, 2017b; Valluru et al., 2017).

It is important to understand yield potential traits to evaluate the effects of heat, drought, and other abiotic stresses (Table 3; Sukumaran et al., 2018). Common QTL for drought and heat stress were identified in a bi-parental mapping study that utilized the Seri/Babax population with controlled phenology (Pinto et al., 2010; Pinto and Reynolds, 2015). Heat tolerance is an important quantitative trait, though only a few QTL mapping studies have focused on heat tolerance (Talukder et al., 2014).

TABLE 3. Genetic control for traits with potential for improving wheat performance under drought, heat, and high salt environments.

Abiotic Stress	Trait	Chromosomal location of genes	Source
Drought			
	Phenology	1A, 1B, 2A, 2B, 2D, 3A, 3B, 4A, 4D, 5A, 5B, 5D, 6A, 6B, 7A, 7B	Snape et al. (2001); Bennett et al. (2012b); Acuña-Galindo et al. (2015)
	Early vigor	2D, 3BL, 4B, 4D, 5A	Rebetzke et al. (2001); Bonneau et al. (2013)
	Staygreen	2B, 2D	Verma et al. (2004)
	Canopy temperature	1B, 2B, 3B, 4A	Pinto et al. (2010); Bennett et al. (2012a)
	Root biomass / penetration ability	1B, 2D, 4A, 6B, 7B	Waines and Ehdaie (2007); Acuña et al. (2014)
	Photosynthetic capacity	1B, 1D, 2B, 2D, 3B, 4A, 4B, 4D, 5A, 5B, 5D, 6A, 6B, 6D, 7A, 7B, 7D	Paulsen (2002); Kumar et al. (2012); Acuña-Galindo (2015); Wang et al. (2015)
	Stomatal conductance	1B, 2A, 2B, 2D, 4A, 4B, 4D, 7A, 7B	Rebetzke et al. (2013); Wang et al. (2015)
	Stem carbohydrates	1A, 2B, 2D, 3B, 4B, 5B, 6B, 7A, 7B	Rebetzke et al. (2008a)
	Glaucousness	2B, 2D, 3A	Tsunewaki and Ebana (1999); Bennett et al. (2012b)
	Coleoptile length	2B, 2D, 4A, 4B, 4D, 5D, 6B	Rebetzke et al. (2007)
	Osmotic adjustment	7A	Morgan and Tan (1996)
	Carbon Isotope Discrimination	1B, 2B, 3B, 4A, 4B, 4D, 5A, 7A	Rebetzke et al. (2008b)

Abiotic Stress	Trait	Chromosomal location of genes	Source
Drought (*cont.*)			
	Transpiration	1A, 1B, 4A, 7B	Parent et al. (2015); Wang et al. (2015)
Heat			
	Phenology	1A, 1B, 2A, 2B, 2D, 3A, 3B, 4A, 4D, 5A, 5B, 5D, 6A, 6B, 7A, 7B	Paulsen (2002); Acuña-Galindo et al. (2015)
	Early vigor	1B, 2D, 4B, 4D, 5A, 7D	Reynolds and Rebetzke (2011); Lopes et al. (2013)
	Staygreen	1As, 2B, 2D, 3A, 3B, 3BS, 6A, 6B, 7A, 7Ds	Kumar et al. (2010); Vijayalakshmi (2010)
	Canopy temperature	1B, 1D, 2B, 3B, 3BL, 4A, 5DL	Reynolds and Rebetzke (2011); Bennett et al. (2012a); Mason et al. (2013); Tiwari et al. (2013)
	Root biomass	1B, 2AL, 7AL, 7BL	Reynolds and Rebetzke (2011)
	Photosynthetic capacity	1A, 1B, 1D, 2B, 2D, 3B, 3D, 4A, 4B, 4D, 5A, 6A, 7A, 7B, 7D	Reynolds and Rebetzke (2011); Azam et al. (2014)
	Stomatal conductance	1B, 2A, 2B, 2D, 4A, 4B, 4D, 5A, 7A, 7B	Reynolds and Rebetzke (2011)
	Stem carbohydrates	1A, 1B, 1D, 2D, 2B, 3D, 4A, 4B, 5B, 6A, 7B, 7D	Rebetzke et al. (2008a)
	Glaucousness	2B, 2D, 3A	Tsunewaki and Ebana (1999); Bennett et al. (2012b)
	Wax	1B, 3D, 5A	Mondal et al., 2015
	Transpiration efficiency (leaf)	1B, 1D, 2D, 3B, 4A, 4B, 4D, 5A, 7A, 7B	Reynolds and Rebetzke (2011)
	Transpiration efficiency (grain)	1D, 2A, 2D, 4B, 4D, 6D, 7B	Reynolds and Rebetzke (2011)
	Membrane thermostability	7A	Ciucă and Petcu (2009)
Salinity			
	Na+ exclusion	2AL, 5AL, 6A	Lindsay et al. (2004); Byrt et al. (2007); Genc et al. (2010)
	Osmotic tolerance	2B, 1B, 5B	Shamaya (2013)
	Chlorophyll content in young leaves	3A	Turki et al. (2015)
	Root length	3A, 6A, 5B	Turki et al. (2015)
	Na+ accumulation in distal leaves	1D, 3B	Shamaya (2013)
	Na+/K+ discrimination	4B	Shamaya (2013)

A recent study in durum wheat identified QTL hotspots for yield and component traits under yield potential, heat stress, and drought stress conditions (Sukumaran et al., 2018). The challenge remains in creating a synthetic line (a cross between durum and *Aegilops tauschii*) where these QTL are incorporated and transferred into an elite line for further use. How much diversity can be incorporated into the elite materials and cultivars depends on knowing the current diversity of yield-contributing QTL, finding alternative QTL sources (landraces, synthetics, etc.) for grain yield, and incorporating them into these lines.

Strategies for Crossing and Selection

PT-based crossing strategies are now within the grasp of breeding programs, complementing extant trait-based approaches that involve genetically simpler characteristics like disease resistance and agronomic traits (Reynolds et al., 2009; Reynolds et al., 2018). As outlined earlier, conceptual models (e.g. Richards et al., 2010; Cossani and Reynolds, 2012; Reynolds et al., 2012) and simple quantitative models help identify potentially complementary traits (Reynolds et al., 2009; Reynolds et al., 2015), while screening genetic resources in an appropriate range of environments can identify actual parental sources. Progeny selection is facilitated by remote sensing (Tattaris et al., 2016). The current crossing and selection scheme in physiological breeding utilizes diverse genetic resources, different conceptual models, and molecular genetic diversity, ultimately delivering improved germplasm to developing countries through the Wheat Yield Collaboration Yield Trial (WYCYT) and Stress Adaptive Trait Yield Nursery (SATYN) international nurseries (Fig. 2).

A comprehensive analysis of results from the WYCYT has shown step changes in yield potential (average 7 percent above elite checks across 26 international sites) in crosses explicitly designed to combine favorable sink-related traits (HI, grains/spike, kernel weight, etc.) with high RUE lines (Reynolds et al., 2018). Biomass was also increased in the same materials by up to 12 percent, depending on genetic background and environment. Considering another example under drought, parents with expression of a cool canopy typically associated with deeper roots were crossed with lines that store and remobilize an above average amount of stem water soluble carbohydrates, resulting in progeny that showed equal or better expression of both traits than either parents as well as superior yield under drought (Reynolds et al., 2015). Other examples come from experimental PT crosses; for example, the wheat cross Bacanora/Weebil—in which both parents achieve high yield by expressing distinct yield components—resulted in progeny with exceptional yield potential (Bustos et al., 2013). In a number of cases, new lines and physiological phenotyping approaches have been adopted by national programs (Table 4; Pask et al., 2014).

TABLE 4. Lines developed using physiological breeding released by the National Wheat Program of Pakistan.

Year	Name	Cross / pedigree	Selection History
2013	Pakistan-13	MEX94.27.1.20/3/SOKOLL//ATTILA/3*BCN	PTSS02B00132T-0TOPY-0B-0Y-0B-38Y-0M-0SY
2016	Borlaug-16	SOKOLL/3/PASTOR//HXL7573/2*BAU	PTSS02B00015S-0Y-0B-0Y-0B-1Y-0M-0SY
2017	Kohat-17	SOKOLL/WBLLI	PTSS02Y00021S-0Y-030ZTM-040SY-040M-5Y-0M-0SY

FIGURE 2. Crossing and selection scheme in physiological breeding leading to international nurseries (WYCYTs and SATYNs).

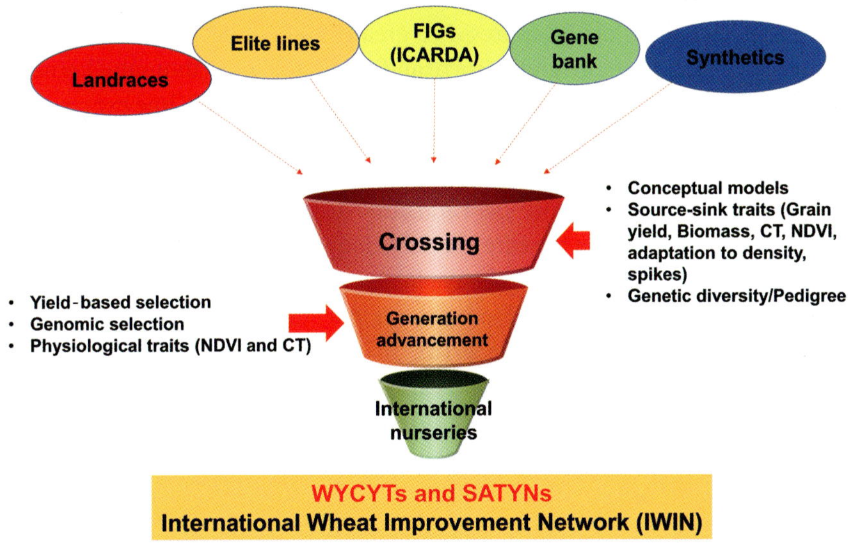

Another promising pre-breeding approach is the use of molecular markers where genetic basis is not known. Instead, breeding and selection decisions are made based on genome-wide markers in genomic selection models (Yu et al., 2016; Crossa et al., 2017). Multiple studies have evaluated the use of genomic prediction models in breeding, and G × E reaction norm models have shown improved prediction accuracy in spring wheat (Sukumaran et al., 2017b, 2017c). Speed breeding-based approaches have the potential to advance six generations of spring wheat per year, potentially in combination with genomic selection (Watson et al., 2018).

Evaluating Genetic Gains

Strategic crossing for PTs incorporates genetic diversity for complex traits into the gene pool and provides an "acid test" for validating hypotheses about trait interactions, genetic background effects, and the value of plant ideotypes across environments. For example, considering source × sink crosses for yield potential, the highest yielding line internationally from the third WYCYT, a PT line with the pedigree MEX94.27.1.20/3/SOKOLL//ATTILA/3*BCN/4/PUB94.15.1.12/WBLL1, was the product of a cross between parents P1 (MEX94.27.1.20/3/SOKOLL//ATTILA/3*BCN) and P2 (PUB94.15.1.12/ WBLL1), which both have Mexican landraces in their pedigree. Parents and progeny sown side by side in breeding environments showed that, while both parents had similar yields (2 percent difference), P1 expressed a greater "source," as indicated by 11 percent more biomass at maturity than P2, while P2 expressed a greater sink, indicated by 8 percent larger HI, 6 percent more grains/spike and 9 percent larger thousand kernel weight (TKW) than P1. In the same environment, the new PT line expressed 11 percent more yield, 4 percent larger HI, and 2 percent greater TKW and biomass at maturity than the best parent. At the selection site, the new PT line expressed 5 percent more yield and almost 20 percent greater TKW and biomass at maturity than the elite check Borlaug 100, which is the best yielding elite CIMMYT line currently available. Its superior performance over Borlaug 100 was also reflected at international sites in terms of yield, biomass, and TKW (Reynolds et al., 2018).

Understanding the genetic bases of trait combinations that result in yield improvement can illustrate successful interactions among alleles. Some of the best progeny of the wheat cross Bacanora/Weebil possessed two loci on chromosomes 1B and 7B that improved grain number and grain weight, respectively (Griffiths et al., 2015), even though these yield components are usually negatively associated. The absence of this trade-off in some progeny resulted in extremely high yields (Bustos et al., 2013) suggesting the loci as targets for marker-assisted selection (Griffiths et al., 2015). Furthermore, genetic dissection of the Seri/Babax wheat population under drought and heat stress (Pinto et al., 2010) showed that yield QTL were co-located with several PT QTL: canopy temperature (three QTL), NDVI (three QTL), and chlorophyll (one QTL), illustrating the genetic correlation of each trait to yield (Fig. 3). Genetic dissection of complex physiological traits can suggest potentially favorable combinations of alleles, even between genotypes with similar phenotypes. In addition, genetic dissection can show that expression of different phenotypes may share a common genetic basis.

FIGURE 3. Genetic dissection of yield by physiological traits. The Venn diagram shows how QTL for yield is shared by several different physiological traits contributing to its expression, including canopy temperature (CT); NDVI, and chlorophyll (CHL). Weaker QTL effects for water-soluble carbohydrates were common to all four (Pinto et al., 2010). Bold fonts denote QTL for yield that are stable across environments, while non-bold fonts denote environment-dependent QTL for yield (from Reynolds and Langridge, 2016).

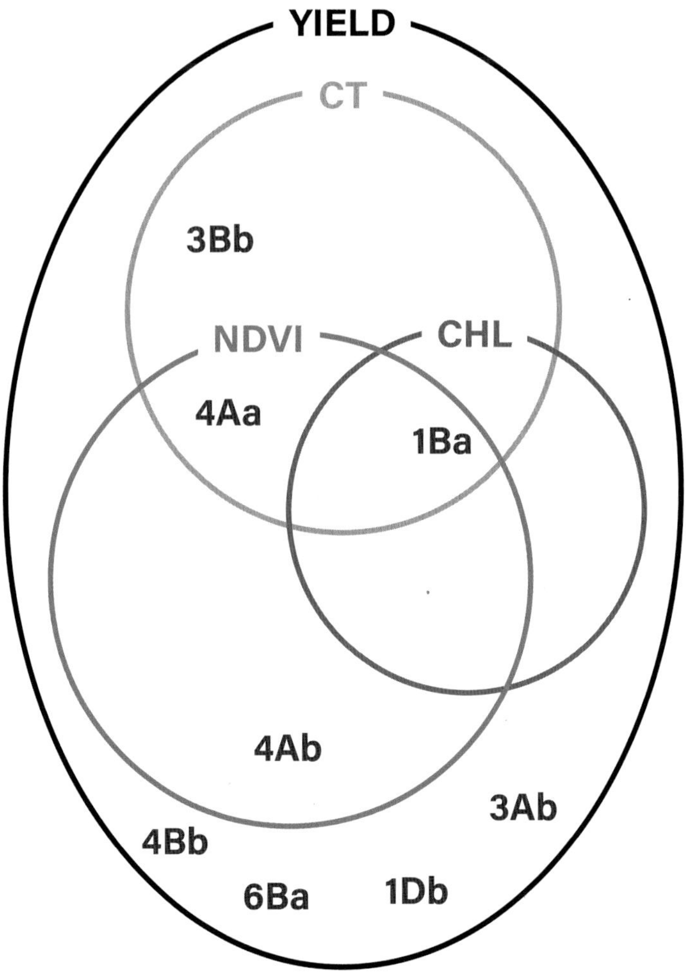

Conclusions

Climate change is a serious threat to food security, but recent advances in genomics and high-throughput phenotyping are enabling new methods of exploring genetic diversity. Physiological breeding is now more feasible as high-throughput methods for evaluating the physiological and genetic bases of yield and adaptation make complex traits more accessible to crossing and progeny selection. Recent proofs of concept outlined in this chapter and elsewhere have shown that genetic gains in wheat can be achieved under yield potential, drought, and heat stress conditions by combing complementary traits in parents and selecting the best progeny. Many successful crosses encompass exotic parents in their pedigrees, which is important for two main reasons. Firstly, it shows that—when selected judiciously as parents—exotic material can contribute benefits to performance in the relatively short term. Secondly, since many of these materials also expressed higher biomass than checks, it is likely that recycling them in breeding gene pools will achieve further genetic gains to yield. Given the recent emphasis on improving crop photosynthesis and understanding the genetic and physiological bases of adaptive traits in crops (e.g., Realizing Increased Photosynthetic Efficiency and the International Wheat Yield Partnership), physiological breeding now offers a tried-and-tested methodology to translate many hypotheses into proofs of concept.

Acknowledgments

The development of physiological breeding as outlined in this chapter has been supported by the Sustainable Modernization of Traditional Agriculture (MasAgro) initiative from the Secretariat of Agriculture, Livestock, Rural Development, Fisheries, and Food (SAGARPA), the International Wheat Yield Partnership (IWYP), the Grains Research and Development Corporation (GRDC), the U.S. Agency for International Development (USAID), and the German Federal Ministry for Economy Cooperation and Development (BMZ). Editorial assistance from Emma Quilligan is also gratefully acknowledged.

References

Acuña, T. B., Rebetzke, G. J., He, X., Maynol, E., and Wade, L. J. (2014). Mapping quantitative trait loci associated with root penetration ability of wheat in contrasting environments. *Mol. Breeding*, *34*(2), 631–642.

Acuña-Galindo, M. A., Mason, R. E., Subramanian, N. K., and Hays, D. B. (2015). Meta-analysis of wheat QTL regions associated with adaptation to drought and heat stress. *Crop Sci.*, *55*(2), 477–492.

Aisawi, K. A. B., Reynolds, M. P., Singh, R. P., and Foulkes, M. J. (2015). The physiological basis of the genetic progress in yield potential of CIMMYT spring wheat cultivars from 1966 to 2009. *Crop Sci.*, *55*(4), 1749–1764.

Araus, J. L., and Cairns, J. E. (2014). Field high-throughput phenotyping: the new crop breeding frontier. *Trends in Plant Sci.*, *19*(1), 52–61.

Asseng, S., Ewert, F., Martre, P., Rötter, R. P., Lobell, D. B., Cammarano, D., Kimball, B. A., Ottman, M. J., Wall, G. W., White J. W., et al. (2015). Rising temperatures reduce global wheat production. *Nat. Clim. Change*, *5*, 143–147.

Asseng, S., Cammarano, D., Basso, B., Chung, U., Alderman, P., Sonder, K., Reynolds, M., and Lobell, D. (2017). Hotspots of wheat yield decline with rising temperatures. *Global Change Biol.*, *23*(6), 2464–2472.

Azam, F. I., Chang, X., and Jing, R. (2015). Mapping QTL for chlorophyll fluorescence kinetics parameters at seedling stage as indicators of heat tolerance in wheat. *Euphytica*, *202*(2), 245–258.

Babar, M. A., Reynolds, M. P., Van Ginkel, M., Klatt, A. R., Raun, W. R., and Stone, M. L. (2006). Spectral reflectance to estimate genetic variation for in-season biomass, leaf chlorophyll, and canopy temperature in wheat. *Crop Sci.*, *46*(3), 1046–1057.

Bateson, A. and Curtiss, B., 1996, A method for manual endmember selection and spectral unmixing. *Remote Sens. Environ.*, *55*(3), 229–243.

Bennett, D., Reynolds, M., Mullan, D., Izanloo, A., Kuchel, H., Langridge, P., and Schnurbusch, T. (2012a). Detection of two major grain yield QTL in bread wheat (Triticum aestivum L.) under heat, drought and high yield potential environments. *Theor. Appl. Genet.*, *125*(7), 1473–1485.

Bennett, D., Izanloo, A., Edwards, J., Kuchel, H., Chalmers, K., Tester, M., Reynolds, M., Schnurbusch, T., and Langridge, P. (2012b). Identification of novel quantitative trait loci for days to ear emergence and flag leaf glaucousness in a bread wheat (Triticum aestivum L.) population adapted to southern Australian conditions. *Theor. Appl. Genet.*, *124*(4), 697–711.

Blum, A. (2015). Towards a conceptual ABA ideotype in plant breeding for water limited environments. *Funct. Plant Biol.*, *42*(6), 502–513.

Bonneau, J., Taylor, J., Parent, B., Bennett, D., Reynolds, M., Feuillet, C., Langridge, P., and Mather, D. (2013). Multi-environment analysis and improved mapping of a yield-related QTL on chromosome 3B of wheat. *Theor. Appl. Genet.*, *126*(3), 747–761.

Braun, H. J., Atlin, G., and Payne, T. (2010). Multi-location testing as a tool to identify plant response to global climate change.. In Reynolds, M. P. (ed.), *Climate change and crop production*, pp. 115–138. Wallingford, UK: CABI Publishers.

Bustos, D. V., Hasan, A. K., Reynolds, M. P., and Calderini, D. F. (2013). Combining high grain number and weight through a DH-population to improve grain yield potential of wheat in high-yielding environments. *Field Crop. Res.*, *145*, 106–115.

Byrt, C. S., Platten, J. D., Spielmeyer, W., James, R. A., Lagudah, E. S., Dennis, E. S., Tester, M., and Munns, R. (2007). HKT1; 5-like cation transporters linked to Na+ exclusion loci in wheat, Nax2 and Kna1. *Plant Physiol.*, *143*(4), 1918–1928.

Calderini, D. F., Savin, R., Abeledo, L. G., Reynolds, M. P., and Slafer, G. A. (2001). The importance of the period immediately preceding anthesis for grain weight determination in wheat. *Euphytica*, *119*, 199–204.

Challinor, A. J., Watson, J., Lobell, D. B., Howden, S. M., Smith, D. R., and Chhetri, N. (2014). A meta-analysis of crop yield under climate change and adaptation. *Nat. Clim. Change*, *4*, 287–291.

Chapman, S. C., Chakraborty, S., Dreccer, M. F., Howden, S. M. (2012). Plant adaptation to climate change—opportunities and priorities in breeding. *Crop Pasture Sci.*, *63*, 251–268.

Chapman, S. C., Merz, T., Chan, A., Jackway, P., Hrabar, S., Dreccer, M. F., Holland, E., Zheng, B., Ling, T. J., and Jimenez-Berni, J. (2014). Pheno-copter: a low-altitude, autonomous remote-sensing robotic helicopter for high-throughput field-based phenotyping. *Agronomy*, *4*(2), 279–301.

Chenu, K., Deihimfard, R., and Chapman, S. C. (2013). Large-scale characterization of drought pattern: a continent-wide modelling approach applied to the Australian wheatbelt–spatial and temporal trends. *New Phytologist*, *198*(3), 801–820.

Ciucă, M., and Petcu, E. (2009). SSR markers associated with membrane stability in wheat (Triticum aestivum L.). *Romanian Agric. Res.*, *26*, 21–24.

Cooper, M., Messina, C. D., Podlich, D., Totir, L. R., Baumgarten, A., Hausmann, N. J., Wright, D., and Graham, G. (2014). Predicting the future of plant breeding: complementing empirical evaluation with genetic prediction. *Crop Pasture Sci.*, *65*, 311–336.

Cossani, C. M., and Reynolds, M. P. (2012). Physiological traits for improving heat tolerance in wheat. *Plant Physiol.*, *160*, 1710–1718.

Cossani, C. M., and Reynolds, M. P. (2015). Heat stress adaptation in elite lines derived from synthetic hexaploid wheat. *Crop Sci.*, *55*, 2719–2735.

Crossa, J., Pérez-Rodríguez, P., Cuevas, J., Montesinos-López, O., Jarquín, D., de los Campos, G., Burgueño, J., Camacho-González, J. M., Pérez-Elizalde, S., Beyene, Y., and Dreisigacker, S. (2017). Genomic selection in plant breeding: methods, models, and perspectives. *Trends Plant Sci.*, *22*(11), 961–975.

Daetwyler, H. D., Calus, M. P., Pong-Wong, R., de los Campos, G., and Hickey, J. M. (2013). Genomic prediction in animals and plants: simulation of data, validation, reporting, and benchmarking. *Genetics*, *193*(2),347–365.

Eagles, H. A., Cane, K., Trevaskis, B., Vallance, N., Eastwood, R. F., Gororo, N. N., Kuchel, H., and Martin, P. J. (2014). Ppd1, Vrn1, ALMT1 and Rht genes and their effects on grain yield in lower rainfall environments in southern Australia. *Crop Pasture Sci.*, *65*, 159–170.

Fahlgren, N., Gehan, M. A., and Baxter, I. (2015). Lights, camera, action: high-throughput plant phenotyping is ready for a close-up. *Curr. Opin. Plant Biol.*, *24*, 93–99.

Fava, F., Colombo, R., Bocchi, S., Meroni, M., Sitzia, M., Fois, N., and Zucca, C. (2009). Identification of hyperspectral vegetation indices for Mediterranean pasture characterization. *Int. J. Appl. Earth Obs. Geoinf.*, *11*(4), 233–243.

Feilhauer, H., Asner, G. P., Martin, R. E., and Schmidtlein, S. (2010). Brightness-normalized partial least squares regression for hyperspectral data. *J. Quant. Spectrosc. RA.*, *111*(12–13), 1947–1957.

Feuillet, C., Langridge, P., and Waugh, R. (2008). Cereal breeding takes a walk on the wild side. *Trends Genet.*, *24*(1), 24–32.

Foulkes, M. J., Slafer, G. A., Davies, W. J., Berry, P. M., Sylvester-Bradley, R., Martre, P., Calderini, D. F., Griffiths, S., and Reynolds, M. P. (2011). Raising yield potential of wheat. III. Optimizing partitioning to grain while maintaining lodging resistance. *J. Exp. Bot.*, *62*(2), 469–486.

Foulkes, M. J., Rivera-Amado, A., Trujillo, E., Sylvester-Bradley, R., and Reynolds, M. P. (2015). Achieving a step-change in harvest index in high biomass wheat cultivars. In Reynolds, M. P., Molero, G., Mollins, J., and Braun, H. J. (eds.) *Proceedings of the International TRIGO (Wheat) Yield Potential Workshop 2015. CENEB, CIMMYT, Cd. Obregón, Sonora, Mexico, 24–26 March 2015*, pp. 31–35.

Füllner, K., Temperton, V. M., Rascher, U., Jahnke, S., Rist, R., Schurr, U., and Kuhn, A. J. (2012). Vertical gradient in soil temperature stimulates development and increases biomass accumulation in barley. *Plant Cell Environ.*, *35*(5), 884–892.

Gallusci, P., Dai, Z., Génard, M., Gauffretau, A., Leblanc-Fournier, N., Richard-Molard, C., Vile, D., and Brunel-Muguet, S. (2017). Epigenetics for plant improvement: current knowledge and modeling avenues. *Trends Plant Sci.*, *22*(7), 610–623.

Gardiner, L. J., Quinton-Tulloch, M., Olohan, L., Price, J., Hall, N., and Hall, A. (2015). A genome-wide survey of DNA methylation in hexaploid wheat. *Genome Biol.*, *16*(1), 273.

Gehan, M. A., Greenham, K., Mockler, T. C., and McClung, C. R. (2015). Transcriptional networks—crops, clocks, and abiotic stress. *Curr. Opin. Plant Biol.*, *24*, 39–46.

Genc, Y., Oldach, K., Verbyla, A. P., Lott, G., Hassan, M., Tester, M., Wallwork, H., and McDonald, G. K. (2010). Sodium exclusion QTL associated with improved seedling growth in bread wheat under salinity stress. *Theor. Appl. Genet.*, *121*(5), 877–894.

Ghanem, M. E., Marrou, H., and Sinclair, T. R. (2015). Physiological phenotyping of plants for crop improvement. *Trends in Plant Sci.*, *20*(3), 139–144.

Griffiths, S., Wingen, L., Pietragalla, J., Garcia, G., Hasan, A., Miralles, D., Calderini, D. F., Ankleshwaria, J. B., Waite, M. L., Simmonds, J., and Snape, J. (2015). Genetic dissection of grain size and grain number trade-offs in CIMMYT wheat germplasm. *PLoS ONE*, *10*(3), e0118847.

Gutierrez, M., Reynolds, M. P., and Klatt, A. R. (2010). Association of water spectral indices with plant and soil water relations in contrasting wheat genotypes. *J. Exp. Bot.*, *61*(12), 3291–3303.

Haboudane, D., Miller, J. R., Tremblay, N., Zarco-Tejada, P. J., and Dextraze, L.(2002). Integrated narrow-band vegetation indices for prediction of crop chlorophyll content for application to precision agriculture. *Remote Sens. Environ.*, *81*(2–3), 416–426.

Hall, A. J., and Richards, R. A. (2013). Prognosis for genetic improvement of yield potential and water-limited yield of major grain crops. *Field Crop. Res.*, *143*, 18–33.

Hansen, P. M., and Schjoerring, J. K. (2003). Reflectance measurement of canopy biomass and nitrogen status in wheat crops using normalized difference vegetation indices and partial least squares regression. *Remote Sens. Environ.*, *86*(4), 542–553.

Harper, J., Armstead, I., Thomas, A., James, C., Gasior, D., Bisaga, M., Roberts, L., King, I., and King, J. (2011). Alien introgression in the grasses Lolium perenne (perennial ryegrass) and Festuca pratensis (meadow fescue): the development of seven monosomic substitution lines and their molecular and cytological characterization. *Ann. Bot.*, *107*(8), 1313–1321.

Hernandez, J., Lobos, G. A., Matus, I., del Pozo, A., Silva, P., and Galleguillos, M. (2015). Using ridge regression models to estimate grain yield from field spectral data in bread wheat (Triticum aestivum L.) grown under three water regimes. *Remote Sens.*, *7*(2), 2109–2126.

Hodson, D., and White, J. (2010). GIS and crop simulation modelling applications in climate change research. In Reynolds, M. P. (ed.), *Climate change and crop production.* Wallingford, UK: CABI Publishers, 245–262.

Hoerl, A. E., and Kennard, R. W. (1970). Ridge regression: applications to nonorthogonal problems. *Technometrics*, *12*(1), 69–82.

Huang, B. E., Verbyla, K. L., Verbyla, A. P., Raghavan, C., Singh, V. K., Gaur, P., Leung, H., Varshney, R. K., and Cavanagh, C. R. (2015). MAGIC populations in crops: current status and future prospects. *Theor. Appl. Genet.*, *128*(6), 999–1017.

Kersting, K., Wahabzada, M., Römer, C., Thurau, C., Ballvora, A., Rascher, U., Léon, J., Bauckhage, C., and Plümer, L. (2012). Simplex distributions for embedding data matrices over time. In *Proceedings of the 2012 SIAM international conference on data mining.* Society for Industrial and Applied Mathematics. doi: 10.1137/1.978161972825.26.

Kover, P. X., Valdar, W., Trakalo, J., Scarcelli, N., Ehrenreich, I. M., Purugganan, M. D., Durrant, C., and Mott, R. (2009). A multiparent advanced generation inter-cross to fine-map quantitative traits in Arabidopsis thaliana. *PLoS Genetics*, *5*(7), e1000551.

Kumar, U., Joshi, A. K., Kumari, M., Paliwal, R., Kumar, S., and Röder, M. S. (2010). Identification of QTLs for stay green trait in wheat (Triticum aestivum L.) in the "Chirya 3" × "Sonalika" population. *Euphytica*, *174*(3), 437–445.

Kumar, S., Sehgal, S. K., Kumar, U., Prasad, P. V., Joshi, A. K., and Gill, B. S. (2012). Genomic characterization of drought tolerance-related traits in spring wheat. *Euphytica*, *186*(1), 265–276.

Langridge, P., and Fleury, D. (2011). Making the most of "omics" for crop breeding. *Trends Biotechnol.*, *29*(1), 33–40.

Langridge, P., and Reynolds, M. P. (2015). Genomic tools to assist breeding for drought tolerance. *Curr. Opin. Biotechnol.*, *32*, 130–135.

Li, X., Yan, W., Agrama, H., Jia, L., Jackson, A., Moldenhauer, K., Yeater, K., McClung, A., and Wu, D. (2012). Unraveling the complex trait of harvest index with association mapping in rice (Oryza sativa L.). *PLoS ONE*, *7*(1), e29350.

Lindsay, M. P., Lagudah, E. S., Hare, R. A., and Munns, R. (2004). A locus for sodium exclusion (Nax1), a trait for salt tolerance, mapped in durum wheat. *Funct. Plant Biol.*, *31*(11), 1105–1114.

Long, S. P., Marshall-Colon, A., and Zhu, X. G. (2015). Meeting the global food demand of the future by engineering crop photosynthesis and yield potential. *Cell*, *161*, 56–66.

Longin, C. F. H., Mi, X., and Würschum, T. (2015). Genomic selection in wheat: optimum allocation of test resources and comparison of breeding strategies for line and hybrid breeding. *Theor. Appl. Genet.*, *128*(7), 1297–1306.

Lopes, M. S., and Reynolds, M. P. (2011). Drought adaptive traits and wide adaptation in elite lines derived from resynthesized hexaploid wheat. *Crop Sci.*, *51*(4), 1617–1626.

Lopes, M. S., Reynolds, M. P., McIntyre, C. L., Mathews, K. L., Kamali, M. J., Mossad, M., Feltaous, Y., Tahir, I. S., Chatrath, R., Ogbonnaya, F., and Baum, M. (2013). QTL for yield and associated traits in the Seri/Babax population grown across several environments in Mexico, in the West Asia, North Africa, and South Asia regions. *Theor. Appl. Genet.*, *126*(4), 971–984.

Lopes, M. S., Dreisigacker, S., Peña, R. J., Sukumaran, S., and Reynolds, M. P. (2015). Genetic characterization of the wheat association mapping initiative (WAMI) panel for dissection of complex traits in spring wheat. *Theor. Appl. Genet.*, *128*(3), 453–464.

Lynch, J. P. (2013). Steep, cheap and deep: an ideotype to optimize water and N acquisition by maize root systems. *Ann. Bot.*, *112*(2), 347–357.

Marx, W. (2012). Tracking historical papers and their citations. *Eur. Sci. Ed.*, *38*(2), 35–37.

Mason, R. E., Hays, D. B., Mondal, S., Ibrahim, A. M., and Basnet, B. R. (2013). QTL for yield, yield components and canopy temperature depression in wheat under late sown field conditions. *Euphytica*, *194*(2), 243–259.

Mir, R. R., Zaman-Allah, M., Sreenivasulu, N., Trethowan, R., and Varshney, R. K. (2012). Integrated genomics, physiology and breeding approaches for improving drought tolerance in crops. *Theor. Appl. Genet.*, *125*, 625–645.

Molero, G., Rivera-Amado, A. C., Pinera-Chavez, F., Trujillo-Negrellos, E., Quintero, A., Gonzalez-Navarro, O. E., Slafer, G. A., Calderini, D. F., Foulkes, J., Berry, P. M., et al. (2016). Evaluation of yield potential traits in elite spring wheat lines with the view to accelerating genetic gains. In Reynolds, M., Molero, G., and Quilligan, E. (eds.), *Proceedings of the 2nd International TRIGO (Wheat) Yield Potential Workshop 2016. CENEB, CIMMYT, Cd. Obregón, Sonora, Mexico, 9–10 March 2016, pp. 25–34.*

Molero, G., Piñera-Chavez, F. J., Rivera-Amado, A. C., Pinto, F., Gimeno, J., Sukumaran, S., and Reynolds, M. P. (2017). Phenotypic characterization of the International Wheat Yield Partnership-Hub (IWYP-HUB) panels, In Reynolds, M. P., Molero, G., and McNab, A. (eds.), *Proceedings of the 3rd International TRIGO (Wheat) Yield Potential Workshop 2017. CENEB, CIMMYT, Cd. Obregón, Sonora, Mexico, 22–23rd March 2017, pp. 64–73.*

Mondal, S., Mason, R. E., Huggins, T., and Hays, D. B. (2015). QTL on wheat (*Triticum aestivum L.*) chromosomes 1B, 3D and 5A are associated with constitutive production of leaf cuticular wax and may contribute to lower leaf temperatures under heat stress. *Euphytica*, *201*(1), 123–130.

Morgan, J. M., and Tan, M. K. (1996). Chromosomal location of a wheat osmoregulation gene using RFLP analysis. *Funct. Plant Biol.*, *23*(6), 803–806.

Ortiz, R., Braun, H. J., Crossa, J., Crouch, J. H., Davenport, G., Dixon, J., Dreisigacker, S., Duveiller, E., He, Z., Huerta, J., and Joshi, A .K. (2008). Wheat genetic resources enhancement by the International Maize and Wheat Improvement Center (CIMMYT). *Genet. Resour. Crop Ev.*, *55*(7), 1095–1140.

Parent, B., Shahinnia, F., Maphosa, L., Berger, B., Rabie, H., Chalmers, K., Kovalchuk, A., Langridge, P., and Fleury, D. (2015). Combining field performance with controlled environment plant imaging to identify the genetic control of growth and transpiration underlying yield response to water-deficit stress in wheat. *J. Exp. Bot.*, *66*(18), 5481–5492.

Pask, A., Joshi, A. K., Manès, Y., Sharma, I., Chatrath, R., Singh, G. P., Sohu, V. S., Mavi, G.S., Sakuru, V.S.P., Kalappanavar, I. K., and Mishra, V. K. (2014). A wheat phenotyping network to incorporate physiological traits for climate change in South Asia. *F. Crop. Res.*, *168*, 156–167.

Paulsen, G. (2002). Application of physiology in wheat breeding. *Crop Sci.*, *42*(6), 2228–2229.

Pinto, R. S., Reynolds, M. P., Mathews, K. L., McIntyre, C. L., Olivares-Villegas, J. J., and Chapman, S. C. (2010). Heat and drought adaptive QTL in a wheat population designed to minimize confounding agronomic effects. *Theor. Appl. Genet.*, *121*(6), 1001–1021.

Pinto, R. S., and Reynolds, M. P. (2015). Common genetic basis for canopy temperature depression under heat and drought stress associated with optimized root distribution in bread wheat. *Theor. Appl. Genet.*, *128*(4), 575–585.

Pinto, R. S., Lopes, M. S., Collins, N. C., and Reynolds, M. P. (2016). Modelling and genetic dissection of staygreen under heat stress. *Theor. Appl. Genet.*, *129*(11), 2055–2074.

Poorter, H., Fiorani, F., Pieruschka, R., Wojciechowski, T., Putten, W. H., Kleyer, M., Schurr, U., and Postma, J. (2016). Pampered inside, pestered outside? Differences and similarities between plants growing in controlled conditions and in the field. *New Phytol.*, *212*(4), 838–855.

Ray, D. K., Mueller, N. D., West, P. C., and Foley, J. A. (2013). Yield trends are insufficient to double global crop production by 2050. *PLoS ONE*, *8*, e66428.

Rebetzke, G. J., Appels, R., Morrison, A. D., Richards, R. A., McDonald, G., Ellis, M. H., Spielmeyer, W., and Bonnett, D. G. (2001). Quantitative trait loci on chromosome 4B for coleoptile length and early vigour in wheat (*Triticum aestivum L.*). *Aust. J. Agric. Res.*, *52*(12), 1221–1234.

Rebetzke, G. J., Richards, R. A., Fettell, N. A., Long, M., Condon, A. G., Forrester, R. I., and Botwright, T. L. (2007). Genotypic increases in coleoptile length improves stand establishment, vigour and grain yield of deep-sown wheat. *F. Crop. Res.*, *100*(1), 10–23.

Rebetzke, G. J., Van Herwaarden, A. F., Jenkins, C., Weiss, M., Lewis, D., Ruuska, S., Tabe, L., Fettell, N. A., and Richards, R. A. (2008a). Quantitative trait loci for water-soluble carbohydrates and associations with agronomic traits in wheat. *Aust. J. Agric. Res.*, *59*(10), 891–905.

Rebetzke, G. J., Condon, A. G., Farquhar, G. D., Appels, R., and Richards, R. A. (2008b). Quantitative trait loci for carbon isotope discrimination are repeatable across environments and wheat mapping populations. *Theor. Appl. Genet.*, *118*(1), 123–137.

Rebetzke, G. J., Rattey, A. R., Farquhar, G. D., Richards, R. A., and Condon, A. T. G. (2013). Genomic regions for canopy temperature and their genetic association with stomatal conductance and grain yield in wheat. *Funct. Plant Biol.*, *40*(1), 14–33.

Reynolds, M. P., Calderini, D. F., Condon, A. G., and Rajaram, S. (2001), Physiological basis of yield gains in wheat associated with the LR19 translocation from Agropyron elongatum. *Euphytica*, *119*, 139–144.

Reynolds, M. P., Pellegrineschi, A., and Skovmand, B. (2005). Sink-limitation to yield and biomass: a summary of some investigations in spring wheat, *Ann. Appl. Biol.*, *146*(1), 39–49.

Reynolds, M., Dreccer, F., and Trethowan, R. (2006). Drought-adaptive traits derived from wheat wild relatives and landraces. *J. Exp. Bot.*, *58*(2), 177–186.

Reynolds, M., Manes, Y., Izanloo, A., and Langridge, P. (2009). Phenotyping approaches for physiological breeding and gene discovery in wheat. *Ann. Appl. Biol.*, *155*, 309–320.

Reynolds, M. P. and Rebetzke, G. (2011). Application of plant physiology in wheat breeding. In Angus, W., Bonjean A., and van Ginkel, M. (eds.), *The world wheat book: a history of wheat breeding*, *2*, pp. 877–906.

Reynolds, M., Foulkes, J., Furbank, R., Griffiths, S., King, J., Murchie, E., Parry, M., and Slafer, G. (2012). Achieving yield gains in wheat. *Plant Cell Environ.*, *35*, 1799–1823.

Reynolds, M., Tattaris, M., Cossani, C. M., Ellis, M., Yamaguchi-Shinozaki, K., and Saint Pierre, C. (2015). Exploring genetic resources to increase adaptation of wheat to climate change. In Ogihara, Y., Takumi, S., and Handa, H. (eds.), *Advances in wheat genetics: from genome to field*, pp. 355–368. Tokyo, Japan: Springer.

Reynolds, M., and Langridge, P. (2016). Physiological breeding. *Curr. Op. Plant Biol.*, *31*, 162–171.

Reynolds, M. P., Pask, A. J., Hoppitt, W. J., Sonder, K., Sukumaran, S., Molero, G., Saint Pierre, C., Payne, T., Singh, R. P., Braun, H. J., and Gonzalez, F. G. (2018). Correction to: strategic crossing of biomass and harvest index—source and sink—achieves genetic gains in wheat. *Euphytica*, *214*(1), 9.

Richards, R. A., Rebetzke, G. J., Watt, M., Condon, A. G., Spielmeyer, W., and Dolferus, R. (2010). Breeding for improved water productivity in temperate cereals: Phenotyping, quantitative trait loci, markers and the selection environment. *Funct. Plant Biol.*, *37*, 85–97.

Rivera-Amado, C. (2016). *Optimizing assimilate partitioning in wheat for genetic enhancement of spike fertility and yield potential.* PhD Thesis, Univ Nottingham Sch Biosci Sutt Bonningt Campus, Leistershire, UK.

Rivera-Amado, A., Trujillo-Negrellos, E., Sylvester-Bradley, R., Molero, G., Sierra-Gonzalez, A., Reynolds, M. P., and Foulkes, M. (2016). Achieving increases in spike growth, fruiting efficiency, and harvest index in high biomass wheat cultivars, In Reynolds, M., Molero, G., and Quilligan,

E. (eds.), *Proceedings of the 2nd International TRIGO (Wheat) Yield Potential Workshop 2016. CENEB, CIMMYT, Cd. Obregón, Sonora, Mexico, 9–10 March 2016, pp. 70–76.*

Römer, C., Wahabzada, M., Ballvora, A., Pinto, F., Rossini, M., Panigada, C., Behmann, J., Léon, J., Thurau, C., Bauckhage, C., and Kersting, K. (2012). Early drought stress detection in cereals: simplex volume maximisation for hyperspectral image analysis. *Funct. Plant Biol.*, *39*(11), 878–890.

Rutkoski, J., Poland, J., Mondal, S., Autrique, E., Pérez, L. G., Crossa, J., Reynolds, M., and Singh, R. (2016). Canopy temperature and vegetation indices from high-throughput phenotyping improve accuracy of pedigree and genomic selection for grain yield in wheat. *G3: Genes Genomes Genet.*, *6*, 2799–2808.

Sadras, V. O. (2007). Evolutionary aspects of the trade-off between seed size and number in crops. *F. Crop. Res.*, *100*(2–3), 125–138.

Serbin, S. P., Dillaway, D. N., Kruger, E. L., and Townsend, P. A. (2011). Leaf optical properties reflect variation in photosynthetic metabolism and its sensitivity to temperature. *J. Exp. Bot.*, *63*(1), 489–502.

Shamaya, N. (2013). *Genetic studies of salinity tolerance in wheat.* PhD Thesis, School of Agriculture, Food and Wine, Australian Centre for Plant Functional Genomics, University of Adelaide, Australia.

Shearman, V. J., Scott, R. K., and Foulkes, M. J. (2005). Physiological Processes Associated with Wheat Yield Progress in the UK. *Crop Sci.*, *45*, 175–185.

Silva-Perez V., Molero G., Serbin S. P., Condon A. G., Reynolds M. P., Furbank R. T., and Evans J. R. (2018). Hyperspectral reflectance as a tool to measure biochemical and physiological traits in wheat. *J. Exp. Bot.*, *69*(3), 483–496.

Slafer, G. A., and Savin, R. (1994). Source-sink relationships and grain mass at different positions within the spike in wheat. *F. Crop. Res.*, *37*(1), 39–49.

Slafer, G. A., Elia, M., Savin, R., García, G. A., Terrile, I. I., Ferrante, A., Miralles, D. J., and González, F. G. (2015). Fruiting efficiency: an alternative trait to further rise wheat yield. *Food and Energy Security*, *4*(2), 92–109.

Smale, M., Reynolds, M. P., Warburton, M., Skovmand, B., Trethowan, R., Singh, R. P., Ortiz-Monasterio, I., and Crossa, J. (2002). Dimensions of diversity in modern spring bread wheat in developing countries from 1965. *Crop Sci.*, *42*(6), 1766–1779.

Snape, J. W., Butterworth, K., Whitechurch, E., and Worland, A. J.(2001). Waiting for fine times: genetics of flowering time in wheat. *Euphytica*, 119, 185–190.

Sukumaran, S., and Yu, J. (2014). Association mapping of genetic resources: achievements and future perspectives. *Genom. Plant Genet. Resour.*, 1, 207–235.

Sukumaran, S., Dreisigacker, S., Lopes, M., Chavez, P., and Reynolds, M. P. (2015a). Genome-wide association study for grain yield and related traits in an elite spring wheat population grown in temperate irrigated environments. *Theor. Appl. Genet.*, *128*(2), 353–363.

Sukumaran, S., Reynolds, M. P., Lopes, M. S., and Crossa, J. (2015b). Genome-wide association study for adaptation to agronomic plant density: A component of high yield potential in spring wheat. *Crop Sci.*, *55*(6), 2609–2619.

Sukumaran, S., Lopes, M. S., Dreisigacker, S., Dixon, L. E., Zikhali, M., Griffiths, S., Zheng, B., Chapman, S., and Reynolds, M. P. (2016). Identification of earliness per se flowering time locus in spring wheat through a genome-wide association study. *Crop Sci.*, *56*(6), 2962–2672.

Sukumaran, S., Lopes, M., Dreisigacker, S., and Reynolds, M. (2017a). Genetic analysis of multi-environmental spring wheat trials identifies genomic regions for locus-specific trade-offs for grain weight and grain number. *Theor. Appl. Genet.* doi: 10.1007/s00122-017-3037-7.

Sukumaran, S., Crossa, J., Jarquin, D., Lopes, M. and Reynolds, M. P. (2017b). Genomic prediction with pedigree and genotype × environment interaction in spring wheat grown in South and West Asia, North Africa, and Mexico. *G3: Genes Genomes Genet.*, 3–116.

Sukumaran, S., Crossa, J., Jarquín, D., and Reynolds, M. (2017c). Pedigree-based prediction models with genotype × environment interaction in multienvironment trials of CIMMYT wheat. *Crop Sci.*, *57*(4), 1865–1880.

Sukumaran, S., Reynolds, M. P., and Sansaloni, C. (2018). Genome-wide association analyses identify QTL hotspots for yield and component traits in durum wheat grown under yield potential, drought, and heat stress environments. *Front. Plant Sci.*, *9*, 81.

Talukder, S. K., Babar, M. A., Vijayalakshmi, K., Poland, J., Prasad, P. V. V., Bowden, R., and Fritz, A. (2014). Mapping QTL for the traits associated with heat tolerance in wheat (Triticum aestivum L.). *BMC Genet.*, *15*(1), 97.

Tardieu, F., Parent, B., Caldeira, C. F., and Welcker, C. (2014). Genetic and physiological controls of growth under water deficit. *Plant Physiol.*, *164*(4), 1628–1635.

Tattaris, M., Reynolds, M. P., and Chapman, S. C. (2016). A direct comparison of remote sensing approaches for high-throughput phenotyping in plant breeding. *Front. Plant Sci.*, *7*, 1131.

Tiwari, C., Wallwork, H., Kumar, U., Dhari, R., Arun, B., Mishra, V. K., Reynolds, M. P., and Joshi, A. K. (2013). Molecular mapping of high temperature tolerance in bread wheat adapted to the Eastern Gangetic Plain region of India. *F. Crop. Res.*, *154*, 201–210.

Trethowan, R. M., and Mujeeb-Kazi, A. (2008). Novel germplasm resources for improving environmental stress tolerance of hexaploid wheat, *Crop Sci.*, *48*(4), 1255–1265.

Tsunewaki, K., and Ebana, K. (1999). Production of near-isogenic lines of common wheat for glaucousness and genetic basis of this trait clarified by their use. *Genes Genet. Syst.*, *74*(2), 33–41.

Turki, N., Shehzad, T., Harrabi, M., and Okuno, K. (2015). Detection of QTLs associated with salinity tolerance in durum wheat based on association analysis. *Euphytica*, *201*(1), 29–41.

Vadez, V., Kholova, J., Zaman-Allah, M., and Belko, N. (2013). Water: the most important "molecular" component of water stress tolerance research. *Funct. Plant Biol.*, *40*(12), 1310–1322.

Valluru, R., Reynolds, M. P., Davies, W. J., and Sukumaran, S. (2017). Phenotypic and genome-wide association analysis of spike ethylene in diverse wheat genotypes under heat stress. *New Phytol.*, *214*(1), 271–283.

van Ginkel, M., and Ortiz, R. (2018). Cross the best with the best, and select the best: HELP in breeding selfing crops. *Crop Sci.*, *58*(1), 17–30.

van Ittersum, M. K. (2016). Crop yields and global food security. Will yield increase continue to feed the world? *Eur. Rev. Agric. Econ.*, *43*(1), 191–192.

Verma V, Foulkes M. J., Worland A. J., Sylvester-Bradley R., Caligari P. D. S., Snape J. W. (2004). Mapping quantitative trait loci for flag leaf senescence as a yield determinant in winter wheat under optimal and drought-stressed environments. *Euphytica*, *135*, 255–263.

Vijayalakshmi, K., Fritz, A. K., Paulsen, G. M., Bai, G., Pandravada, S., and Gill, B. S. (2010). Modeling and mapping QTL for senescence-related traits in winter wheat under high temperature. *Mol. Breed.*, *26*(2), 163–175.

Waines, J. G., and Ehdaie, B. (2007). Domestication and crop physiology: roots of green-revolution wheat. *Ann. Bot.*, *100*(5), 991–998.

Walter, J., Edwards, J., McDonald, G., and Kuchel, H. (2018). Photogrammetry for the estimation of wheat biomass and harvest index. *F. Crops Res.*, *216*, 165–174.

Wang, S. G., Jia, S. S., Sun, D. Z., Wang, H. Y., Dong, F. F., Ma, H. X., Jing, R. L., and Ma, G. (2015). Genetic basis of traits related to stomatal conductance in wheat cultivars in response to drought stress. *Photosynthetica*, *53*(2), 299–305.

Watson, A., Ghosh, S., Williams, M. J., Cuddy, W. S., Simmonds, J., Rey, M. D., Hatta, M. A. M., Hinchliffe, A., Steed, A., Reynolds, D., and Adamski, N. M. (2018). Speed breeding is a powerful tool to accelerate crop research and breeding. *Nature Plants*, *4*, 23–29.

Wilkinson, S., Kudoyarova, G. R., Veselov, D. S., Arkhipova, T. N., and Davies, W. J. (2012). Plant hormone interactions: innovative targets for crop breeding and management. *J. Exp. Bot.*, *63*(9), 3499–3509.

Wingen, L. U., Orford, S., Goram, R., Leverington-Waite, M., Bilham, L., Patsiou, T. S., Ambrose, M., Dicks, J., and Griffiths, S. (2014). Establishing the AE Watkins landrace cultivar collection as a resource for systematic gene discovery in bread wheat. *Theor. App. Genet.*, *127*(8), 1831–1842.

Yu, J., Pressoir, G., Briggs, W. H., Bi, I. V., Yamasaki, M., Doebley, J. F., McMullen, M. D., Gaut, B. S., Nielsen, D. M., Holland, J. B., and Kresovich, S. (2006). A unified mixed-model method for association mapping that accounts for multiple levels of relatedness. *Nat. Genet.*, *38*(2), 203.

Yu, J., Holland, J. B., McMullen, M. D., and Buckler, E. S. (2008). Genetic design and statistical power of nested association mapping in maize, *Genet.*, *178*(1), 539–551.

Yu, X., Li, X., Guo, T., Zhu, C., Wu, Y., Mitchell, S. E., Roozeboom, K. L., Wang, D., Wang, M. L., Pederson, G. A., and Tesso, T. T. (2016). Genomic prediction contributing to a promising global strategy to turbocharge gene banks, *Nat. Plants*, *2*(10), 16150.

Zarco-Tejada, P.J., Catalina, A., González, M. R., and Martín, P. (2013). Relationships between net photosynthesis and steady-state chlorophyll fluorescence retrieved from airborne hyperspectral imagery, *Remote Sens. Environ.*, *136*, 247–258.

Zhang, Z., Ersoz, E., Lai, C.Q., Todhunter, R.J., Tiwari, H. K., Gore, M. A., Bradbury, P. J., Yu, J., Arnett, D. K., Ordovas, J. M., and Buckler, E. S. (2010). Mixed linear model approach adapted for genome-wide association studies, *Nat. Genet.*, *42*(4), 355.

Zhao, K., Tung, C. W., Eizenga, G. C., Wright, M. H., Ali, M. L., Price, A. H., Norton, G. J., Islam, M. R., Reynolds, A., Mezey, J., and McClung, A. M. (2011). Genome-wide association mapping reveals a rich genetic architecture of complex traits in Oryza sativa. *Nature Communications*, 2, 467.

Zhu, C., Gore, M., Buckler, E. S., and Yu, J. (2008). Status and prospects of association mapping in plants. *The Plant Genome*, 1(1), 5–20.

ADAPTIVE WILD CEREAL GENOMIC RESOURCES FOR CROP IMPROVEMENT

Eviatar Nevo
Institute of Evolution, University of Haifa

Introduction

Genetic diversity is the basic variable enabling evolutionary processes and the origin of adaptations either by natural selection (Darwin, 1859) or by natural genetic engineering (Shapiro, 2011), or both. Human civilization is based on domestication of plants and animals, with wild cereals among the foremost domesticates. However, domestication has in many cases led to the erosion of genetic diversity (Nevo, 2004; Fu et al., 2005, 2006; Fu and Somers, 2009; see Avni et al., 2017) partly adapted to resist both *abiotic* (solar radiation, heat, drought, salinization, mineral shortage, and global warming) and *biotic* (pathogens and parasites) stresses. For these reasons, most cultivars have over time become vulnerable and endangered to environmental stresses. This threatens future food production in a world where the population is projected to increase from 7.6 to 9.8 billion people by 2050 and 11.2 billion in 2100 (UN DESA, 2017). This occurred because historically farmers, in the domestication process, focused on higher yield and other characters they selected for, such as non-brittle rachis to keep the seeds on spikes without shattering, and big grain size. During this process, the cultivars often became vulnerable to environmental stresses. At the same time, the ongoing process of global warming (IPCC, 2014) has already caused a depletion of genetic diversity in the wild progenitors of wheat and barley (Nevo et al., 2012). Those progenitors are the best hope for restoring and improving adaptive genetic diversity in cultivars (Feldman and Sears, 1981; Nevo et al., 2012; Henry and Nevo 2014).

Importantly, tetraploid cultivar durum pasta wheat, *Triticum durum*, the major commercial tetraploid wheat today, is derived from domesticated emmer wheat (*T. dicoccum*, which in turn was domesticated from the wild *T. dicoccoides*). Pasta wheat has revealed by wide crossing good disease resistance, yield stability, and industrial quality across Mediterranean sites where it is cultivated (Zaim et al., 2017). It confers, through its S13 gene, resistance to the new immensely devastating virulent race group Ug99 stem rust pathogen *Blumeria* (formerly *Puccinia*) *graminis* f.sp *tritici*, of which the strain *Pgf* appeared in Africa at the beginning of this century (Zhang, 2017). Unfortunately, this strain remains virulent in the presence of most stem rust resistance genes currently deployed in wheat, and thus poses a serious threat to global wheat production. This situation highlights the unique importance of the wild genetic resources in saving mankind from a global disaster, since genetic improvements in wheat, the foremost staple food, could save mankind from a global hunger predicament. Therefore, wild progenitors, land races, and domesticates need to be conserved both *in situ* and *ex situ* (Nevo, 1998) to save their highly important genetic diversity for crop improvement and to guarantee future food production. Here I will briefly overview *some* of the rich genetic adaptive complexes of wild cereals, which should be conserved to save existing cultivars from their current predicament of genetic diversity erosion, which threatens agriculture and food production in the face of a constantly increasing human population (Nevo, 2016).

The Wild Progenitors of Wheat and Barley

Domestication of cereals in the Old World appear first in the Early Pre-Pottery Neolithic B (PPNB) farming villages that developed in southwest Asia about 10,500–11,000 years before present (BP) (see Map 1 in Zohary et al., 2012, and a recent review by Rotenberg, 2017). Pre-agriculture collection, started before domestication, was found for wild barley in Kebara cave, Mount Carmel, Israel, 50,000 years ago (Lev et al., 2005), and later at Ohalo II, 23,000 years ago, south of the Lake of Galilee (Weiss, 2002). The assemblage of eight founder crops, three cereals, four legumes, and flax characterize the domestication of plants in the Old World (Zohary et al., 2012). The first three domesticated cereals, which gave rise to those developed subsequently, are: emmer wheat (*T. dicoccum*), einkorn wheat (*T. monococum*), and barley (*Hordeum vulgare*). Emmer wheat and barley appear to have been the more common cultivated crops. Cultivated emmer wheat was domesticated from its hulled wild progenitors (*T. dicoccoides*) very early, giving rise both to cultivated emmer (*T. dicoccum*) and developed into the free-threshing pasta wheats (*T. durum*). In turn, cultivated tetraploid emmer wheat became one of the parents of hexaploid bread wheat (*T. aestivum*). As for barley, cultivated barley (*Hordeum vulgare*) was selected from its wild progenitor *H. spontaneum* (Zohary et al., 1969). Both cultivated wheat and barley were domesticated in the eastern Mediterranean (Zohary et al., 2012).

FIGURE 1. Wild emmer wheat, *T. dicoccoides*, the progenitor of cultivated wheat. (A) Spiking wild emmer wheat, *T. dicoccoides* in a greenhouse, immature, (B) Lush natural field of wild emmer wheat, *T. dicoccoides*, in the region of its origin and greatest diversity in Upper Galilee, Israel, several kilometers east of Zefat (Safed), (C) Color polymorphism of *T. dicoccoides* black, yellow and green morphs, and (D) *TD* in Yehudiyya: Sun and shade micro-scale genetic divergence in the shade under oak canopy and in sun nearby (Nevo, original photos).

Wild Emmer Wheat, *T. dicoccoides*, the Progenitor of Bread Wheat

Allotetraploid wild emmer wheat, *T. dicoccoides* (TD) 2n=28, the progenitor of most cultivated wheat, is an ecological specialist, and an excellent model organism for advancing evolutionary theory, wheat evolution, and a major genetic resource for wheat improvement (Fig. 1) (Aaronsohn, 1910; Nevo et al., 2002; Peng et al., 2003; Gustafson et al., 2009; Peng et al., 2011a, 2011b; Nevo, 2014). The 10.1 gigabase genome assembly of wild emmer genome architecture and diversity elucidate wheat evolution and domestication (Avni et al., 2017). The coupling of population genetics within and between wild emmer populations and the recent genome assembly and analysis of gene content, and intergenic diversity (e.g., noncoding regulatory DNA) will accelerate the genome-assisted improvement of cultivated wheat as the major food provider worldwide (Nevo, 2016). The center of origin and diversity of TD is northeastern Upper Galilee (Figure 1b) and the Golan Heights. Elsewhere in the Near East Fertile Crescent, it occurs in semi-isolated and isolated populations. The population genetic structures are generally an "archipelago," i.e., neighboring populations may have distinct divergent allele frequencies. Regional and local genetic patterns are partly or largely adaptive at the protein and DNA levels, both at *coding* and *noncoding* portions of the wheat genome, correlated with, and predictable by, environmental *abiotic* and *biotic* stresses. TD is rich in mostly untapped genetic resources for improving cultivated wheat, harboring drought, salt, mineral, and disease resistances, grain proteins, and high variation in photosynthetic yield (Nevo, 2011). TD was chromosomally mapped with 549 molecular markers and 70 QTLs (quantity trait loci) for 11 traits of agricultural importance and domestication (Peng et al., 2003; Peng et al., 2011a). TD is affected by global warming (Nevo et al., 2012) both *phenotypically* (advancing flowering time) and *genotypically* (genetic erosion, allelic turn-over, and novel drought-resistance alleles). Hence, it should be conserved *in-situ* and *ex-situ* (Nevo, 1998) to safeguard its great adaptive genetic potential for wheat improvement and increased food production in a world with an exploding population (Nevo, 2016).

FIGURE 2. Wild barley, *H. spontaneum*. A lush field of wild barley, *H. spontaneum*, growing on the Natufian terrace and cemetery of the Oren cave at "Evolution Canyon" lower Nahal Oren, Mount Carmel, Israel (Nevo, original photo).

Wild Barley, *H. spontaneum* (HS), the Progenitor of Barley

Wild barley, *H. spontaneum*, the progenitor of cultivated barley, *Hordeum vulgare*, originated 5.5 million years ago in southwest Asia and is distributed in the eastern Mediterranean, Balkans, North Africa, central Asia, and Tibet (Nevo, 2012; Zohary et al., 2012). *H. vulgare*, the fourth most important world crop, currently used for animal feed, beer, and human food, was domesticated by humans at least 10,000 years ago, in the Neolithic revolution or earlier. Preagricultural wild barley seed gathering by humans occurred as early as 23,000 years ago in Ohalo II (Weiss, 2002), and 50,000 years ago in Kebara cave in Mount Carmel (Lev et al., 2005). Barley domestication from the progenitor, *H. spontaneum*, to the cultivated *H. vulgare* started in the Near East Fertile Center, proceeded to central Asia and reached Tibet (Dai et al., 2014). *H. vulgare*, comprised of thousands of land races and cultivars, is widespread at higher altitudes and other places to which most crops can't adapt easily. However, it has deteriorated genetically due to domestication, and primarily due to breeding (i.e., cultivar improvement), and needs genetic reinforcement. Wild barley, *H. spontaneum*, the modern barley progenitor (Zohary, 1969), the best hope for barley improvement, is a hardy ecological generalist, adapted to a wide range of extreme latitudes, altitudes, climates (warm and cold), and multiple soil types. Adaptation occurs at all levels: *genomically, proteomically*, and *phenomically*, both *regionally* and *locally*. It displays an "archipelago" genetic structure, rich genetically, and harbors large adaptive *abiotic* and *biotic* resistances precious to barley and cereal improvement. Sequencing the cultivated barley, *H. vulgare* genome (Mascher et al., 2017) , and its progenitor *H. spontaneum* genome (Nevo, in preparation) will therefore reveal huge, mostly untapped, genetic resources. The current global warming stresses affects *H. spontaneum* (Nevo et al., 2012), hence it is of critical importance to conserve it *in situ* and *ex situ* (Nevo, 1998) to safeguard its future immense contribution to barley improvement. I will discuss adaptations of wild emmer wheat, *T. dicoccoides* (TD), and wild barley, *H. spontaneaum* (HS), which have been model organisms for studying their adaptive complexes against environmental stresses in natural populations within the Near East Fertile Crescent, at the Institute of Evolution, University of Haifa since 1975 (HS) and since 1980 (TD) (see Nevo list of wild cereal publications at http://evolution.haifa.ac.il). I will present some examples of genetic resources to combat abiotic and biotic stresses of wild cereals studied primarily at the Institute of Evolution, University of Haifa, based on the extensive gene bank of wild progenitors of wheat and barley collected throughout their range in the Fertile Crescent (Israel, Jordan, Turkey, and Iran) since 1975.

Genetic Resources of Wild Cereals against Abiotic Stresses

Salt and Drought Resistance in TD and HS

Drought and salinity are the major abiotic stresses that threaten the world food supply. Wild emmer wheat, TD, and wild barley, HS, together with other Triticeae, possess high genetic potential for drought and salt tolerance that have been practically identified, tested and transferred to wheat cultivars with proven expression of improved tolerance levels in experimental trials (Nevo et al. 1984a, 1984b; Nevo et al., 1993; Nevo, 2002; Zhang et al., 2002; Chen et al., 2004a, 2004b; Suprunova et al., 2004, 2007; Peleg et al., 2005; Zhang et al., 2005; Peleg et al., 2008; Chen et al., 2009, 2010, 2011, 2013, 2015; Yang et al., 2009; Nevo and Chen 2010; Yang et al., 2011; Nevo, 2012; Nevo et al., 2012;Yang et al., 2013; Bedada et al., 2014; Nevo, 2014; Chen et al., 2002).

We first compared and contrasted agronomically important phenotypic traits in populations involving both TD and HS, originating from diverse habitats across Israel. The field tests were conducted in two sharply contrasting environments, *xeric desert* (Avedat farm and Sede Boker) and *mesic Mediterranean* (Mount Carmel, Haifa) (Nevo et al., 1984a, 1984b). These populations showed widespread divergence for allozymes (Nevo et al., 1982) and disease resistance polymorphisms (Moseman et al., 1983; Moseman et al., 1984, 1985, 1990; Nevo et al., 1984a, 1984b, 1986, 1991; Jana and Nevo, 1991; Fetch Jr. et al., 2003). The agronomic traits compared involved ten variables including germination, earliness, biomass, and yield parameters. The results indicated dramatic phenotypic differences between and within populations in strains from the two tested environments, *xeric* and *mesic.* This experiment illustrated that wild emmer and wild barley populations in Israel varied not only in genetic polymorphism of allozymes and disease resistance but also in quantitative traits of agronomic importance correlated with their drought resistance. Populations from xeric environments in the Negev Desert, where drought adaptation is of cardinal importance for survival, proved superior to populations from Mt. Carmel and Haifa in their drought resistance capabilities.

A series of studies on drought resistance in wild emmer wheat and wild barley have been conducted since 1975 at the Institute of Evolution, University of Haifa, cited above. These studies deciphered that both TD and HS possess immense potential for drought and salt tolerance that has been extensively and practically identified, tested, and some were transferred to wheat cultivars with proven expression of tolerance in experimental trials (Nevo and Chen, 2010). Both TD and particularly HS have adapted to a broad range of environments and have developed rich genetic diversities for drought and salt tolerance. Drought- and salt-tolerant genes and QTLs have been identified in TD and HS and have great potential in wheat and barley improvement. Advanced backcross QTL analysis, utilizing the introgression libraries based on TD and HS as donors, and positional cloning of natural QTLs, will play important roles in elucidating the molecular role of drought and salt tolerance. In the near future wild genetic resistances to drought and salinity will be shifted from field experiments to the farmer (Nevo and Chen, 2010).

Phenotypic diversity for drought response in TD populations was described by Peleg et al. (2005). Later, Peleg et al. (2008) described the patterns of spatiotemporal variation of allelic diversity of SSR loci and their ecological correlates. They found 56 percent of allelic diversity observed in 54 microsatellite loci analyzed in 145 genotypes was within populations and 44 percent was between populations. Ecology was more important than geographic separation in explaining both diversity and divergence patterns. The allelic diversity of SSRs was correlated with annual rainfall ($r = 0.74$, $p < 0.0002$), indicating that the allelic diversity was higher in populations growing under conditions of intermediate stress (rainfall 350–550 mm).

Multilevel regulation and adaptive signaling processes to terminal drought were described by transcriptome analysis in TD (Krugman et al., 2010). A total of 5892 differentially regulated transcripts were identified between drought *resistant* (R) and drought *susceptible* (S) genotypes. Functional enrichment analysis revealed that regulatory and signaling processes were significantly enriched among the drought induced transcripts, particularly in the R genotypes. Annotation of 221 genes revealed that 26 percent were involved in multilevel regulation including transcriptional regulation, RNA binding, kinase activity, and calcium and abscisic acid (ABA) signaling related to stomatal closure. Differential expression patterns included in drought adaptive genes, cell wall adjustment, cuticular wax deposition, lignification, osmo-regulation, redox homeostasis, dehydration protection, and drought-induced senescence, suggested that the TD genotypes analyzed contained candidate genes for improving drought resistance. Drought stress also characterized TD roots of R and S genotypes (Krugman et al., 2011) by differences in expression of hormone-related genes (gibberilins, ABA, and auxin), biosynthesis, RNA binding, and calcium and phosphatidylinositol signaling. ABA in the R genotype roots increased in response to drought, whereas it was invariant in the S genotype roots. The R genotypes expressed higher drought including glucose, trehalose, proline, and glycine. Adaptive transcriptomics and metabolomics were displayed by the R genotypes, optimizing bioenergetics, carbon metabolism, and cell homeostasis. Complementarily, both roots and shoots of TD involved candidate genes improving drought tolerance in wheat cultivars.

Many transcriptional factors (TFs) including DREB1/CBF, DREB 2, AREB/ABF, and NAC (Uauy et al., 2006) can be used to improve tolerance to abiotic stresses. For example, xeric wild barley, *H. spontaneum* transcriptional factor HsDREB 1A under transcriptional control of the stress-inducible barley HVA1 promoter was introduced into the apomictic bahiagrass (*Paspalum notatum*). Transgenic bahiagrass tolerated severe salt stress and repeated cycles of severe dehydration stress in comparison with nontransgenic plants. The TFs cloned from wild barley and wild emmer wheat should also be useful for barley and wheat, respectively, improvement of drought and salt tolerances. Most of the transformed genes were from model plants such as *Arabidopsis* and rice (*Oryza sativa*) or from wheat and barley cultivars. These approaches could be applied to wild relatives whose genes and networks of adaptive complexes may have stronger functions. This suggestion awaits experimental confirmation and field testing as was substantiated in natural populations under severe drought and salt stresses (Nevo, 2012, 2014).

Wild Barley eibi1 Mutation Identifies a Gene Essential for Leaf Water Conservation

Land plants have developed a cuticle preventing uncontrolled water loss. Recently, we reported that an ATP-binding cassette (ABC) subfamily G (ABCG) full transporter is required for leaf water conservation in both wild barley and rice (Chen et al., 2004–2011). A spontaneous mutation, eibi1.b, in wild barley has a low capacity to retain leaf water, a phenotype associated with reduced cutin deposition and a thin cuticle (Fig. 3). Map-based cloning revealed that Eibi1 encodes an HvABCG31 full transporter (Fig. 4). The gene was highly expressed in the elongation zone of a growing leaf (the site of cutin synthesis), and its gene product also was localized in developing, but not in mature tissue. A de novo wild barley mutant named "eibi1.c," along with two transposon insertion lines of rice mutated in the ortholog of HvABCG31 also were unable to restrict water loss from detached leaves. HvABCG31 is hypothesized to function as a transporter involved in cutin formation. Homologs of HvABCG31 have been found in green algae, mosses, and lycopods, indicating that this full transporter is highly conserved in the evolution of land plants and supported their land colonization (Chen et al., 2011).

• • •

FIGURE 3. The drought sensitivity of the eibi1.b wild barley mutant and its cuticle structure. (A) Wild-type (Left) and eibi1.b (Right) seedlings. The leaf comprises the three regions: the EmBL, the NEZ below the point of emergence (POE), and the EZ above the point of leaf insertion (POLI). (B) Water loss from detached leaves of eibi1 and wild-type plants. (C) Electron micrographs of the cuticle in EmBL, NEZ, and EZ of wild-type plants and the eibi1. b mutant. Arrows indicate the cuticle thickness. (D) Major cutin monomers and wax components in the EmBL of the eibi1.b mutant and wild type. All bars represent mean ± SD (n = 5). Asterisks denote significant differences (P < 0.0001) between wild type and mutant as determined by Student's t-tests. (E) Epidermis cells with (eibi1.b) or without (wild type) cytoplasmic protrusions into the vacuole. c, cytoplasm; p, protrusions; v, vacuole. (From Chen et al., 2011).

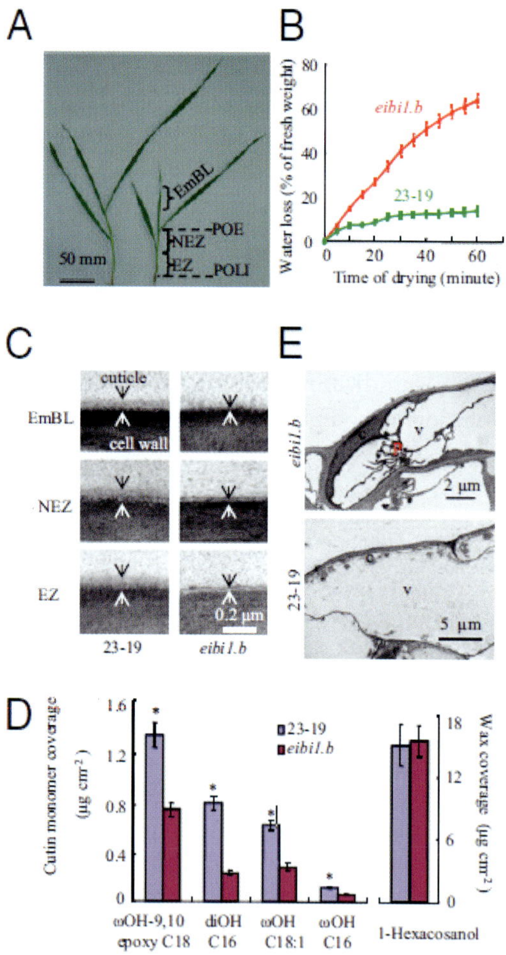

Figure 4. Positional cloning of eibi1 and its functional assignment. (A) The Eibi1 locus maps to barley chromosome 3H and its ortholog to rice chromosome 1. Numbers to the left of the barley map indicate the number of observed recombination events in a population of 9,070 gametes. The rice and barley orthologs are connected by dashed lines. (B) Exon/intron structure of barley Eibi1. The single-nucleotide difference between the wild-type and the eibi1.b mutant sequence is indicated. The red dashed line indicates the 9-bp deletion in eibi1.c exon 10. Untranslated regions are indicated by an empty box at each end. (C) The ⊠ irradiation-induced eibi1.c mutant and its wild type at the flowering stage. (D) Water-loss test of detached leaves from eibi1.c and OUH602. (E) Exon/intron structure of rice OsABCG31 indicating Tos17 insertions (arrows). (F) Wild-type (one tall seedling) and osabcg31.b mutants (four dwarf seedlings) at the three leaf stage. (G) Enlarged view of dwarf seedlings in F. (H) Water loss in the detached leaf of the wild type and the mutant. (I) Mutant and wild-type leaves of barley air dried for 1 h and rice air dried for 0.5 h. (J) Toluidine blue staining of leaf segments of mutants and wild types. (Scale bar: 20 mm for wild barley; 5 mm for rice.). (From Chen et al., 2011).

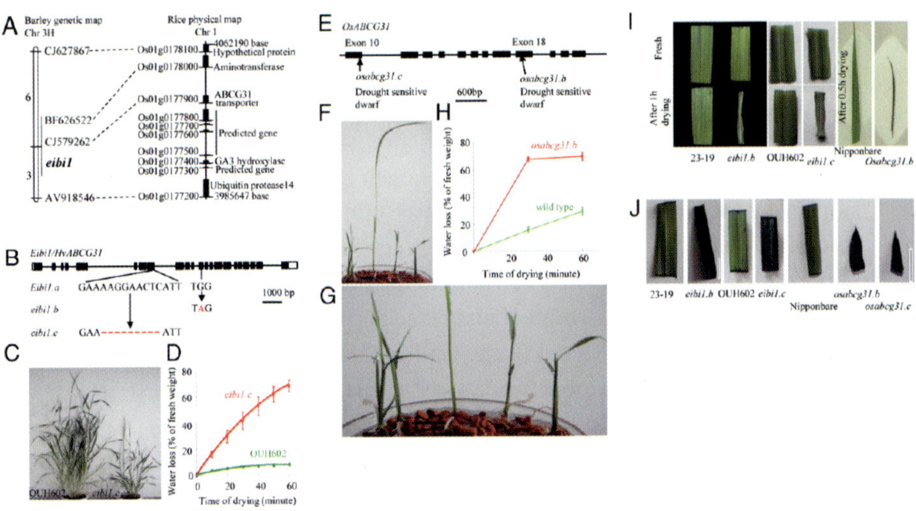

Genetic Resources for Disease Resistance in Wild Wheat and Barley

The Near East Fertile Crescent extending from Iran to Israel is the primary center of origin and diversity of cultivated barley (Nevo, 2012) and wheat (Nevo, 2014). The following are examples of this rich diversity not only for *abiotic* but also *biotic* stresses, involving smuts in Israel (Savchenko et al., 2015), rusts in Israel (Savchenko et al., 2014a, 2014b), and powdery mildew (Moseman et al., 2003–2005; Nevo et al., 2002). Utilization of these resources in cultivars is summarized for rusts in Xie and Nevo (2008). Following is an overview a few of our relevant studies conducted at the Institute of Evolution on disease resistance in wild cereals primarily in Israel.

Resistance of *Hordeum spontaneum* Collected in Israel to Infection with Blumeria graminis f. sp. hordei (Moseman et al., 1983)

The progenitor of barley, *H. spontaneum*, is used extensively for developing improved barley (*H. vulgare*) cultivars. Sites in Israel were identified with H. spontaneum plants resistant or moderately resistant to *Blumeria graminis* f. sp. *hordei*, which causes the disease barley powdery mildew (Moseman et al., 1983). The sites represent the range of ecological and environmental conditions and of genetic diversity and morphology of plants in Israel. The three cultures of Blumeria graminis f. sp. *hordei*, 64.54, 59.11, and R1, used to screen the accessions, possessed the virulence genes corresponding to most of the known resistance genes to that pathogen in barley. Eighty-nine accessions were identified as resistant to all three cultures, and 180 accessions as at least moderately resistant. Sixty-three accessions were resistant to one set of the cultures and moderately resistant to the other cultures, suggesting that those accessions may have more than one gene conditioning their resistance. All of the accessions collected in the Galilee, Northern Coastal Plains, and Judean Highlands sites, where the pathogen is usually present, were resistant or moderately resistant to all three cultures of the pathogen. No resistant or moderately resistant accessions were obtained from sites in the Negev or on Mt. Hermon, where the pathogen is seldom observed. Relationships between resistance to *Blumeria graminis* f. sp. *hordei*, genetic diversity and morphology of *H. spontaneum* plants, and the ecological and environmental conditions at specific sites were analyzed to identify sites with plants having the maximum disease resistance and genetic diversity. The information was used in planning future plant germplasm explorations (see below the analysis of

Yin et al., 2015) powdery mildew at "Evolution Canyon" as a micro-model for the evolution of host–pathogen relationships.

Resistance of *Triticum dicoccoides* to Infection with Blumeria graminis f. sp. tritici (Moseman et al., 1984).

The reactions of 233 *T. dicoccoides* accessions collected at ten sites in Israel and Turkey to infection with cultures of *Blumeria graminis* f. sp. *tritici* were determined (Moseman et al., 1984). The reactions indicated that the number of sources of resistance to *Blumeria graminis* f. sp. *tritici*, which can be obtained from *T. dicoccoides* populations growing in Israel and Turkey, is almost unlimited. One hundred and fourteen, or 49 percent, of the accessions were resistant, and 137, or 59 percent, of the accessions were resistant or moderately resistant to infection with four cultures of B. *graminis* f. sp. *tritici*, which possessed the virulence genes corresponding to most of the identified resistance genes in wheat. Accessions collected at sites with marginal habitats where *T. dicoccoides* grows poorly and has lower grain weight, were more susceptible than were accessions collected at sites with an optimal habitat for growth of *T. dicoccoides*. The results agreed with those in a previous study with *H. spontaneum*, and indicate that to obtain *H. spontaneum* or *T. dicoccoides* accessions with the highest level of resistance to the powdery mildew pathogens, plants should be collected at sites in ecological and geographic regions where those two species occupy optimum habitats and are exposed to the powdery mildew pathogens. A similar pattern for powdery mildew at the "Evolution Canyon" microscale model will be described below (Yin et al., 2015).

Resistance of *T. dicoccoides* Collected in Israel to Infection with *Puccinia recondita tritici* (Moseman et al., 1985)

Triticum dicoccoides, a progenitor of hexaploid wheat (*T. aestivum*) and tetraploid wheat (*T. durum*), grows wild in many sites in Israel, primarily in northern Israel and the Golan Heights, its center of origin and divergence (Nevo, 2014). Reactions were determined utilizing 687 accessions of *T. dicoccoides* from Israel to infection with culture PRTUS6 of *Puccinia recondite* f. sp. *tritici*, which incites the disease wheat leaf rust. Fourteen percent (or ninety-eight accessions) were at least moderately resistant in the first-leaf stage to infection with culture PRTUS6. Nineteen of thirty-four accessions collected at site 12 were moderately resistant as were ten of thirty-five from

Site 2 and four of nine from Site 8. Resistant accessions were obtained in low frequencies from Sites 6 and 7. Resistant and moderately resistant accessions were obtained among 353 accessions collected at approximately 180 sites throughout the region from Upper and Lower Galilee, Mt. Gilboa and the Judean Mountains, and Mt. Hermon and the Golan Heights. Additional resistant and moderately resistant accessions of *T. dicoccoides* could probably be obtained by collecting from additional sites. The resistant accessions identified in this study were used to develop enhanced hexaploid and tetraploid wheat germplasm resistant to the rusts *Puccinia recondita* f. sp. *tritici*, *Blumeria graminis* f. sp. tritici, and *Puccinia. Striiformis* West.

Reactions of *H. spontaneum* to Infection with Two Cultures of *Puccinia hordei* from Israel and United States (Moseman et al., 1990)

The reactions to infection with two cultures of *Puccinia hordei* were determined for 292 *H. spontaneum* accessions collected at 16 sites encompassing the ecological range of *H. spontaneum* in Israel (Moseman et al., 1990). Culture Tel-Aviv was from Israel and culture 57.19 was from the United States. Fifty-two percent of the accessions were resistant to culture Tel-Aviv and 67 percent were resistant to culture 57.19. Forty-three percent of the accessions were resistant to both cultures. The average infection type (IT) of accessions within sites ranged from 2.7 to 7.5 on a 0–9 rating scale. The results showed that the presence of *Ornithogalum* species, the alternate hosts of *P. hordei*, may increase the percentage of *H. spontaneoum* accessions resistant to *P. hordei*. More accessions were resistant at sites where humidity at 14:00 was higher, the annual evaporation was lower, and where the glumes were shorter. In a previous study a lower percentage of *H. spontaneum* accessions also was found to be resistant to a culture of *E. graminis hordei* from Israel than to cultures from other countries. Previous studies also have shown that cultures of *Puccinia hordei* and *Blumeria graminis* f. sp. *hordei* from Israel contain many genes for virulence on barley, and that *H. spontaneum* accessions from Israel contain many genes for resistance to these two pathogens. Previous results and the results reported in this paper support the hypothesis of co-evolution of resistant host genes and virulent pathogen genes where hosts and pathogens have coexisted for thousands of years.

Resistance of Wild Wheat to Stripe and Stem Rusts: Predictive Method by Ecology and Allozyme Genotypes (Nevo et al., 1986a, 1986b, 1991)

From 114 accessions of wild emmer wheat from eleven sites in Israel, known for their allozymic variation (Nevo et al., 1982), individual genotypes were tested for resistance to one isolate of stripe rust (*Puccinia striiformis*) both in the seedling stage in a growth chamber and in the adult plant stage in the field (Nevo et al., 1986b). The results indicated that resistance to stripe rust in seedlings and adults were significantly correlated ($rs = 0.40$, $p < 0.001$). Genetic polymorphisms of resistance to stripe rust vary geographically and are predictable by climatic as well as allozymic markers (Nevo, 1987). Three variable combinations of rainfall, evaporation, and temperature explain significantly 0.40–0.53 of the spatial variance in disease resistance to stripe rust, suggesting the operation of natural selection. Several allozyme genotypes were significantly associated with disease resistance. Similarly, Nevo et al. (1991) described the resistance of wild emmer wheat to stem rust (*Blumeria graminis*) predicted by ecological, pathological, and allozyme associations. They concluded that natural populations of wild emmer wheat in Israel contain large amounts of disease resistance genes. These populations could be effectively screened by phytopathologists for resistant genotypes that will be used by breeders to produce new resistant cultivars. Prediction of resistant genetic resources for crop improvement by isozyme markers and ecology was described by Nevo (1987). Genome associations with disease resistances in humans has been reviewed, primarily of the major histocompatibility complex (Trowsdale and Knight, 2013). Israel as an optimal natural laboratory for identifying genetic resources of wild cereals for crop improvement was described by Nevo (1986).

"Evolution Canyon": A Microsite Model for the Origin of Disease Resistance in Wild Cereals

"Evolution Canyon" (ECI) at Lower Nahal Oren, Mount Carmel, Israel (Figs. 5 and 6) is an optimal natural microscale model for unraveling cereal evolution in action, highlighting the basic evolutionary processes of biodiversity evolution, adaptation, and speciation (Nevo, 1995, 2009, 2012, 2014, and Nevo list of "Evolution Canyon" at http://evolution.haifa.ac.il). Recently, we have shown that it is also a remarkable model for the origin of disease resistances in wild emmer wheat, *T. dicoccoides,* and wild barley, *H. spontaneum,* imitating locally the regional pattern across Israel from the mesic north to the xeric south. Major model organisms in ECI, among others, are wild emmer, *T. dicoccoides*, the progenitor of cultivated wheat, and wild barley, *H. spontaneum*, which display dramatic inter-slope adaptive and species divergence on the tropical-hot and xeric "African" slope (AS, or South Facing Slope, SFS) and the abutting temperate-mesic "European" slope (ES, or North Facing Slope, NFS), separated on average by 250 meters (Figs. 5 and 6) (see Nevo, 2006 for wild barley sympatric speciation of HS). Recently, it was also found that TD and HS both undergo adaptive ecological incipient sympatric speciation at EC I (Nevo, in preparation).

Evolution Canyon Micro-Scale Model for the Evolution of Host-Pathogen Coevolution and the Origin of Disease Resistance Genes against Powdery Mildew TD and Rust in *T. dicoccoides* and *H. spontaneum*

Yin et al. (2015) examined 278 single sequence repeats (SSRs) and the phenotype diversity of resistance to powdery mildew in *T. dicoccoides* between the opposite slopes. Polymorphic products between the phenotypes of AS and ES were found in 148 pairs of SSR primers (53.23 percent, Fig. 7). Remarkably, all *T. dicoccoides* growing on the hot and arid AS=SFS were *susceptible* to a composite of powdery mildew (*Blumeria graminis*). By contrast, all plants of wild emmer growing on the ES=NFS were *highly resistant* to powdery mildew at both seedling and adult stages, despite the short average distance of 250 meters between the opposite slopes. The plants on the hot and dry AS did not develop resistance to powdery mildew, whereas the plants on the ES, a cool and humid slope, did develop resistance since the disease stress was strong there, in contrast to that on AS. Thus, the "Evolution Canyon" model represents an especially interesting model of host-pathogen interaction demonstrating how resistance develops when pathogen stress causes an adaptive result at a *micro-scale* model, paralleling locally the regional pattern in Israel (Moseman et al., 1984). A similar, yet unpublished result was demonstrated locally in "Evolution Canyon" in powdery mildew and stem rust attacking wild barley, *H. spontaneum* (Rints and Nevo, unpublished data). Likewise, with the secondary metabolites, Dimboa and Diboa in roots and shoots of wild emmer wheat at EC I (Ben Abu et al., 2018). This pattern of EC I micro-scale model imitates the disease resistance pattern regionally across Israel, with primarily resistance at the mesic North and susceptibility at the hot-dry desert South, described earlier for both HS and TD (Moseman et al., 1983 and 1984, respectively).

FIGURE 5. The "Evolution Canyon" (EC I) at Lower Nahal Oren, Mount Carmel, Israel. Note the tropical savannoid, park-forest, hot and dry "African" slope, on the right hand side (AS = SFS slope) versus the temperate, densely forested, cool and humid ES = NFS slope, on the left hand side, separated, on average, by only 250 meters. Higher terrestrial species richness occurs on the more stressful tropical AS. Aquatic species richness is higher on the milder, temperate cool and humid "European" Slope (ES). This model has been studied for biodiversity evolution, adaptation, and adaptive ecological incipient sympatric speciation across life from viruses and bacteria to mammals, as well as for global warming, and now also for host-pathogen coevolution, and the origin of disease resistances in wild cereals (original photo by Nevo).

• • •

FIGURE 6. The "Evolution Canyon" (EC I) model at Mount Carmel (EC I), sampling stations (A-C) and air view. (A) Schematic diagram, showing the cause of microclimatic interslope divergence, (B) Cross section view of "Evolution Canyon" I (EC I), Lower Nahal Oren, Mount Carmel, and (C) Air view of EC I. Note the distinct divergent plant formations on the opposite slopes. The temperate, forested, cool, humid, temperate slope, dubbed "European Slope" (ES), is also dubbed north facing Slope (NFS), sharply contrasting with the tropical, savannoid, hot and dry open park forest, dubbed the "African slope" (AS), also dubbed the south facing slope (SFS). In EC I at Mount Carmel, as in other three "Evolution Canyons" studied in Israel, seven sampling stations (populations) are designated: three on the AS = SFS (nos. 1–3), one at the valley bottom (no. 4) and three on the NFS (nos. 5–7) (From Sikorski and Nevo, 2006).

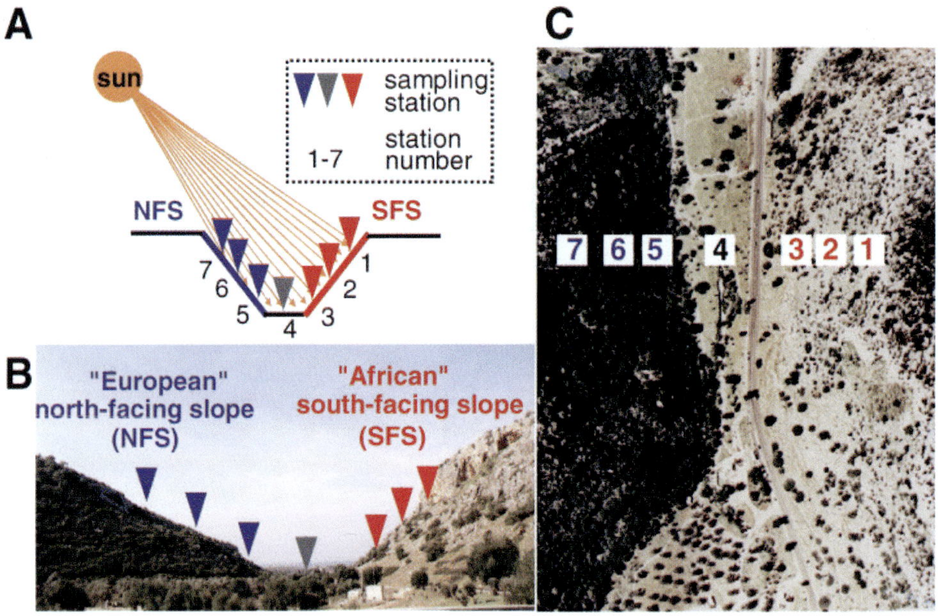

FIGURE 7. SSR polymorphism divergence in the A and B genomes of wild emmer wheat *T. dicoccoides* on the opposite slopes at "Evolution Canyon." (A) Tested SSR markers between AS and ES on genomes A and B. (B) SSR frequency polymorphism between AS and ES on the two genomes. Red color indicates high frequency and blue color low frequency. Very little *intra-slope* polymorphism is displayed versus *inter-slope* polymorphism (From Yin et al., 2015). Some inter-slope polymorphism is seen between genomes A and B. Interestingly, all wild emmer genotypes on AS-SFS were susceptible, and all genotypes on the abutting ES=NFS were resistant at both seedlings and adult stages (From Yin et al., 2015). Clearly, the AS, a dry and hot slope, did not develop resistance to powdery mildew, whereas the ES, a cool and humid slope, did develop resistance since disease stress was strong there. This pattern parallels the regional Israeli pattern and is a remarkable demonstration in host-pathogen interaction on how disease resistance to powdery mildew develops when pathogen stress, prevalent on ES=NFS, causes an adaptive disease resistance result at a micro-scale distance, paralleling the evolution of disease resistance result in humid and cold Upper Galilee (From Yin et al., 2015).

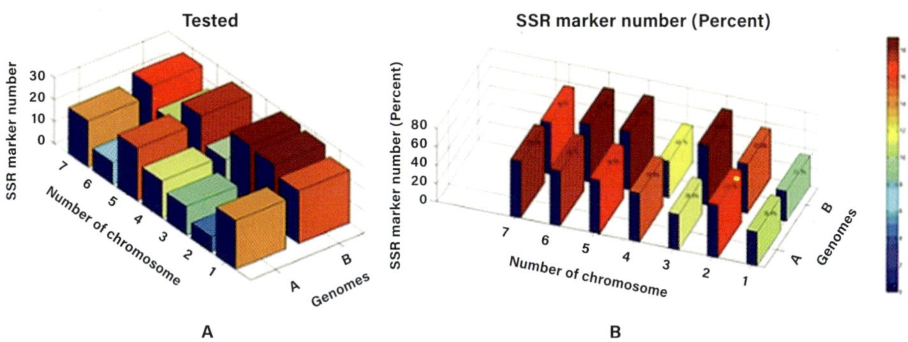

Wild Emmer: Genetic Resources, Gene Mapping, and Potential for Wheat Improvement (Xie, 2008)

Wild emmer, *T. dicoccoides*, the progenitor of cultivated wheat, harbors rich genetic resources in Israel and the Near East Fertile Crescent for wheat improvement (Nevo et al. 2002; Nevo, 2014). They include many agronomic traits such as *abiotic* stress tolerances (salt, drought and heat), *biotic* stress tolerances (powdery mildew, rusts, and *Fusarium* head blight, mosaic viruses), grain protein quality and quantity, and micronutrient concentrations (Zn, Fe, and Mn). In the review of Xie and Nevo (2008), we summarized (1) traits and controlling genes identified and mapped in *T. dicoccoides* (See Table 1 in Xie and Nevo, 2008); and (2) the genes transferred to improve cultivated wheat from *T. dicoccoides* (Table 2 in Xie and Nevo, 2008). These genes control important agronomic traits such as disease resistance, high protein (Nevo et al., 1986; Pagnotta et al., 1995; Nevo et al., 1995) and micronutrient content (Jun et al., 2011, in HS; Cakmak et al., in TD), should contribute to wheat production and food nutrition. However, most of the rich genetic reservoir in wild emmer wheat remains *untapped*, highlighting the need for further exploration and utilization for long-term wheat breeding programs, especially when the reference genome of wild emmer wheat is available (Avni et al., 2017), as well as that of cultivated barley (Mascher et al., 2017). Barley genes could be used in the wheat and vice versa.

A significant number of genes have been identified and mapped (Table 1 in Xie and Nevo, 2008). Markers closely linked to many of the traits will be useful for transferring them to cultivated wheat and subsequently for their incorporation into improved cultivars. The cloning of *Gpc-B1* and of other genes may also permit molecular improvement of wheat through transgenic strategies. Until now, several *T. dicoccoides*-derived genes, namely, *Pm16, Pm 30, Pm 31, PM41, Lr53, Yr15, Yr35, Yr36,* Qfhs.ndsu-3AS, and Gpc-B1 (Table 2 in Xie and Nevo, 2008) have been transferred to cultivated wheat lines, and some of them should contribute to wheat production and end-product nutrition in the near future. Wild emmer has not been exploited very widely, despite considerable knowledge of its potential for many years. There are several reasons for this. Foremost, innovative ideas and resources take time to become established. Secondly, it takes many years to introgress traits derived from wild relatives to cultivated wheat and to exploit them in improved cultivars. Thirdly, short-term and easier solutions are usually preferred by breeders. However, risk aversions may be unjustified over the long term. Domestication and modern breeding practices have made wheat vulnerable to diverse *abiotic* and *biotic* stresses. For example, race Ug99 of *Blumeria graminis* f. sp. *tritici*, discovered in Uganda in

1999 and subsequently spread to Kenya, Ethiopia, Yemen, and Iran, is virulent for a number of previously effective resistance genes and is predicted to spread rapidly through these regions and possibly further afield, potentially causing a wheat production disaster that would affect food production (http://www.biologynews.net/ archives/2007/01/17/). Therefore, wild emmer, along with other wheat progenitors and relatives, is a vital genetic resource that will be required to meet this threat and others that may arise in the future, certainly when the reference genome is available (Avni et al., 2017), and could be effectively used to solve pending disasters.

Variation in Response to Infection with *Blumeria graminis* f. sp. *hordei* and *Puccinia hordei* in Wild Barley Populations in a Center of Diversity (Jana and Nevo, 1991)

The Near East Fertile Crescent extending from Iran to Israel is the primary center of origin of cultivated barley, and the primary region of origin and diversity of the wild barley, *H. spontneum*, the progenitor of cultivated barley. Genetic diversity of wild barley accessions collected from different parts of this region were evaluated for their reactions to infection with three isolates of *Blumeria graminis* f. sp. *hordei* and two isolates of the rust pathogen, *P. hordei* (Jana and Nevo, 1991). One culture of each pathogen was isolated in Israel and the others, either in Japan or the United States. Out of a total of 330 wild barley accessions collected from 14 sites in Iran, Turkey and Syria, only 18.8 percent were resistant to the Israeli culture, and 14.8 percent were resistant to a composite of the Japanese and American cultures of *E. graminis* f. sp. *hordei*. Out of 105 accessions collected from six sites in Iran and Turkey, none was found to be resistant to the Israeli culture and 34.3 percent were resistant to the American culture of *P. hordei*. Considerable variation was observed both within and among sites for reactions to infection with different cultures of each of the two pathogens. These results were compared with those of an earlier study involving wild barley accessions from Israel to illustrate the relative importance of different sub-regions in the Near East Fertile Crescent as sources of new genes for resistance to *E. graminis* f. sp. *hordei* and *P. hordei*.

The discovery of association of ecological and allozymic variation with the variation for resistance to powdery mildew, leaf rust, and stripe rust in wild emmer wheat (Nevo et al., 1985, 1986b, respectively), underscores the importance of *in situ* conservation of genetically heterogeneous wild or landrace populations (Nevo, 1998). The results of powdery mildew and leaf rust indicated that the maximum number of new resistant genes to the pathogens or their combination to the two pathogens occurred more in Israel than in the neighboring countries. Maximizing collection sites, even at microsites with divergent ecologies, such as in Evolution Canyon, may increase the discovery of resistant genes (e.g. Yin et al., 2015). Israel appears particularly important in identifying novel resistant genes to the two pathogens, as well as other pathogens due to its outstanding ecological diversity, geological, edaphic, and climatic regions (Nevo, 1987).

Identification of Molecular Markers Linked to the Yr15 Stripe Rust Resistance Gene of Wheat Originated in Wild Emmer Wheat, *T. dicoccoides*

The *Yr15* gene of wheat confers resistance to the stripe rust pathogen *P. striiformis* West, which is one of the most devastating diseases of wheat throughout the world. Sun et al. (1997) identified molecular markers flanking the *Yr15* gene of wheat using the near-isogenic-lines approach. RFLP screening of 76 probe-enzyme combinations revealed one polymorphic marker (Nor/^{1}aqI) between the susceptible and the resistant lines. In addition, out of 340 RAPD primers tested, six produced polymorphic RAPD bands between the susceptible and the resistant lines. The genetic linkage of the polymorphic markers was tested on segregating F2 population (123 plants) derived from crosses between stripe rust-susceptible *T. durum* wheat, cv D447, and a BC3 F9 resistant line carrying *Yr15* in a D447 background. A 2.8-kb fragment produced by the Nor RFLP probe and a 1420-bp PCR product generated by the RAPD primer OPB13 showed linkage, in coupling, with the *Yr15* gene. Employing the standard maximum-likelihood technique it was found that the order OPB131420—Yr15—Nor1 on chromosome 1B appeared to be no less than 1,000 times more probable than the closest alternative. The map distances between OPB131420—Yr15—Nor1 are 27.1 cM and 11.0 cM for the first and second intervals, respectively. The application of marker-assisted selection for the breeding of new wheat cultivars with the stripe rust resistance gene is of cardinal importance. Additional studies were conducted on *Yr 15* (Fahima et al., 1998; Peng et al., 1998, 1999, 2000a, 2000b; Chague et al., 1999). Later we identified a close locus to *YR15*, designated *YrH52*.

Two stripe-rust resistance genes, *YrH52* and *Yr15*, derived from the Israeli wild emmer wheat, *T. dicoccoides*, have been located on chromosome 1B. The main objectives of the study by Peng (2000b) were to increase marker density in the vicinity of YrH52 gene by means of AFLP, RAPD and microsatellite markers, to improve the map of another *T. dicoccoides*-derived stripe-rust resistance gene *Yr15* using microsatellite markers, and to preliminarily discriminate these two genes. Additional 26 marker loci comprising 20 AFLPs, three RAPDs, and three microsatellites were found to be linked to *YrH52*. An updated genetic map consisting of 45 marker loci, in the region of *YrH52* gene was constructed with a total map length of 107.7 cm. The mean interval length was 0.96 cm in the region Xgwm359b–P55M53b carrying the *YrH52* gene. *YrH52* was bracketed by Xgwm413 (*Nor1* and UBC212a) and Xgwm273a (Xgwm273d) with map distance of 1.3 and 2.7 cm from either side, respectively. Eight additional microsatellite markers were found to be linked to *Yr15*, and the linkage map of *Yr15* gene was thus obviously improved. In the *YrH52*-mapping population, no crossover was detected in the interval UBC212a (Xgwm413)–*Yr15*–*Nor1*, and *YrH52* was located distally outside this interval. It may suggest that *YrH52* is different from *Yr15* even though both of them are derived from *T. dicoccoides* and are mapped on chromosome 1BS. The large number of molecular markers revealed in Peng et al (2000b) would be helpful for the marker-assisted introgression of the *T. dicoccoides*-derived *YrH52* and *Yr15* stripe-rust resistance genes into elite cultivars of wheat, and the high-density map would accelerate the map-based cloning of the two genes.

Microsatellite Tagging of the Stripe-rust Resistance Gene YrH52 Derived from Wild Emmer Wheat, *T. dicoccoides*, and Suggestive Negative Crossover Interference on Chromosome 1B

Stripe rust (caused by *P. striifomis*) is one of the most devastating diseases relating to wheat production. Wild emmer wheat, *T. dicoccoides*, the tetraploid progenitor of cultivated wheat, has proven to be a valuable source of novel striperust resistance genes for wheat breeding. For example, *T. dicoccoides* accessions from Mt. Hermon, Israel, are uniformly and highly resistant to stripe rust. Peng et al. (1999) mapped the stripe-rust resistance gene, derived from the unique Mt. Hermon population of wild emmer, using microsatellite markers. An F2 mapping population was established by crossing stripe-rust resistant *T. dicoccoides* accession H52 from Mt. Hermon with the *T. durum* cultivar "Langdon." The stripe-rust resistance derived from accession H52 was found to be controlled by a single dominant gene, which was temporarily designated as YrH52. Out of 120 microsatellite markers tested,

109 (91 percent) showed polymorphism between the parental lines. Among 79 segregating microsatellite loci generated from 56 microsatellite primer pairs, nine were linked to YrH52 with recombination frequencies of 0.02–0.35, and LOD scores of 3.56–54.22. A genetic map of chromosome 1B, consisting of ten microsatellite loci and the stripe-rust resistance gene YrH52, was constructed with a total map length of 101.5 cM. YrH52 is also closely linked to RFLP marker Nor1 with a map distance of 1.4 cM and a LOD value of 29.62. A negative crossover interference was observed in chromosome 1B, especially in the region spanning the centromere. Negative crossover interference may be a common characteristic of gene-rich regions or gene clusters in special chromosomes.

Diversity and Sources of Multiple Disease Resistance in *H. spontaneum*

Hordeum spontaneum, the progenitor of cultivated barley, is known to be a rich source of disease resistance genes. The diversity of *H. spontaneum* accessions from Israel and Jordan was assessed for their reaction to six fungal pathogens of importance to cultivated barley in the United States and Canada. Overall, a high level of *macro*-scale (across collection sites) and *micro*-scale (within a collection site) diversity for disease reaction was found in the 116 accessions of *H. spontaneum* evaluated at the seedling stage (Fetch et al., 2003). Additionally, genetic heterozygosity for resistance loci was common in *H. spontaneum*. The frequency of resistance in accessions from Jordan and Israel was high for *Septoria* speckled leaf blotch (77 and 98 percent, respectively), leaf rust (70 and 90 percent), net blotch (72 and 68 percent), and powdery mildew (58 and 70 percent); intermediate for spot blotch (*Cochliolobus sativus*) (53 and 46 percent); and low for stem rust (2 and 26 percent). The level of disease resistance in *H. spontaneum* was not strongly correlated with any of the weather variables (temperature, precipitation, and humidity) monitored near the collection sites. However, in general, resistance was more often found in germplasm from mesic (e.g., Mediterranean coast) than in xeric (e.g., Negev Desert) areas. Two *H. spontaneum* accessions (Shechem 12-32 and Damon 11-11) were resistant to all six pathogens and may be useful in programs breeding barley for multiple disease resistance. The high level of diversity and heterozygosity for disease reaction found in this study indicated that *H. spontaneum* is an extraordinarily rich and largely untapped source of unique disease resistance alleles for cultivated barley improvement. Importantly, rest resistance in *H. spontaneum* was also identified on the mesic European slope (ES) of Evolution Canyon (Rints and Nevo, unpublished).

Storage Proteins in Wild Emmer Wheat, *T. dicoccoides*: Diversity of HMW Subunits from Israel. 1. Geographical Patterns and Ecological Predictability

The diversity of High Molecular Weight HMW glutenin subunits in the tetraploid wild progenitor of wheat, *T. dicoccoides,* was studied electrophoretically in 231 individuals representing 11 populations of wild emmer from Israel (Nevo and Payne, 1987). The results showed that: (a) The two HMW glutenin loci, *Glu-A1* and *Glu-B1*, were rich in variation, having 11 and 15 alleles, respectively (Fig. 8); (b) Genetic variation in HMW glutenin subunits was often severely restricted in individual populations, supporting an island population genetic model; and (c) Significant correlations were found between glutenin diversity and the frequencies of specific glutenin alleles and physical (climate and soil) and biotic (vegetation) variables (Fig. 8). The results suggested that (a) at least part of the glutenin polymorphisms in wild emmer can be accounted for by environmental factors; and (b) the endosperm of wild emmer contains many allelic variants of glutenin storage proteins that are not present in bread wheat and could be utilized in breeding varieties with improved breadmaking qualities.

• • •

FIGURE 8. (A) SDS-PAGE of grain proteins found in a population of wild emmer from Kokhav Hashahar, near Jerusalem (Nevo and Payne, 1987). The region of the gel containing the HMW glutenin subunits is marked by brackets. The proteins thought to be coded by the *Glu-A1* and *Glu-B1* loci are prefixed with letters A and B, respectively. The HMW glutenin subunits of bread-wheat cultivar Chinese Spring are separated in slot 1. They are, with increasing mobility, 2 (chromosome 1D), 7+ 8 (1B), and 12 (1D). (B) SDS-PAGE of grain proteins in two wild emmer populations: slots 1–15: Qazrin; slots 16–23: Yehudiyya. (C) SDS-PAGE of a range of wild emmers to show the extent of allelic variation in HMW glutenin subunits. Grains were taken from populations Qazrin (slot 1). Yehudiyya (slots 2, 7, 8, and 10). Tabigha (slot 12). Bet-Meir (slot 14). Rosh Pinna (slot 11). Bat Shelomo (slot 9), Mt Hermon (slot 15), Sanhedriyya (slots 3–6, and 13), and Taliyiba (slot 16). This HMW glutenin diversity is potential diversity for improving bread quality.

A

B

C

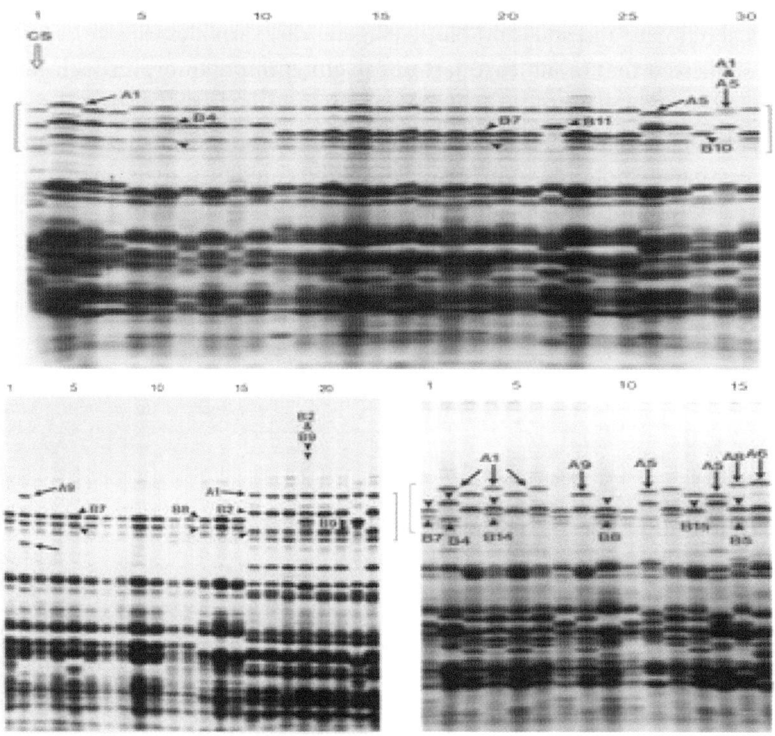

Wheat Storage Proteins: Glutenin Diversity in Wild Emmer, *T. dicoccoides*, in Israel and Turkey. 2. DNA Diversity Detected by PCR

Germplasm of *T. dicoccoides* collected from different environments in Israel was evaluated by using the PCR as a molecular marker (Pagnotta et al., 1995). Two pairs of primers were used in the PCR in amplifying the DNA tracts coding the high- and low-molecular-weight glutenin subunits. Analyses reveal great variability within and between populations indicating the high values of this germplasm for future breeding programs to improve the protein quality in wheat.

Wheat Storage Proteins: Glutenin DNA Diversity in Wild Emmer Wheat, *T. dicoccoides*, in Israel and Turkey. 3. Environmental Correlates and Allozymic Associations

The diversity of high-molecular-weight (HMW) and low-molecular-weight (LMW) glutenin subunits in the tetraploid wild progenitor of wheat, *T. dicoccoides*, was studied at the DNA level by the polymerase chain reaction (PCR) (Nevo et al., 1995). The DNA diversity of HMW and LMW glutenins was shown to be correlated to environmental physical and biotic factors (climate, soil and pathogen resistance) and to allozyme variation. We concluded that glutenin DNA diversity is *nonrandomly* distributed and could be optimally sampled in nature for future breeding programs to improve bread quality.

Wild Emmer Wheat an Important Source for Improving Zinc and Iron in Bread Wheat

Deficiency in key micronutrients and proteins is prevalent in human nutrition. TD is an important genetic resource for increasing zinc and iron in bread wheat (Cakmak et al., 2004). TD shows impressive variation and the highest concentrations of micronutrients, especially in homoeologous chromosomes 5 significantly exceeding those of cultivated wheat, especially in chromosomes 6A, and 5B. Clearly TD is an important genetic resource for increasing concentration and content of Zn and Fe in modern cultivated wheat, thereby improving human nutrition and health. The discovery of a NAC gene in *T. dicoccoides*, regulating senescence and associated with increased grain protein, zinc, and iron content (Uauy et al., 2006) is an excellent source for improving grain protein, zinc and iron content in cultivated wheat, thus increasing its human nutrition and health (Uauy et al., 2006).

Conclusions

The few examples reviewed here primarily from the research program at the Institute of Evolution, University of Haifa, conducted in the period 1975–2018, and ongoing, unfold the rich genetic resource of wild cereals for crop improvement. The Near East Fertile Crescent, and primarily small Israel, harbor immense biodiversity resources for wheat and barley improvement (Nevo, 1987). These include rich abiotic (e.g., drought, cold, heat, salt, tolerances) and biotic (viral, bacterial, fungal and herbicide resistances) resistances, both regionally and locally; high quantity and quality of storage proteins (glutenins, gliadins, hordeins), amylase, photosynthetic yields, and micronutrient minerals (e.g., Zn, Fe, Se resources). Most of these resources are yet untapped and provide precious sources for cereal improvement and therefore to increases both the quantity and quality of human nutrition. Marker-assisted selection could provide a solid basis for introgression, cloning, and transformation of agronomically important genes and QTLs from the wild to the cultivars reinforcing them genetically to resist climatic change, desertification, salinization, and pollution. The cultivars have deteriorated genetically during domestication, and both cultivars and wild progenitors are now deteriorating genetically due to global warming (Nevo et al., 2012), and urbanization. This ongoing deterioration of the genetic resources is especially crucial in view of the still rising human population—anticipated to reach 9.8 billion people in 2050—pollution, and climate change (Nevo, 2016). The wild cereals are the best source of additional genes for crop improvement and therefore need to be preserved both *in situ* and *ex situ* (Nevo, 1998) to safeguard their future immense potential for crop improvement. The recently published reference genomes of cultivated barley, *H. vulgare* (Mascher et al., 2017), and wild emmer wheat, *T. dicoccoides* (Avni et al., 2017) will be of paramount importance in isolating gene and regulatory genomic elements for crop improvement and safeguarding future food production. This is especially true for wild emmer wheat and wild barley whose origin and diversity is the Near East Fertile Crescent, particularly north Israel and the Golan Mountains.

Prospects

Generating genetically resistant cultivars is the imperative of future breeding, hence the importance of genetics, genomics, and epigenomics: i.e., genes and regulation. The genome sequencing of cultivated barley (Mascher et al., 2017), wild barley, *H. spontaneum* (Nevo, in preparation), wild emmer wheat (Avni et al., 2017) and bread wheat (to be completed in the foreseeable future, by the wheat consortium), are all either completed or intensively progressing (Chapman et al., 2015). This will unfold the rich wild and cultivated cereal genomes for crop improvement. It will be of cardinal importance to utilize in crop improvement both *coding* and *noncoding* genomic elements. The latter have recently been exposed in humans (ENCODE). To understand how cells interpret the information within the genome much more is needed to be learned about the regulatory functions of the *noncoding genome*. This became the task of ENCODE, the Encyclopedia of DNA Elements, the aim of which was to describe all functional elements encoded in the human genome. Nine years after launch (2012), its main efforts culminate in the publication of thirty coordinated papers, six of which are in the issue of *Nature* (check in the internet *Nature* ENCODE explorer). ENCODE and other genomic programs are unraveling genome function and regulation, revealing complex genomic networks that underlie adaptive complexes that could be of utmost importance in crop improvement. Wheat is grown on 17 percent of all crop area and represents the staple food for more than 40 percent of the world's population. Barley ranks fourth in the world production behind wheat, rice, and maize (*Zea mays* L.) and is widely used for animal feed and food industry. The progenitors of cereals could be the best source for the crop improvement based on the new genomics and epigenomics for increasing and improving crop production.

Crop Improvement by Dead Sea Filamentous Fungi

As a very intriguing epilogue, crops like cultivated wheat and barley can be improved by the genetic resources of Dead Sea filamentous fungi (Figure 9). The Dead Sea (DS) is a uniquely important natural laboratory model for evolution under extreme hypersaline environment, ten times saltier than the oceans (34.8 gL^{-1}), and is *still* a potentially important gene pool for crop improvement and future agricultural genetic engineering prospects. Following the uniquely important findings of life in the hypersaline DS by Elazar Volcani in the early 1930s, including archea, bacteria, and one alga, we identified in the DS the first records of filamentous fungi (Buchallo et al., 1998).

Our studies of their biodiversity, including seventy-seven species, both permanent and ephemeral, involved their biology and genetics in the context of the evolution of the geology, climate, and limnology of the Dead Sea (Nevo, Oren, and Waser, eds., 2003), highlighting their great potential for crop improvement (Jin et al., 2005).

FIGURE 9. Two filamentous fungi from the Dead Sea, *Aspergillus ruber* (red) and *Aspergillus amstelodami* (yellow) (Kis Papo et al, 2014; Perl et al, 2017). These filamentous fungi harbor genetic resources for crop improvement against multiple stresses including salinity, drought, temperatures, hot and cold, freezing-thawing (Jin et al., 2005).

Our first demonstration of their importance for agriculture was by cloning the *EhHOG* gene, a mitogen-activated protein kinase (MAPK) from the DS fungus *Eurothium herbariorum* (Jin et al., 2005). The gene is an osmoregulator in yeasts and other eukaryotes. When expressed in *Saccharomyces cerevisiae,* the growth and aberrant morphology of the *hog 1* mutant was restored under high osmolarity. Glycerol content in the transformant outperformed the wild type and had higher fitness under high lithium and freezing-thawing conditions. The study revealed the putative presence of high osmolarity glycerol response pathway in *E. herbariorum,* and the adaptive significance of *EhHOG* pathway in osmotic regulation, heat stress, freezing stress, and oxidative stress. Later (Kis papo et al., 2004), we sequenced the whole genome of the DS fungus *Aspergillus rubrum (*Figure 9*)*

and identified its adaptive strategies based on genome analysis and transcriptome profiling. The 26.2 Mb genome showed gains in gene families and loss of transport processes. Transcriptome analyses under different growth conditions revealed differentially expressed genes encoding ion and metabolite transporters. *E. rubrum* actively counteracts the salinity stress. Many of the genes encode for proteins with significantly higher proportion of acidic amino acid residues, a trait characteristic of the halophilic prokaryotes, supporting the theory of convergent evolution under extreme hypersalinity stress.

Fungal species diversity steadily decreased, highly and significantly correlated with the constant declining water level and increasing density and salinity of DS water, revealing both species *extinction* and *evolution* (Perl et al., 2017). During the period from 2000 to 2015, fungal species declined from thirty-four species to two frequent species in summer 2015: *Aspergillus amstelodami*, *Aspergillus rubrum* (Figure 9); later, *Aspergillus niger* and *Penicillium chrysogenum* were also found. Remarkably, most species became extinct, while the few remaining survived by evolving adaptive survival strategies. Hog transformed *Arabidopsis* survived high salinity (W. Song, in preparation). Even though the surviving *Aspergilli* are dormant now in the DS, as spores, they grow in the lab and develop mycelium under diluted DS water. However, these unique survivors harbor networks of salinity adapted gene pathways that could transform saline agriculture for crop improvement. They could be used in crop improvement, including wheat and barley, against multiple stresses, adapted to the increasingly salinized, warmer, and drier planet Earth, especially under the predicament of global warming. The Dead Sea could save human life from severe hunger in the future.

Acknowledgments

This paper is dedicated to all my students and collaborators, in Israel and the world. I thank the Ancell-Teicher Research Foundation of Genetics and Molecular Evolution for continuous financial support throughout this study. I also thank Ruth Ben David, Esther Kabuliansky, Robin Permut, and Kexin Li for their great help in advancing my research programs.

References

Aaronsohn, A. (1910) *Agricultural and botanical explorations in Palestine*. U.S. Government Printing Office.

Avni, R., et al., (2017). Wild emmer genome architecture and diversity elucidate wheat evolution and domestication. *Science*, *357*, 93–97.

Bedada, G., Westerbergh, A., Nevo, E., Korol, A., and Schmid, K. J. (2014) DNA sequence variation of wild barley *Hordeum spontaneum* (L.) across environmental gradients in Israel. *Heredity, 112*, 646–55.

Beb Abu, Y., Beiles, A., Flom, D., and Nevo, E. (2018). Adaptive evolution of bezoxazinoids in wild emmer wheat, *Triticum dicoccoides*,at "Evolution Canyon," Mount Carmel, Israel. *PLoS ONE, 13*(2), e0190424

Chague, V., Fahima, T., Dahan, A., Sun, G. L., Korol, A. B., Ronin, Y. I., Grama, A., Roder, M. S., and Nevo, E. (1999). Isolation of microsatellite and RAPD markers flanking the *Yr15* gene of wheat using NILs and bulked segregant analysis. *Genome, 42*, 1050–1056.

Chapman, J. A., Mascher, M., Buluç, A., Barry, K., Georganas, E., Session, A., and Schmutz, J. (2015). A whole-genome shotgun approach for assembling and anchoring the hexaploid bread wheat genome. *Genome biology, 16*(1), 26.

Chen, G., Komatsuda, T., Ma, J. F., Li, C., Yamaji, N., and Nevo, E. (2011). A functional cutin matrix is required for plant protection against water loss. *Plant Signal Behav., 6*, 1297–1299.

Chen, G., Komatsuda, T., Pourkheirandish, M., Sameri, M., Sato, K., Krugman, T., Fahima, T., Korol, A. B., and Nevo, E. (2009). Mapping of the eibi1 gene responsible for the drought hypersensitive cuticle in wild barley (*Hordeum spontaneum*). *Breed Sci., 59*, 21–26.

Chen, G., Krugman, T., Fahima, T., Chen, K., Hu, Y., Röder, M., Nevo, E., and Korol A. (2010). Chromosomal regions controlling seedling drought resistance in Israeli wild barley, *Hordeum spontaneum* C. Koch. *Genet. Resour. Crop Evol., 57*, 85–99.

Chen, G., Sagi, M., Weining, S., Krugman, T., Fahima, T., Korol, A. B., and Nevo, E. (2004a). Wild barley eibi1 mutation identifies a gene essential for leaf water conservation. *Planta, 219*, 684–93.

Chen, G., Suprunova, T., Krugman, T., Fahima, T., and Nevo, E. (2004b). Ecogeographic and genetic determinants of kernel weight and colour of wild barley (*Hordeum spontaneum*) populations in Israel. *Seed Sci. Res., 14*, 137–46.

Chen, L., Ren, J., Shi, H., Chen, X., Zhang, M., Pan, Y., Fan, J., Nevo, E., Sun, D., and Fu, J. (2013). Physiological and molecular responses to salt stress in wild emmer and cultivated wheat. *Plant Mol. Biol., 31*, 1212–1219.

Chen, L., Ren, J., Shi, H., Zhang, Y., You, Y., Fan, J., Chen, K., Liu, S., Nevo, E., and Fu, J. (2015). TdCBL6, a calcineurin B-like gene from wild emmer wheat (*Triticum dicoccoides*), is involved in response to salt and low-K+ stresses. *Mol. Breed., 35*, 1–12.

Chen, G., Krugman, T., Fahima, T., Korol, A. B., and Nevo, E. (2002). Comparative study of morphological and physiological traits related to drought resistance between xeric and mesic *Hordeum spontaneum* lines in Israel. *Barley Genet. Newslett., 32*, 22–33.

Dai F., Chen, Z.-H., Wang, X., Li, Z., Jin, G., Wu, D., Cai, S., Wang, N., Wu, F., and Nevo, E. (2014). Transcriptome profiling reveals mosaic genomic origins of modern cultivated barley. *PNAS, 111*, 13403–13408.

Darwin, C. (1859). *On the origins of species by means of natural selection*. John Murray.

Fahima, T., Röder, M. S., Grama, A., and Nevo, E. (1998). Microsatellite DNA polymorphism divergence in *Triticum dicoccoides* accessions highly resistant to yellow rust. *Theor. Appl. Genet., 96*, 187–195.

Feldman, M., and Sears, E. R. (1981). The wild gene resources of wheat. *Scientific American, 244*, 102–12.

Fetch Jr., T., Steffenson, B., and Nevo, E. (2003). Diversity and sources of multiple disease resistance in *Hordeum spontaneum*. *Plant Dis, 87*, 1439–48.

Fu, Y.-B., Peterson, G. W., Richards, K. W., Somers, D., DePauw, R. M., and Clarke J. M. (2005). Allelic reduction and genetic shift in the Canadian hard red spring wheat germplasm released from 1845 to 2004. *Theor. Appl. Genet., 110*, 1505–16.

Fu, Y.B., Peterson, G. W., Yu, J.-K., Gao, L., Jia, J., and Richards K. W. (2006). Impact of plant breeding on genetic diversity of the Canadian hard red spring wheat germplasm as revealed by EST-derived SSR markers. *Theor. Appl. Genet., 112*, 1239–47.

Fu, Y.-B., and Somers, D. J. (2009). Genome-wide reduction of genetic diversity in wheat breeding. *Crop Sci., 49*, 161–68.

GuoXiong, C., Pourkheirandish, M., Sameri, M., Ning, W., Nair, S., Shi, Y., Chao, L., Nevo, E., and Komatsuda, T. (2009). Genetic targeting of candidate genes for drought sensitive gene eibi1 of wild barley (*Hordeum spontaneum*). In *Breeding science*, pp. 637–44. Japanese Society of Breeding.

Gustafson, J., Raskina, O., Ma, X., and Nevo, E. (2009). Evolutionary genomics of wheat. In Carver, B. F. (ed.), *Wheat: science and trade*, pp. 5–30. New Jersey: Wiley-Blackwell.

Henry, R. J., and Nevo, E. (2014). Exploring natural selection to guide breeding for agriculture. *Plant Biotech J, 12*, 655–62.

IPCC (2014). *Intergovernmental Panel on Climate Change 2014. Impacts, adaptation, and vulnerability*. Part B Regional Aspects. Cambridge: Cambridge University Press.

Jana, S., and Nevo, E. (1991). Variation in response to infection with Erysiphe graminis hordei and Puccinia hordei in some wild barley populations in a centre of diversity. *Euphytica, 57*, 133–40.

Jin, Y., Weining, S., and Nevo, E. (2005). A MAPK gene from Dead Sea fungus confers stress tolerance to lithium salt and freezing–thawing: prospects for saline agriculture. *Proceedings of the National Academy of Sciences of the United States of America, 102*(52), 18992–18997.

Jun, Y., Wang, F., Qin, H., Chen, G., Nevo, E., Fahima, T., and Cheng, J. (2011). Natural variation in grain selenium concentration of wild barley, *Hordeum spontaneum*, populations from Israel. *Biol. Trace Elem. Res., 142*, 773–786.

Krugman, T., Chagué, V., Peleg, Z., Balzergue, S., Just, J., Korol, A. B., Nevo, E., Saranga, Y., Chalhoub, B., and Fahima, T. (2010). Multilevel regulation and signalling processes associated with adaptation to terminal drought in wild emmer wheat. *Funct. Integr. Genomics., 10*, 167–86.

Krugman, T., Peleg, Z., Quansah, L., Chagué, V., Korol, A. B., Nevo, E., Saranga, Y., Fait, A., Chalhoub, B., and Fahima, T. (2011). Alteration in expression of hormone-related genes in wild emmer wheat roots associated with drought adaptation mechanisms. *Funct. Integr. Genomics., 11*, 565–83.

Lev, E., Kislev, M. E., and Bar-Yosef, O. (2005) Mousterian vegetal food in Kebara cave, Mt. Carmel. *Journal of Archaeological Science, 32*(3), 475–484.

Mascher, M., et al. (2017). A chromosome conformation capture ordered sequence of the barley genome. *Nature, 544*, 427–433.

Moseman, J., Nevo, E., and El-Morshidy, M. (1990). Reactions of Hordeum spontaneum to infection with two cultures of *Puccinia hordei* from Israel and United States. *Euphytica, 49*, 169–175.

Moseman, J., Nevo, E., Gerechter-Amitai, Z., El-Morshidy, M., and Zohary, D. (1985). Resistance of *Triticum dicoccoides* collected in Israel to infection with Puccinia recondita tritici. *Crop Sci., 25*, 262–265.

Moseman, J., Nevo, E., Morshidy, M. E., and Zohary, D. (1984). Resistance of *Triticum dicoccoides* to infection with *Erysiphe graminis tritici. Euphytica, 33*, 41–47.

Moseman, J., Nevo, E., and Zohary, D. (1983) Resistance of *Hordeum spontaneum* collected in Israel to infection with *Erysiphe graminis hordei. Crop Sci., 23*, 1115–1119.

Nevo, E. (1986). Genetic resources of wild cereals and crop improvement: Israel, a natural laboratory. *Isr. J. Bot., 35*, 255–278.

Nevo, E. (1987). Plant genetic resources: prediction by isozyme markers and ecology. In Rattazi, M. C., Scandalios, J. G., and Whitt, G. S. (eds.), *Isozymes: current topics in biological research*. Volume 16: Agriculture, physiology and medicine, pp. 247–267. New York: Alan R. Liss.

Nevo, E. (1995). Asian, African and European biota meet at "Evolution Canyon," Israel: local tests of global biodiversity and genetic diversity patterns. *Proc. Roy. Soc. Lond. B, 262*,149–155.

Nevo, E. (1998). Genetic diversity in wild cereals: regional and local studies and their bearing on conservation *ex situ* and *in situ*. *Genet. Resour. Crop Evol., 45*, 355–70.

Nevo, E. (2002). *Evolution of wild emmer and wheat improvement: population genetics, genetic resources, and genome organization of wheat's progenitor, Triticum dicoccoides*. Springer Science & Business Media.

Nevo, E. (2004). Genomic diversity in nature and domestication. In Henry, R. J. (ed.), *Plant Diversity and evolution: genotypic and phenotypic variation in higher plants*, pp. 287–315. Cambridge, MA: CABI Publishing.

Nevo, E. (2006). "Evolution Canyon": a microcosm of life's evolution focusing on adaptation and speciation. *Isr. J. Ecol. Evol., 52*(3–4), 485–506.

Nevo, E. (2009). Evolution in action across life at "Evolution Canyon," Israel. *Trends in Evol. Biol., 1*, e3.
Nevo, E. (2011). *Triticum*. In Kole, C. (ed.), *Wild crop relatives: genomic and breeding resources*. Cereals, pp. 407–456. Berlin: Springer-Verlag.
Nevo, E. (2012). "Evolution Canyon," a potential microscale monitor of global warming across life. *PNAS, 109*, 2960–2965.
Nevo, E. (2014). Evolution of wild emmer wheat and crop improvement. *J. Syst. Evol., 52*, 673–696.
Nevo, E. (2016). Global warming and future food production. *Geoinfor.Geostat. An overview*, 4, 4.
Nevo, E., Oren, A., and Wasser S. P. (eds.) (2003). *Fungal life in the Dead Sea*. Germany: Verlag.
Nevo, E., Beiles, A., Gutterman, Y., Storch, N., and Kaplan, D. (1984a). Genetic resources of wild cereals in Israel and vicinity. I. Phenotypic variation within and between populations of wild wheat, *Triticum dicoccoides. Euphytica, 33*, 717–35.
Nevo, E., Beiles, A., Gutterman, Y., Storch, N., and Kaplan, D. (1984b). Genetic resources of wild cereals in Israel and vicinity. II. Phenotypic variation within and between populations of wild barley, *Hordeum spontaneum. Euphytica, 33*, 737–56.
Nevo, E., Moseman, J. G., Beiles, A., and Zohary, D. (1985). Patterns of resistance of Israeli wild emmer wheat to pathogens. I. Predictive method by ecology and allozyme genotypes for powdery mildew and leaf rust. *Genetica, 67*, 209–222.
Nevo, E., Grama, A. Beiles, A., and Golenberg, E. M. (1986). Resources of high protein genotypes in wild emmer, Triticum dicoccoides, in Israel: predictive method by geology and allozyme markers. *Genetica, 68*, 215–227.
Nevo, E., and Payne, P. I. (1987). Wheat storage proteins: diversity of HMW glutenin subunits in wild emmer from Israel. I. Geographical patterns and ecological predictability. *Theor. Appl. Genet., 74*, 827–836.
Nevo, E., and Chen, G. (2010). Drought and salt tolerances in wild relatives for wheat and barley improvement. *Plant Cell Environ., 33*, 670–685.
Nevo, E., Fu, Y.-B., Pavlicek, T., Khalifa, S., Tavasi, M., and Beiles, A. (2012). Evolution of wild cereals during 28 years of global warming in Israel. *PNAS, 109*, 3412–3415.
Nevo, E., Gerechter-Amitai, Z., and Beiles, A. (1991). Resistance of wild emmer wheat to stem rust: ecological, pathological and allozyme associations. *Euphytica, 53*, 121–130.
Nevo, E., Gerechter-Amitai, Z., Beiles, A., and Golenberg, E. M. (1986). Resistance of wild wheat to stripe rust: predictive method by ecology and allozyme genotypes. *Plant Syst. Evol., 153*, 13–30.
Nevo, E., Golenberg, E., Beiles, A., Brown, A., and Zohary, D. (1982). Genetic diversity and environmental associations of wild wheat, *Triticum dicoccoides*, in Israel. *Theor. Appl. Genet., 62*, 241–254.
Nevo, E., Krugman, T., and Beiles, A. (1993). Genetic resources for salt tolerance in the wild progenitors of wheat (*Triticum dicoccoides*) and barley (*Hordeum spontaneum*) in Israel. *Plant Breed., 110*, 338–341.
Nevo, E., Pagnotta, M. A., Beiles, A., and Porceddu, E. (1995). Wheat storage proteins: glutenin DNA diversity in wild emmer wheat, *Triticum dicoccoides*, in Israel and Turkey. 3. Environmental correlates and allozymic associations. *Theor. Appl. Genet., 91*: 415–420.
Pagnotta, M. A., Nevo, E., Beiles, A., and Porceddu, E. (1995). Wheat storage proteins: glutenin diversity in wild emmer wheat, *Triticum dicoccoides*, in Israel and Turkey. 2. DNA diversity detected by OCR. *Theor. Appl. Genet., 91*: 409–414.
Peleg, Z., Fahima, T., Abbo, S., Krugman, T., Nevo, E., Yakir, D., and Saranga, Y. (2005). Genetic diversity for drought resistance in wild emmer wheat and its ecogeographical associations. *Plant, Cell & Environment, 28*, 176–91.
Peleg, Z., Saranga, Y., Krugman, T., Abbo, S., Nevo, E., and Fahima T. (2008). Allelic diversity associated with aridity gradient in wild emmer wheat populations. *Plant Cell Environ., 31*, 39–49.
Peng, J. H., Fahima, T., Röder, M. S., Li, Y. C., Dahan, A., Grama, A., Ronin, Y. I., Korol, A. B., and Nevo, E. (1999). Microsatellite tagging of stripe-rust resistance gene *YrH52* derived from wild emmer wheat, *Triticum dicoccoides*, and suggestive negative crossover interference on chromosome 1B. *Theor. Appl. Genet., 98*, 862–872.

Peng, J. H., Fahima, T., Röder, M. S., Li, Y. C., Grama, A., and Nevo, E. (2000). Microsatellite high-density mapping of the stripe rust resistance gene *YrH*52 region on chromosome 1B and evaluation of its marker-assisted selection in the F2 generation in wild emmer wheat. *New Phytologist, 146*, 141–154.

Peng, J. H., Korol, A. B., Fahima, T., Roder, M. S., Ronin, Y. I., Li, Y. C. and Nevo, E. (2000). Molecular genetic maps in wild emmer wheat, *Triticum dicoccoides*: genome-wide coverage, massive negative interference, and putative quasi-linkage. *Genome Res., 10*, 1509–1531.

Peng J., Ronin, Y., Fahima, T., Röder, M. S., Li, Y., Nevo, E., and Korol, A. (2003). Domestication quantitative trait loci in *Triticum dicoccoides*, the progenitor of wheat. *PNAS*, 100, 2489–2494.

Peng J., Sun, D., and Nevo, E. (2011a). Wild emmer wheat, *Triticum dicoccoides*, occupies a pivotal position in wheat domestication process. *Australian Journal of Crop Science.*

Peng, J. H., Sun, D., and Nevo, E. (2011b). Domestication evolution, genetics and genomics in wheat. *Mol Breed., 28*, 281–301.

Perl, T., Kis-Papo, T., and Nevo, E. (2017). Fungal biodiversity in the hypersaline Dead Sea: extinction and evolution. *Biological Journal of the Linnean Society, 121*(1), 122–132.

Rottenberg, A. (2017). Has agriculture dispersed worldwide from a single origin? *Genetic Resources and Crop Evolution, 64*(6), 1–7.

Savchenko, K. G., Heluta, V. P., Wasser, S. P., and Nevo, E. (2014a). Rust fungi (Puccinales) of Israel I. All genera except *Puccinia* and *Uromyces* with *Caeoma origani*, sp. *Nova Hedwigia, 98*, 163–178.

Savchenko, K. G.,, Heluta, V. P., Wasser, S. P., and Nevo, E. (2014b). Rust Fungi in Israel. II. The genus *Uromyces. Nova Hedwigia, 98*, 393–407.

Savchenko, K. G., , Wasser, S. P., Heluta, V. P., and Nevo, E. (2015). Smut fungi of Israel. In P.A. Voltz (ed.), *Biodiversity of Cyanoprokaryotes, Algae and Fungi of Israel*, pp. 111–116. Germany: Koeltz Scientific Books.

Shapiro J. A. (2011). *Evolution: a view from the 21st century*. New Jersey: FT Press Science.

Sikorski, J., and Nevo, E. (2006). On the necessity to study natural bacterial populations—the model of *Bacillus simplex* from "Evolution Canyon" I and II, Israel. *Isr. J. Ecol. Evol., 52*(3–4), 527–542.

Sun, G. L., Fahima, T., Korol, A. B., Turpeinen, T., Grama, A., Ronin, Y. I., and Nevo, E. (1997). Identification of molecular markers linked to the *Yr15* stripe rust resistance gene of wheat originated in wild emmer wheat, *Triticum dicoccoides. Theor. Appl. Genet., 95*, 622–628

Suprunova, T., Krugman, T., Distelfeld, A., Fahima, T., Nevo, E., and Korol, A. (2007). Identification of a novel gene (*Hsdr4*) involved in water-stress tolerance in wild barley. *Plant Mol. Biol., 64*, 17–34.

Suprunova T., Krugman, T., Fahima, T., Chen, G., Shams, I., Korol, A., and Nevo, E. (2004). Differential expression of dehydrin genes in wild barley, *Hordeum spontaneum*, associated with resistance to water deficit. *Plant Cell Environ., 27*, 1297–1308.

Trowsdale, J., and Knight, J. C. (2013). Major histocompatibility complex genomics and human disease. *Ann. Rev. Genomics Hum Genet, 14*, 301–323.

Uauy, C., Distelfeld, A., Fahima, T., Blechl, A., and Dubcovsky, J. (2006). A NAC gene regulating senescence improves grain protein, zinc, and iron content in wheat. *Science, 314*, 1298–1301.

Weiss, E. (2002) *Issues in reconstructing the human economy and society of Ohalo II inhabitants by the macrofossil botanical remains*. PhD, Bar-Ilan University.

Xie, W., and Nevo, E. (2008). Wild emmer: genetic resources, gene mapping and potential for wheat improvement. *Euphytica, 164*(3), 603–614.

Yang, Z., Zhang, T., Bolshoy, A., Beharav, A., and Nevo, E. (2009). Adaptive microclimatic structural and expressional dehydrin 1 evolution in wild barley, *Hordeum spontaneum*, at "Evolution Canyon," Mount Carmel, Israel. *Mol. Ecol., 18*, 2063–2075.

Yang, Z., Zhang, T., Lang, T., Li, G., Chen, G., and Nevo, E. (2013). Transcriptome comparative profiling of barley eibi1 mutant reveals pleiotropic effects of *HvABCG31* gene on cuticle biogenesis and stress responsive pathways. *Int. J. Mol. Sci., 14*, 20478–20491

Yang, Z., Zhang, T., Li, G., and Nevo, E. (2011). Adaptive microclimatic evolution of the dehydrin 6 gene in wild barley at "Evolution Canyon," Israel. *Genetica*, 139, 1429–1438

Yin, H. Ben Abu, Y., Wang, H., Li, A., Nevo, E., and Kong, L. (2015). Natural selection causes adaptive genetic resistance in wild emmer against powdery mildew at "Evolution Canyon" microsite, Mt. Carmel, Israel. *PLoS ONE, 10*, e0122344.

Zaim, et al. (2017). Wide crosses of durum wheat (*Triticum durum* Desf.) reveal good disease resistance, yield stability, and industrial quality across Mediterranean sites. *Field crops research 2017, 2014*, 219–227.

Zhang, F., Chen, G., Huang, Q., Orion, O., Krugman, T., Fahima ,T., Korol, A., Nevo, E., and Gutterman, Y. (2005). Genetic basis of barley caryopsis dormancy and seedling desiccation tolerance at the germination stage. *Theor. Appl. Genet., 110*, 445–453.

Zhang, F., Gutterman, Y., Krugman, T., Fahima, T., and Nevo, E. (2002). Differences in primary dormancy and seedling revival ability for some Hordeum spontaneum genotypes of Israel. *Isr. J. Plant Sci., 50*, 271–6.

Zhang, W., Chen, S., Abate, Z., Nirmala, J., Rouse, M., and Dubcovscy, J. (2017). Identification and characterization of S13, a tetraploid wheat gene that confers resistanceto the Ug99 stem rust race group. *Proc. Nat. Acad. Sci USA*. doi: 10.1073/pnas.1706277114

Zohary, D. (1969). The progenitors of wheat and barley in relation to domestication and agricultural dispersal in the Old World. In Ucko, P.J., and Dimbelby, G.W. (eds.) *The domestication and exploitation of plants and animals*, pp. 47–66 (London: Gerald Duckworth & Co., 1969).

Zohary, D., Hopf, M., and Weiss, E. (2012). *Domestication of plants in the Old World: the origin and spread of domesticated plants in Southwest Asia, Europe, and the Mediterranean Basin*. Oxford University Press on Demand.

GENOME ENGINEERING TECHNIQUES

Opportunities and Challenges

Wendy Harwood
John Innes Centre, Norwich Research Park

Introduction

A stroll through supermarket vegetable aisles will reveal an array of different *Brassica* vegetables all derived from an ancestral wild cabbage type plant. This diversity of *Brassica* food crops is only possible because breeders have been able to select plants with mutations that have led to this range of different crop forms. Many of these mutations have occurred naturally, but others have been induced.

Mutation breeding is an approach for producing induced mutations that dates back to the 1930s. It followed the discovery by Lewis John Stadler that radiation led to various aberrations in maize (*Zea mays*) and barley (*Hordeum vulgare*) including yellowing due to chlorophyll related mutations (Stadler and Sprague, 1936). This work provided the first demonstration that mutations could be artificially induced and formed the basis of the mutation breeding approach. A little later, in the 1940s it was found that chemical mutagens could also cause mutations. One common mutagenic chemical used was ethyl methanesulfonate (EMS), and it is still employed today to produce random mutations in plant DNA.

A joint FAO/IAEA mutant variety database demonstrates the very widespread use of mutation breeding techniques (https://mvd.iaea.org/). The database records mutant lines developed through a range of induced mutagenesis techniques including physical or chemical mutagenesis as well as mutants induced as a result of passage through tissue culture. Currently there are almost 3,300 entries to the database, with new entries still being added. These mutant cultivars cover the whole range of crop plants including the major cereals, wheat (*Triticum* ssp.), rice (*Oryza sativa*) and maize, vegetable and fruit crops, as well as flowers, trees, and root crops. Many of our current food crops have mutation breeding steps in their pedigrees and the method has led to the development of many successful varieties. One example is the successful malting barley Golden Promise that made a very large contribution to the production of grain for malting and today is the model cultivar for modern genome engineering technologies.

TILLING (Targeting Induced Local Lesions IN Genomes) is another breeding strategy for harnessing mutations caused by chemical mutagenesis. Resources for TILLING have been developed in a range of crop species including wheat, and recently combining this original technology with high-throughput sequencing capacity has increased the power of the approach (Uauy et al., 2017). The technique has had past success in crop improvement, for example in wheat where many mutant alleles of the *waxy* gene were identified (Chen et al., 2014). A bread wheat cultivar with an almost null waxy phenotype was developed utilizing these mutations (Slade et al., 2005). Despite the success of both mutation breeding and TILLING methods, they both suffer from the drawback that they result in very large numbers of randomly induced mutations throughout the genome. It is well known that deleterious mutations will be created as well as the desired ones, leading to the need for extensive backcrossing to remove the unwanted mutations.

As described above, conventional induced mutagenesis methods have enjoyed considerable success. These methods are also exempt from genetically modified organism (GMO) regulations. However, due to the random nature of the mutations generated, the methods rely on chance to produce the desired outcome. The serious and multiple challenges facing us in providing sufficient, safe, and nutritious food for the future mean that alternative more targeted approaches are needed; in addition to the technologies based on random mutagenesis. Plant breeders urgently require new sources of variation, for example new sources of resistance genes to help tackle existing and emerging disease threats.

The first genetically modified food crop, a tomato (*Solanum lycopersicum*) with delayed ripening, was made available in the U.S. in 1994. The first commercial field planting of a GM crop was in 1996 when herbicide tolerant soybean was grown for the first time. From 1996 to 2017 the area on which GM crops were grown increased from 1.7 million hectares to 189 million hectares, a 110-fold increase (ISAAA Brief 53). The majority of these GM crops contained traits for herbicide tolerance and/or insect resistance although new traits including drought tolerance are now being added. Genetic modification technology has greatly expanded the genetic diversity available to breeders, as they were no longer restricted to genes from cross-compatible species only. The impact of this technology has been huge; however, GM methods are unable to insert new genes into precise locations in the plant genome.

Genome editing technologies are currently causing great excitement because for the first time precise changes can be made to DNA sequences at a precise location within a plant genome. The type of change that can be made varies from the induction of a small mutation at a precise location through to the precise insertion of a new gene at a chosen target site. Induction of a small mutation, sufficient to knock out the function of the target gene, is often indistinguishable from a similar mutation occurring naturally or from one produced by mutation breeding. Genome editing, when used to introduce a new gene at a precise genomic location, offers advantages over the random insertion associated with traditional GM approaches. Recently, genome editing methods have further expanded to allow precise DNA base changes and specific up or down regulation of a target gene opening multiple new opportunities for sophisticated and precise genome engineering.

Genetic Modification in Crop Improvement

The very first GM plants were produced in the early 1980s and included tobacco, petunia, and sunflower. Genetic modification of our important crop plants followed from these first successes with wheat being transformed for the first time in 1992 (Vasil et al., 1992) using particle bombardment as the DNA delivery system. Agrobacterium-mediated transformation is now often the preferred delivery system as it leads to lower transgene copy numbers and fewer rearrangements of the DNA (Travella et al., 2005). Despite it now being possible to transform almost all crop plants, efficiency is still low for some species and most transformation systems are genotype dependent, with some genotypes being highly transformable and others being recalcitrant to transformation. Despite these difficulties, certain GM crops including soybeans (*Glycine max*), cotton (*Gossypium*), maize, and oilseed rape (canola) (*Brassica napus*) have enjoyed enormous success.

Current GM Crops and Their Impact to Date

Genetically modified crops were grown on over 189 million hectares in 2017 in twenty-four countries. In addition to the countries cultivating GM crops, a further forty-three countries are importing GM crops for food, feed, and processing (ISAAA brief 53). The most widely grown GM crop is soybeans, making up approximately 50 percent of total GM crop area. This is followed by maize (31 percent), cotton (13 percent), and oilseed rape (5 percent). Overall 77 percent of all soybeans grown are GM and 80 percent of all cotton. Despite these four main crops still dominating, the range of available GM crops has expanded to include alfalfa (*Medicago sativa*), sugar beet (*Beta vulgaris*), papaya (*Carica papaya*), squash (*Cucurbita*), eggplant (*Solanum melongena*), potatoes (*Solanum tuberosum*) and apples (*Malus*). The range of GM traits included in these crops has also expanded from the original herbicide tolerance and/or insect resistance to include non-browning traits, modified oil content, drought tolerance, and new disease resistances.

There is no doubt that GM crops have had a significant impact since they were first grown commercially in 1996. A recent review by Brookes and Barfoot (2017) looked at the impact on farm income and production from 1996 to 2015. The key findings were that significant economic benefits had been achieved at the farm level due to cultivation of the four main GM crops, soybeans, maize, cotton, and oilseed rape. In 2015 alone, the benefits amounted to $15.4 billion split between farmers in developed countries (49 percent) and in developing countries (51 percent). Most of these benefits were due to increased yield gains and lower production costs because of improvements to weed control and reduced levels of pest damage. In some regions an overreliance on the herbicide glyphosate, that many GM crops are resistant to, combined with sub-optimal agricultural practices has resulted in the development of resistant weeds. This has led to increased costs for producers but even so, the overall picture is one of very significant economic benefit.

Safety Assessment of GM Crops

Safety assessment of GM crops has been the subject of a very large number of studies, many of which involved animal feeding trials. A recent review characterized the studies most commonly cited as evidence for adverse effects of GM food or feed (Sanchez and Parrott, 2017). Thirty-five individual articles were examined out of a total of around seven hundred related to GM crop food and feed safety issues (Sanchez, 2015). From all the papers dealing with safety issues, less than 5 percent published negative outcomes. These studies focus on a small number of GM events and close examination revealed flaws in methodology sufficient to invalidate the conclusions (Sanchez and Parrott, 2017). In contrast, forty-four peer reviewed articles have described ninety-day feeding studies for a range of crops, and none of them showed any safety concerns (Ricroch et al., 2014). This is in addition to twenty-three longer-term studies and nineteen multi-generational studies that also showed no adverse outcomes (Sanchez and Parrott, 2017). These findings should be considered alongside the fact that all GM crops undergo a comprehensive safety assessment before approval. In Europe, the European Food Safety Authority (EFSA) evaluates an extensive dossier of scientific information on any new GM crop or event and issues an opinion on safety. In addition, billions of animals and humans have been consuming GM crops for twenty-two years with no evidence of any adverse effects.

Environmental Impacts of GM Crops

Agricultural crops, including GM crops, have a major impact on the environment. It has been reported that GM technology has led to cuts in fuel use and to changes in agricultural practices by facilitating no or low-till cultivation. Together, this has resulted in a significant reduction in greenhouse gas emissions that in 2015 was equivalent to taking 11.9 million cars off the roads (Brookes and Barfoot, 2017). GM crops have certainly led to changes in the spectrum of herbicides used by farmers. Most obvious is the replacement of more selective herbicides with broad-spectrum herbicides, principally glyphosate, generally considered to be a more environmentally benign product. This has led to the development of herbicide resistant weeds so that farmers have had to incorporate additional herbicides alongside glyphosate, meaning that the level of environmental benefit has decreased (Brookes and Barfoot, 2017). The main environmental impacts of agriculture are independent of whether the crops grown are GM or not. Of far greater importance is the type of crop grown, how it is cultivated, and the traits that it contains. Other considerations include the impact of GM crops on biodiversity. A monoculture of any crop is not conducive to biodiversity, but again this is an issue with any agricultural system, not just GM. The practice of no-till farming, where the soil is not turned over and more of the previous crop remains on the field, offers environmental advantages. This practice is facilitated by growing herbicide tolerant crops and has expanded following the availability of GM herbicide tolerant crops. GM crops, like conventional crops, can have positive and negative environmental impacts thus requiring environmental risk assessments, to ensure that the negative impacts are not greater than those for conventional crops, a crucial part of the total risk assessment process.

• • •

Regulation of GM Crops

GM crops are regulated more stringently than any other type of crop or agricultural technology. The process of regulation for GM crops varies between countries and in some cases is greatly influenced by the political environment leading to considerable costs and extensive delays in approvals being issued. In the United States, the same laws that apply to conventional agricultural technologies are applied to GM crops. In the European Union, however, laws specific to GM crops are applied and the regulatory system is not functioning as originally intended. Approvals for imports of GM crops to the EU can take a very long time with delays of over ten years in some cases. There have only been two approvals of GM varieties for cultivation in the EU, an insect resistant (Bt) maize and a potato with modified starch for industrial use, although only the Bt maize is currently grown. An additional problem is that the EU regulatory system has influenced GMO regulatory systems in Africa (Qaim, 2016). This negatively impacts technology development in countries that potentially have much to gain by adopting specific GM crops.

The high regulatory costs associated with GM crops mean that large biotech companies with large resources are the only ones able to move crops through to commercialization. They understandably focus efforts on the key commercial crops such as soybeans, maize, and cotton while other crops, only important in localized areas, are neglected. Variation in regulatory requirements between countries has a detrimental effect on international trade. Delays in approvals, so called asynchronous authorizations, are a problem with some of the commonly traded crops such as soybeans. A GM event approved in some countries might not have been approved in the EU or other countries, meaning that any contamination of imports with the unauthorized event would cause serious trade disruption.

Future Opportunities and Challenges

Despite the problems with regulation outlined above, GM crops may be the best way forward to tackle some of the problems that will face us in the future. One example is the case of existing and new plant diseases. Late blight of potato is a familiar disease that requires multiple pesticide applications in a season to control it. However, wild relatives of potato are a rich source of resistance genes (Witek et al., 2016). Using conventional breeding to introduce such genes is difficult and very slow but a GM approach allows the introduction of multiple resistance genes meaning that the resistance is likely to be more durable than that conferred by a single resistance gene alone. New races of existing diseases also pose a considerable threat and the Ug99 race group of wheat stem rust is a specific example. This disease is a major threat to global wheat (*Triticum* ssp.) production but, again, it is possible to find resistance genes in wild wheat relatives (Yu et al., 2017), and a GM approach to introduce several such genes is the most rapid way to incorporate these genes into breeding programs and to respond in a timely way to this serious threat.

Climate change threatens our ability to produce sufficient food for an expanding population. Extreme weather events, including extreme rainfall and flooding, heatwaves, and drought are now more common (EASAC, 2018). Considering the example of drought, GM approaches offer opportunities to enhance the stress tolerance pathways that enable plants to better cope with periods of water limitation (for example, see Al Abdallat et al., 2013). The first drought tolerance traits are now being incorporated into GM crops including maize, soybeans, and sugar cane (ISAAA GM approval database).

Another recent exciting GM application is the enhancement of photosynthetic capacity and therefore crop productivity. This was first demonstrated in tobacco (Kromdijk et al., 2016), where plant dry matter was increased by 15 percent. It has also been shown that manipulating photosystem II subunit S can lead to increases in water use efficiency with a 25 percent reduction in water loss compared to controls (Glowacka et al., 2018). These achievements, if applied to other major crop plants, could lead to significant yield increases even under conditions of lower water availability. Some of the world's big challenges for crop improvement could only be realized by using a GM approach. Our leguminous crops, including peas and beans, can fix atmospheric nitrogen and therefore do not require nitrogen fertilizers. Cereal crops do not have this ability, but if they could be engineered to fix even a small amount of nitrogen, then this would have a major impact on cereal production, in particular in regions of the world where nitrogen fertilizers are unavailable or prohibitively expensive. Engineering nitrogen fixation into cereal crops is a future goal that can currently only be realized using a GM approach.

The regulatory challenges facing GM crops have been mentioned above, but it is also important to consider socio-economic issues and ethical concerns over the use of GM technology. Socio-economic issues include farm-level impacts, impacts due to coexistence regulations, impacts along the supply chain, impacts on consumers, impacts on food security, and environmental economic impacts. All these issues are extensively considered by Garcia-Yi et al. (2014) who point out that GM crops are increasingly included as part of a wider approach to food security and that they could help mitigate food shortages related to population growth and the effects of climate change in specific regions of the world. Ricroch et al. (2018) considered the conflicting ethical views surrounding GM crops. One general objection to GM crops is the opposition to patenting living objects, as this is often linked to restrictions placed on farmers growing GM crops, preventing them from saving seed for use in subsequent years. After consideration of a range of ethical issues, Ricroch et al. concluded that refocusing the debate from considering only the technology toward considering universal values such as freedom, justice, and truth might be a way forward.

Genome Editing in Crop Improvement

The technique of genome editing allows, for the first time, specific changes to a plant's DNA at a precise genomic location. This contrasts with GM methodology that inserts DNA at a random location within a plant genome. Genome editing can be used to introduce small mutations, the insertion or deletion of one or a few nucleotides that are sufficient to stop a target gene from functioning. This type of mutation is indistinguishable from the multiple mutations that occur naturally within plants and that form the basis of genetic variability exploited in plant breeding. The mutations induced by genome editing are also indistinguishable from those induced by other methods of mutation breeding. In addition, genome editing can allow the precise insertion or replacement of a gene as well as having many other applications for the precise modification, or control of expression of target genes.

Methods for Precise Genome Editing

The discovery that double strand DNA breaks could be induced at specific sites by endonucleases linked to either a guiding protein complex or RNA molecule forms the basis of all new genome editing technologies. The first genome editing technologies included Zinc Finger Nucleases (ZFNs) and Transcription Activator-Like Effector Nucleases (TALENs). Both systems rely on rather complex protein assemblies to guide the nuclease to its target site. This was followed by the application of Clustered Regularly Interspaced Short Palindromic Repeats (CRISPR) / CRISPR associated (Cas) systems that have led to unprecedented excitement over potential biotechnological applications (Ricroch et al., 2017).

CRISPR/Cas9 based genome editing originates from an adaptive bacterial defense strategy that has been reprogrammed to allow genome editing in eukaryotic cells. The CRISPR/Cas9 system requires just two components, the Cas9 nuclease and a guide RNA (sgRNA). The guide RNA has a variable region of twenty base pairs that is complementary to a target site in the plant genome. The rest of the guide RNA, the backbone or scaffold region, does not vary. The target site within the plant also needs to be twenty base pairs, followed by a PAM motif that has the form NGG, where N can be any nucleotide. This means that the identification of a target site within the target gene is quite straightforward, as is the design of the guide RNA. The Cas9 nuclease will cut the DNA target 3–4 nucleotides upstream of the PAM motif if there is sufficient homology between the sgRNA and the target sequence. The seed region, 8–10 bases at the 3' end of the sgRNA, usually needs to be a perfect match. Some mismatches toward the 5' end, further away from the PAM, may be tolerated. This tolerance can lead to so called "off-target" mutations in genes with sufficient homology to the target gene, as well as "on-target" mutations. This tolerance of limited mismatches can be beneficial if the aim is to mutate multiple members of a closely related gene family (Lawrenson et al., 2016).

Once the Cas9 nuclease has cleaved the target gene DNA to leave a double strand break, one of the plant cell's repair pathways will repair the break. The most common repair pathway is non-homologous end joining (NHEJ), an error-prone process that sometimes leads to mistakes being made, causing a mutation in the target gene. The mistake is often just a single, or small number of, nucleotides inserted or deleted at the break point which can knock-out the function of the target gene. The alternative repair pathway is homology-directed repair that is much less prevalent in plants. If a repair template consisting of a region or gene to be inserted, flanked by homology arms corresponding to the regions either side of the predicted break point, is introduced along with the Cas9 and guide RNA, then precise insertion of the donor fragment can occasionally occur. This process is referred to as "knock-in" as opposed to the "knock-out" resulting from NHEJ.

All applications of genome editing require the editing components to be delivered to plant cells. The Cas9 and guide RNA may not need to be inserted into the plant genome to cause a required mutation; however, it is common to introduce the components as transgenes using GM methods. Once the required edit has occurred, the inserted transgenes can be removed during segregation in the following generation, leaving plants with the mutation but no foreign DNA. Genome editing does require, as a minimum, an efficient system to regenerate plants from edited cells and as such can suffer from the same limitations as GM, in that the methods may be genotype dependent.

Current Examples of Gene Editing in Crops

The principal use of genome editing to date has been as a research tool to aid in determination of gene function. The ability to produce targeted mutations in any plant gene is extremely valuable in identifying those genes that may be useful for crop improvement. There have been many studies using CRISPR based genome editing for a range of potentially useful traits (for review see Jaganathan et al., 2018). These traits include resistance to a range of plant diseases such as wheat powdery mildew (Wang et al., 2014) and rice blast (Wang et al., 2016). Taking the example of rice blast, this is a major rice disease causing typical yield losses of 10–30 percent (Dean et al., 2012). Wang et al. (2016) used CRISPR / Cas9 to mutate the *OsERF922* gene and recovered homozygous mutant plants that contained no transgenes. These plants showed significantly reduced infection with rice blast while agronomic traits were unchanged from the wildtype controls. This example demonstrates how the technology could be a valuable approach for tackling this devastating disease.

Other traits engineered using CRISPR / Cas9-based genome editing in crops include tolerance to a range of abiotic stresses such as drought (Shi et al., 2017). In this study CRISPR technology was used to generate a range of novel variants of the gene *ARGOS8*. Some of the gene variants had increased grain yield under conditions of limited water but had no yield loss under well-watered conditions. This is an example of the use of CRISPR / Cas9 to generate novel allelic variation that could contribute to the production of drought tolerant crops.

There are also opportunities to produce crops with improved nutritional quality characteristics using CRISPR / Cas9 methods. The reduction of allergens or antinutritional compounds in crops is a realistic prospect that could be achieved by knock-out of specific genes. An example is the work toward the production of gluten free wheat (Sánchez-León et al., 2018). In this study CRISPR /Cas9 was used to mutate up to thirty-five different α-gliadin genes thus reducing the amount of these immunoreactive proteins in the grain. This finding is of importance for the 7 percent of the western population affected by celiac disease or gluten intolerance.

Despite the promising examples given above, there are only a few gene-edited crops reported to be close to market. These include a gene edited non-browning mushroom that the U.S. Department of Agriculture (USDA) decided will not be regulated so can be freely cultivated (Waltz 2016). Dupont Pioneer has developed an improved waxy corn variety using CRISPR / Cas9 that they aim to launch around 2020. There are many gene-editing startup companies working to market a range of edited crops with diverse traits, so the situation is likely to rapidly change over the next few years.

Regulation of Gene-Edited Crops

The regulation of gene-edited crops has been a much-debated topic focusing on whether they should fall under GM regulatory frameworks or not. One difficulty is that the technology can be used to insert new genes or sequences in a precise location, an application that clearly leads to a GM product. Or, it can be used to create a mutation that is indistinguishable from natural mutations or those induced by mutation breeding, in which case it has been argued that the product is not GM. The multiple applications of genome editing highlight a second difficulty, related to the tendency to consider the technology used in the regulatory process. If regulation only considered the modified trait in the final product, rather than the technology used to create it, then much of this difficulty would be removed.

Different countries have approached the regulation of gene-edited crops in different ways. Considering only the crops containing mutations introduced by gene editing and no foreign DNA, the USDA announced that it will not regulate such gene-edited crops as it considered the method just a more rapid form of traditional breeding. Canada regulates based on whether novel traits have been introduced and on a case-by-case basis, so gene-edited crops do not require any amendment to the regulations. Other countries including Argentina, Chile, and Brazil regulate on a case-by-case basis and consider whether transgenes are present. Therefore, gene-edited crops without transgenes are not subject to GMO regulations.

In Europe the situation is different. The question of whether certain classes of plants derived using new breeding techniques, very similar to genome editing, should be exempt from the EU GMO regulation was recently considered by the European Court of Justice. Their ruling was that "Organisms obtained by mutagenesis are GMOs and are, in principle, subject to the obligations laid down by the GMO Directive" (CJEU press release No 111/18). The only exemption was for mutagenesis techniques conventionally used and with a long history of safe use; therefore, mutation breeding was exempted as previously, but mutagenesis using genome editing was included under the GMO regulations. This ruling isolates Europe in terms of regulation of gene-edited crops, and there is concern about possible consequences due to the prohibitively large costs of moving products through the GMO approvals process.

• • •

Future Opportunities and Challenges

Regulation of gene-edited crops is a major challenge in Europe. As mentioned when considering GM crops, the EU regulatory process is not functioning as originally intended leading to excessive delays in approvals. If gene edited crops are caught up in this system, then it is likely that there will be little appetite for commercialization in Europe. Elsewhere, however, the technology is expected to be fully embraced, leading to benefits for economies, farmers, and consumers.

Another challenge, but also opportunity, concerns the very rapid development of CRISPR-based genome editing technologies. Simple use of CRISPR / Cas9 to induce mutations is just the beginning, with many other applications already demonstrated or in the pipeline. Deletion of a specific gene or fragment is possible if a pair of guide RNAs are introduced together and cut at the same time, leading to a perfect deletion between the guides. This strategy has application, for example, in manipulation of transcription factors to remove a specific motif or binding site. There are different versions of the Cas9 protein, and one referred to as dCas9 can be fused to a range of other proteins to allow up or down regulation of target genes or to allow specific base changes within a target gene (Zong et al., 2017). Recently, dCas9 was fused to a DNA methyl transferase to alter the methylation status of target genes in mammalian cells (Liu et al., 2016). This epigenetic editing is also a technology of great future interest in plant systems. These rapid advances do however present challenges for regulatory systems that are not flexible enough to be able to respond quickly to new technologies.

The ability to insert genes or regulatory sequences in a precise genomic location is also set to revolutionize crop breeding. This is of particular importance for developing crops resistant to current and emerging diseases, where it is often desirable to insert multiple disease resistance genes at the same genomic location. Indeed, this is just one of several applications of genome editing for disease resistance reviewed by Langner et al. (2018). This technology, referred to as gene targeting, is still challenging in plants because of the low frequency of homologous recombination. However, progress has been made in wheat using a viral replicon to increase the number of copies of the repair template available for insertion at the break point (Gil-Humanes et al. 2017). It is likely that this, together with other technology developments, will make gene targeting more widely available in the near future.

Conclusion

Genome engineering, using both genetic modification and genome editing technology, is set to play an increasingly important role in the development of the crops that will supply sufficient, nutritious food to the world's present and future population. These technologies follow in the footsteps of conventional techniques, such as mutation breeding, that have increased the genetic diversity available to breeders. The challenges now facing us mean that new sources of genetic diversity are urgently required. Genome engineering techniques are increasingly able to meet this need with greater precision and flexibility than previously possible. If regulatory systems can respond in an appropriate and timely manner, then there is real hope that we can deliver the crops that will provide the yields required under challenging future conditions.

Acknowledgment

This work was supported by the UK Biotechnology and Biological Research Council (BBSRC) through Institute Strategic Grant BB/P013511/1 to the John Innes Centre.

References

Al Abdallat, A. M., Ayed, J. Y., Abu Elenein, J. M., Al Ajlouni, Z., and Harwood, W. A. (2013). Over-expression of the transcription factor *HvSNAC1* improves drought tolerance in barley (*Hordeum Vulgare* L.) *Molecular Breeding, 33*, 401–414.

Brookes, G., and Barfoot, P. (2017). Farm income and production impacts of using GM crop technology 1996–2015, *GM Crops & Food, 8*(3), 156–193.

Chen, L., and Hao, L. (2014). Progress in TILLING as a tool for functional genomics and improvement of crops. *J. of Integrative Plant Biology, 56*, 425–443.

Court of Justice of the European Union (CJEU) press release No 111/18, Luxembourg, July 25, 2018.

Dean, R., Van Kan, J. A., Pretorius, Z. A., Hammond-Kosack, K. E., Di Pietro, A., Spanu, P. D., Rudd, J. J., Dickman, M., Kahmann, R., Ellis, J., Foster, G. D. (2012), The top 10 fungal pathogens in molecular plant pathology. *Molecular Plant Pathology, 13*, 414–430.

European Academies Science Advisory Council (2018). Extreme weather events in Europe, preparing for climate change adaptation: an update on EASAC's 2013 study.

FAO/IAEA mutant variety database. Retrieved from https://mvd.iaea.org/.

Garcia-Yi, J., Lapikanonth, T., Vionita, H., Vu, H., Yang, S., Zhong, Y., Li, Y., Nagelschneider, V., Schlindwein, B., Wesseler, J. (2014). What are the socio-economic impacts of genetically modified crops worldwide? A systematic map protocol. *Environmental Evidence, 3*, 24.

Gil-Humanes, J., Wang, Y., Liang, Z., Shan, Q., Ozuna, C. V., Sanchez-Leon, S., Baltes, N. J., Starker, C., Barro, F., Gao, C., Voytas, D. F. (2017). High-efficiency gene targeting in hexaploid wheat using DNA replicons and CRISPR/Cas9. *Plant J., 89*, 1251–1262.

Glowacka, K., Kromdijk, J., Kucera, K., Xie, J., Cavanagh, A. P., Leonelli, L., Leakey, A. D. B. Ort, D. R., Niyogi, K. K., and Long, S. (2018). Photosystem II Subunit S overexpression increases the efficiency of water use in a field-grown crop. *Nature Communications, 9*, 868.

ISAAA Brief 53-2017: Executive Summary. Retrieved from http://www.isaaa.org/resources/publications/briefs/53/executivesummary/default.asp

ISAAA GM approval database. Retrieved from http://www.isaaa.org/gmapprovaldatabase/gmtraitslist.

Jaganathan, D., Ramasamy, K., Sellamuthu, G., Jayabalan, S., Venkataraman, G. (2018). CRISPR for Crop Improvement: An update review. *Frontiers in Plant Science, 9*, 985. doi: 10.3389/fpls.2018.00985

Kromdijk, J., Glowacka, K., Leonelli, L., Gabilly, S. T., Iwai, M., Niyogi, K. K., and Long, S.P. (2016). Improving photosynthesis and crop productivity by accelerating recovery from photoprotection. *Science, 354*, 857–861.

Langner, T., Kamoun, S., and Belhaj, K. (2018). CRISPR crops: plant genome editing toward disease resistance. *Annual Review of Phytopathology 56*, 479–512.

Lawrenson, T., Shorinola, O., Stacey, N., Liu, C., Østergaard, L., Patron, N., Uauy, C., and Harwood, W. (2015). Induction of targeted, heritable mutations in barley and *Brassica oleracea* using RNA-guided Cas9 nuclease. *Genome Biology, 16*, 258.

Liu, X. S., Wu, H., Stelzer, Y., Wu, X., Czauderna, S., Shu, J., Dadon, D., Young, R. A., and Jaenisch, R. (2016). Editing DNA methylation in the mammalian genome. *Cell, 167*, 233–247.

Qaim, M. (2016). GM crop regulation. In *Genetically modified crops and agricultural development*. Palgrave Studies in Agricultural Economics and Food Policy. Palgrave Macmillan, New York.

Ricroch, A. E., Boisron, A., and Kuntz, M. (2014). Looking back at safety assessment of GM food/feed: an exhaustive review of 90-day animal feeding studies. *Int. J. Biotechnology, 13*, 230–256.

Ricroch, A. E., Guillaume-Hofnung, M., and Kuntz, M. (2018). The ethical concerns about transgenic crops. *Biochem J., 475*, 803–811.

Ricroch, A., Clairand, P., and Harwood, W. (2017). Use of CRISPR systems in plant genome editing: towards new opportunities in agriculture. *Emerging Topics in Life Sciences, 1*, 169–182.

Sanchez, M. A. (2015). Conflict of interests and evidence base for GM crops food/feed safety research. *Nature Biotechnology, 33*, 135–137.

Sanchez, M. A., and Parrott, W. A. (2017). Characterization of scientific studies usually cited as evidence of adverse effects of GM food/feed. *Plant Biotechnology Journal, 15*, 1227–1234.

Sánchez-León, S., Gil-Humanes, J., Ozuna, C. V., Giménez, M. J., Sousa, C., Voytas, D. F., and Barro, F. (2018). Low-gluten, nontransgenic wheat engineered with CRISPR/Cas9. *Plant Biotechnology Journal, 16:* 902–910.

Shi, J., Gao, H., Wang, H., Lafitte, H. R., Archibald, R. L., Yang, M., et al. (2017). ARGOS8 variants generated by CRISPR-Cas9 improve maize grain yield under field drought stress conditions. *Plant Biotechnol. J., 15*, 207–216.

Slade, A. J., McGuire, C., Loeffler, D., Mullenberg, J., Skinner, W., Fazio, G., Holm, A., Brandt, K. M., Steine, M. N., Goodstal, J. F., and Knauf, V. C. (2012). Development of high amylose wheat through TILLING. *BMC Plant Biol, 12*, 69.

Stadler, L. J., and Sprague, G. F. (1936). Genetic effects of ultra-violet radiation in Maize. II. Filtered radiations. *Proc. Natl. Acad. Sci., 22*, 579–583.

Travella, S., Ross, S. M., Harden, J., Everett, C., Snape, J. W., and Harwood, W. A. (2005). A comparison of transgenic barley lines produced by particle bombardment and *Agrobacterium*-mediated techniques. *Plant Cell Reports, 23*, 780–789.

Uauy, C., Wulff, B., and Dubcovsky, J. (2017). Combining traditional mutagenesis with new high-throughput sequencing and genome editing to reveal hidden variation in polyploid wheat. *Annual Review of Genetics, 51*, 435.

Vasil, V., Castillo, A. M., Fromm, M. E., and Vasil, I. K. (1992). Herbicide resistant fertile transgenic wheat plants obtained by microprojectile bombardment of regenerable embryogenic callus. *Biotechnology, 10*, 667–674.

Waltz, E. (2016). Gene-edited CRISPR mushroom escapes US regulation. *Nature, 532*, 293.

Wang, F., Wang, C., Liu, P., Lei, C., Hao, W., Gao, Y., et al. (2016). Enhanced rice blast resistance by CRISPR/Cas9-targeted mutagenesis of the ERF transcription factor gene OsERF922. *PLoS ONE, 11*, e0154027.

Wang, Y., Cheng, X., Shan, Q., Zhang, Y., Liu, J., Gao, C., et al. (2014). Simultaneous editing of three homoeoalleles in hexaploid bread wheat confers heritable resistance to powdery mildew. *Nat. Biotechnol.*, *32*, 947–951.

Witek, K., Jupe, F., Witek, A. I., Baker, D., Clark, M. D., and Jones, J. D.G. (2016). Accelerated cloning of a potato late blight-resistance gene using RenSeq and SMRT sequencing. *Nature Biotechnology*, *34*, 656–660.

Yu, G., Champouret, N., Steuernagel, B., Olivera, P. D., Simmons, J., Williams, C., Johnson, R., Moscou, M. J., Hernandez-Pinzon, I., Green, P., Sela, H., Millet, E., Jones, J. D., Ward, E. R., Steffenson, B. J., and Wulff, B. B. (2017). Discovery and characterization of two new stem rust resistance genes in *Aegilops sharonensis*. *Theor Appl Genet*, *130*(6) 1207–1222.

Zong, Y., Wang, Y., Li, C., Zhang, R., Chen, K., Ran, Y., Qiu, J-L., Wang, D., and Gao, C. (2017). Precise base editing in rice, wheat and maize with a Cas9-cytidine deaminase fusion. *Nature Biotechnology*, *35*, 438–440.

THE PROBLEM OF INSECTICIDE, FUNGICIDE, AND HERBICIDE RESISTANCE IN AGRICULTURAL CROP PESTS

Michael J. Christoffers and Robert P. Sabba
North Dakota State University

Introduction

The cultivation of land plants for food and fiber is an ancient practice. Ever since the earliest forms of crop production, undesirable animals, plants, and pathogens have competed and interfered with the quality and quantity of desired agricultural products. While a variety of biological organisms may be classified as pests because they damage the growth and yield of crops, some of the most troublesome are insects, weeds, and fungi. Humans have used a variety of methods throughout history to control these pests, with the availability of technology, labor, money, and time being just some of the considerations when implementing pest management.

Among available tools for the control of pests, chemical pesticides have played a central role in agricultural modernization and productivity gains. At the same time, however, environmental and health consequences of pesticide use have come under increased scrutiny, and the long-term reliability of specific pesticide-based management practices have been called into question. In particular, the ability of unwelcome insects, fungi, and weeds to evolve resistance to the chemicals meant to control them has become a critical problem in agricultural crop production. The biological, economic, and socioeconomic aspects of pesticide resistance in agriculture are quite complex and are often case-specific. Thus, an exhaustive analysis of the topic is beyond the scope of this chapter. Rather, this primer is focused on demonstrating the importance of pesticide resistance as a problem; providing a general overview of the biological aspects of resistance to insecticides, fungicides, and herbicides; and covering basic resistance management strategies as well as potential alternative technologies for the future.

Crop Losses Due to Pests

The varied ways in which pests negatively affect crops are not yet fully understood. But in general, pests may reduce crop quality and yield by 1) reducing initial crop germination and establishment, 2) reducing access to and/or uptake and translocation of water and nutrients, 3) directly reducing photosynthesis by shading or damaging photosynthetic tissue, 4) damaging supporting structures such as roots and stems, 5) increasing susceptibility to damage from other pests, 6) damaging reproductive parts before harvest, such as flowers, seeds, fruits, and/or tubers, 7) damaging, spoiling, or contaminating harvested products, or 8) interfering with harvest, cultivation, storage, product processing, or other operations (list based on Walker, 1983).

Infamous examples of pests causing substantial crop losses include the oomycete *Phytophthora infestans*, the cause of late blight and the Irish Potato Famine of the 1840s (see Bourke, 1964); locusts of the family Acrididae, whose devastating swarms have been documented throughout recorded history (see Simpson and Sword, 2008); and the recent rise of herbicide-resistant Palmer amaranth (*Amaranthus palmeri*) as a significant agronomic weed in the U.S. (Ward, Webster, and Steckel, 2013). Oerke et al. (1994) estimated that without pest control efforts, worldwide crop losses for eight major crops during 1988–1999 would have approached 70 percent of the total potential production attainable without pests, based on crop financial value. About 30 percent of this estimated potential loss would have been due to weeds, about 23 percent due to animal pests including insects, and about 17 percent due to diseases. Even with pest control, actual losses due to pests were still estimated at 42 percent of worldwide potential production. For 2001–2003, estimates of potential worldwide crop losses in the absence of pest control ranged from about 50 percent for wheat (*Triticum* spp.) to about 80 percent for cotton (*Gossypium* spp.) (Oerke, 2006), with weeds again accounting for the highest potential loss (34 percent loss across crops). Although worldwide potential crop losses due to weeds are estimated to be higher than for insect pests and diseases, greater control of weeds compared to other pests can often be achieved due to the availability of mechanical weed control practices in addition to chemical and other measures (Oerke and Dehne, 2004).

The Early History of Pesticides in Crop Production

The use of chemical materials to control pests is a practice that dates to ancient times. Sulfur was used by the Sumerians to control insects and mites as early as 2500 BC (Dent, 2000), and in Homer's *The Odyssey* (Greek literature, c. 800 BC), Ulysses burns sulfur to "cleanse from pollution" the hall, house, and court, in an apparent reference to early fumigation (Homer, Murray, and Dimock, 1995). The ancient Romans also used sulfur for purification (Kutney, 2013), and in perhaps an unsubstantiated claim (see Ridley, 1986), are reported to have used salt to destroy soil fertility during their defeat of Carthage in c. 149 BC.

The ancient Greeks and Romans also attempted to use biological materials to kill unwanted trees (see Smith and Secoy, 1976). The Greek philosopher Democritus (c. 460–c. 370 BC) suggested soaking lupin (*Lupinus* spp.) flowers in hemlock juice and applying the solution to tree roots as a method of clearing forests. Another Greek philosopher, Theophrastus (c. 371–c. 287 BC), wrote that oil—most likely olive (*Olea europaea*) oil—could kill young trees if poured over their roots. According to the Roman scholar Varro (c. 116–c. 27 BC), amurca—a watery sediment from olive oil—was also used to kill trees and other weeds in fields. Wine was used as a fungicidal seed treatment as early as c. 60 CE (Russell, 2005).

During the following centuries, recommendations for weed control included salt and human feces (see Smith and Secoy, 1976). But by 900 CE the Chinese were using arsenic to combat garden insects (Pavela, 2014), foreshadowing an era of more effective pest control practices. In France, nicotine in the form of tobacco (*Nicotiana* spp.) extract was first used in 1690 to control lacebugs on pears (*Pyrus* spp.), and was a commonly used insecticide for a variety of insect pests in England by 1800 (McIndoo, 1943).

The 1700s also saw a variety of substances being used as seed treatment fungicides, including arsenic, mercurous chloride, and copper sulfate (Russell, 2005; Zimdahl, 1993). Copper sulfate was used as a cereal seed treatment as early as c. 1760 (Russell, 2005), and in 1807, Prevost scientifically demonstrated copper sulfate's effectiveness on *Tilletia caries*, the fungus causing bunt of wheat (Johnson, 1935; Russell, 2005). Copper sulfate was also proposed as a treatment to kill contaminating weed seeds in wheat, but risk to subsequent wheat germination appears to have prevented its adoption for this purpose (see Smith and Secoy, 1976).

In the early- to mid-1800s, Robertson recommended applying sulfur with soap suds for powdery mildew control in peaches (*Prunus persica*) (Robertson, 1821), and Kenrick recommended a mixture of lime and sulfur together for both insect and powdery mildew control (Kenrick, 1844). Then in 1845, the Belgian botanist Charles Morren proposed using lime with salt and copper sulfate as a soil treatment for potato (*Solanum tuberosum*) late blight, a primary cause of famine at the time (Morren, 1846). Johnson (1935) lamented that had this fungicide been better accepted and alternatively proposed as a foliar treatment, it may have succeeded as a practical control for late blight at a time when such a fungicide was greatly needed. The use of lime in addition to copper sulfate was also found to reduce crop damage during seed and foliar treatments (Russell, 2005).

Lime itself was recommended as an herbicide for the control of horsetail (*Equisetum* spp.) in Germany as early as 1840 (see Mukula and Ruuttunen, 1969). Other chemicals recommended as herbicides by German scientists during the mid-1800s included salt, calcium chloride, sulfuric acid, and iron sulfate. Use of sulfuric acid for weed control expanded in Europe and North America in the late 1800s and early 1900s (see Timmons, 2005).

Destruction of grapevines (*Vitis* spp.) by phylloxera (*Daktulosphaera vitifoliae*), an aphid-like pest native to North America, was first observed in France in 1863 (Menudier, 1879). The resulting devastation to the country's grape and wine production came to be known as The Great French Wine Blight (Smith, 1992; Simms, 2017). Efforts to control phylloxera included Thenard's 1869 recommendation to fumigate soil with carbon disulfide (see Fleming and Baker, 1935). By 1906, carbon disulfide had also been recommended for weed control (Walker, 1906). Another fumigant, hydrogen cyanide, was used to control insects on citrus trees in California as early as 1886 (see Shepard, 1951).

A form of copper and arsenic called Paris green (cuprous acetoarsenite) was reported in 1868 to control the Colorado potato beetle (*Leptinotarsa decemlineata*) in the U.S. (Reynolds, 1869), and quickly become a heavily used insecticide. Paris green was produced from verdigris (cupric acetate), a product of the French wine industry that was applied to grapevines in order to deter thieves from taking ripened fruit (Johnson, 1935). Johnson (1935) reasoned that demand for Paris green as an insecticide may have encouraged the use of copper sulfate mixed with lime as a substitute for verdigris in French vineyards. Copper sulfate and lime was also used as a preservative for vineyard stakes and, along with the introduction of downy mildew into France in 1878, the stage was set for the next advance in fungicides. In 1882, the French botanist Alexis Millardet observed reduced downy mildew growth on grapevines treated with copper sulfate and lime, and in 1885, Millardet reported a refinement of the copper sulfate/lime combination that came to be known as the Bourdeaux mixture (Johnson, 1935; Russell, 2005).

The Bourdeaux mixture became widely used for a variety of plant pathogens, and it was observed that the mixture would also blacken the leaves of wild mustard (*Sinapis arvensis*) (Zimdahl, 1993). In 1896, Bonnett used copper sulfate—a component of the Bordeaux mixture—to selectively control wild mustard and wild radish (*Raphanus raphanistrum*) in an oat (*Avena sativa*) field (see Mukula and Ruuttunen, 1969). Use of copper sulfate and other heavy-metal salts for weed control expanded during the early 1900s (Zimdahl, 1993), although results were not always satisfactory (Timmons, 2005).

The use of arsenical pesticides greatly increased after lead arsenate was used in 1892 for gypsy moth (*Lymantria dispar*) control in Massachusetts (Anonymous, 1895). In the early 1900s, lead arsenate began replacing Paris green as an insecticide (Hood, 2006), and sodium arsenite became one of the first widely used herbicides, especially for aquatic, perennial, and annual forb weeds (Timmons, 2005). Arsenic trichloride was also commercialized as an herbicide for control of morning glory (*Ipomoea* spp.) in the 1920s (Zimdahl, 1993). Other significant inorganic herbicides of the early- to mid-1900s included sodium chlorate, first used in 1923, and sodium borates beginning in the early 1940s (see Timmons, 2005).

Modern Pesticides and Genetically Modified Crops

While comprehensive review of modern agricultural pesticides and crops genetically modified for pesticide tolerance is not possible here, the following summarizes some of the key developments. Beginning in the late 1800s and early 1900s, a major shift in pesticide development came with the commercialization of synthesized organic compounds for pest control. In 1892, the Bayer company introduced one of the first synthetic pesticides, DNOC (4,6-dinitro-*o*-cresol) (Ferguson, 1992). It was originally used to control nun moth (*Lymantria monacha*) larvae in German pine (*Pinus* spp.) forests, and by 1932 became one of the first synthetic herbicides when it was used for weed control in French cereal fields (Cobb and Reade, 2010). While DNOC had some activity as a fungicide (Kataria and Gisi, 1990), the first notable synthetic fungicides were the organomercurial seed treatments, being introduced into Germany in 1913 (Klittich, 2008). The next major advance in fungicide development came in 1934 with the patenting of dithiocarbamates, which were later commercialized as broad-spectrum seed and foliage treatments in the 1940s (Morton and Staub, 2008).

The 1930s were an important decade in insecticide research, culminating in 1939 with discovery of the insecticidal properties of DDT (dichlorodiphenyltrichloroethane) by the Swiss chemist Paul Hermann Müller. DDT was initially used to combat insect vectors of human diseases such as malaria, especially during World War II, and it quickly became a significant agricultural and household insecticide (Jarman and Ballschmiter, 2012). One of its initial uses in crop protection was in the control of the Colorado potato beetle (Gyrisko, Jodka, and Rawlins, 1945). In 1948, Müller was awarded the Nobel Prize in Physiology or Medicine for his discovery of DDT's activity against insects. The widespread use of DDT, however, resulted in negative environmental effects on aquatic animals, birds, and beneficial insects such as bees. Thus, the first decision of the Environmental Protection Agency when it was created in 1972 was to prohibit the use of DDT in the U.S., except for special circumstances (Russell, 1999).

Bacillus thuringiensis (Bt) is a soil bacterium that produces insecticidal proteins during sporulation. These Cry and/or Cyt proteins (also known as d-endotoxins) are narrow in their toxicity target range and target species vary according to the specific strain (Sanahuja et al., 2011). Topical formulations of spores have been available commercially since Sporeine was made available in France in 1938. There are now many different formulations of spores or proteins for different target insects. Although generally of low toxicity to mammals, these formulations have a narrow window of usage and need to be applied directly to the plant organs that are ingested by the target insect in order to be effective (Sanahuja et al., 2011). In the mid-1980s, Bt Cry/Cyt genes were genetically engineered into tobacco and tomato (*Solanum lycopersicum*) plants that expressed the proteins directly in tissues, improving efficacy. The first commercially available transgenic potato, maize (*Zea mays*), and cotton varieties expressing Bt Cry/Cyt proteins were made available in the U.S. in 1995.

Templeman and Marmoy (1940) were the first to report that IAA (indole-3-acetic acid), a natural plant hormone of the auxin class, could be used in cereal crops to kill broadleaf weeds. The synthetic auxin 2,4-D (2,4-dichlorophenoxyacetic acid) was first synthesized a year later in 1941 (Zimdahl, 1993), and was commercialized for selective control of broadleaf weeds in 1945 (Cobb and Reade, 2010). Along with other auxinic herbicides, 2,4-D remains a critical herbicide today.

The first triazine herbicide was discovered in 1952 by Giegy Laboratories in Switzerland. These compounds, such as atrazine and simazine, were initially synthesized in the 1950s for broad spectrum, pre-emergence weed control in maize. They inhibit photosynthesis in target species by binding to the photosystem II complex (Shaner, 2014). Triazine herbicides remain widely used and there are now hundreds of formulations of trazine compounds used to control weeds in over fifty crops all over the world (LeBaron et al.,

2008). Weed resistance to the triazines is widespread, however, and there are concerns over contamination of ground and surface water because of their residual activity (LeBaron et al., 2008).

Glyphosate was first tested as an herbicide in 1970 and was released as a commercial product in 1974 (see Duke and Powles, 2008). While glyphosate provided excellent control of a wide spectrum of weeds, its non-selective activity limited its agricultural role to situations where desirable crops would not be exposed (Duke, 2018). This began to change, however, with the development and introduction of genetically modified crops in the 1990s. The commercial release of glyphosate-resistant soybean (*Glycine max*) in 1996 significantly increased agricultural use of the herbicide (Duke, 2018). While it was first thought to be unlikely that weeds would adapt to glyphosate (Bradshaw et al., 1997), rigid ryegrass (*Lolium rigidum*) from an Australian orchard was first documented as resistant to glyphosate in 1996 (Powles et al., 1998). As of 2018, there are forty-two weed species with confirmed resistance to the herbicide (Heap, 2018).

In part due to evolved glyphosate resistance among weeds, tolerance to other herbicides including 2,4-D, dicamba, and glufosinate have been incorporated into crops such as cotton, maize, and soybean. The introduction of dicamba-tolerant soybean has given growers an additional tool to control weeds resistant to other herbicides. However, widespread use of dicamba, an auxinic herbicide, has increased the frequency of dicamba injury to nearby non-tolerant crops due to drift of the herbicide onto non-tolerant soybean and other sensitive off-target crops (Hartzler, 2017).

Under cool, wet environmental conditions favorable to the growth of *Phytophthora infestans*, potato crops can be devastated by late blight. Multiple applications of fungicides are often required under these conditions to prevent or reduce quality and yield loss (Miller et al., 2006). Attempts to breed resistance through classical means have produced disappointing results. In 2017, three varieties of potato genetically engineered with a gene from a South American wild potato species that provides resistance to certain strains of late blight were approved for use in the United States and Canada (Ridler, 2017a, 2017b). It is claimed that the use of these varieties will reduce the requirement for fungicide applications to control late blight (J. R. Simplot Co., 2017).

The Value of Pesticides

Popp, Pető, and Nágy (2013) estimated that pesticide use in the U.S. prevents about $52 billion in annual crop yield losses. In a study by CropLife America

(Anonymous, 2011), pesticides were estimated to add $51 billion to the annual value of field crops in the U.S., and accounted for 36 percent of the total field crop production value. Pesticide use in nut, fruit, and vegetable crops was estimated to add another $30 billion of value. However, Pimentel (2005) estimated the environmental and societal costs of pesticide use in the U.S. to be about $10 billion annually, including over $1 billion in costs due to increased pesticide resistance.

The ability of modern pesticides to reduce the severity of crop losses due to pests is well known, and has been reviewed elsewhere (see Weller et al., 2014). Some notable examples include the contributions of fungicides and herbicides to long-term increases in U.K. wheat yields (Stokstad and Grullón, 2013), and the results of Brazilian field trials in the mid-2000s showing soybean yields that increased an average of 44 percent due to fungicides for the control of Asian soybean rust (*Phakopsora pachyrhizi*) (Scherm et al., 2009). Results of a study by Rashid et al. (2012) showed up to 116 percent higher net return for rice (*Oryza* spp.) farmers in Bangladesh when herbicidal weed control was used instead of hand-weeding, mostly due to the high labor costs of weeding by hand. Beginning in 2013, emergency treatment of over 1.2 million hectares of land with insecticide successfully reduced the severity and stopped the spread of a locust plague that was destroying up to 40 percent of the rice crop in some parts of Madagascar (Anonymous, 2015).

Looking to the future, Rosegrant et al. (2014) estimated that herbicides, fungicides, and insecticides each have the potential to increase global wheat, maize, and rice yields another 1 to 5 percent by 2050, through increased and more effective use, compared to 2010 production practices. If adopted together, this would reduce the number of food-insecure people in 2050 by an estimated 9 percent in developing countries.

History of Resistance to Agricultural Pesticides

The first recognition of pesticide resistance in agriculture came in 1908 in Clarkston, Washington, when San Jose scale (*Quadraspidiotus perniciosus*) insects were found to be resistant to sulfur-lime treatments used in orchards (Melander, 1914). Although unconfirmed, earlier 1897 reports of control difficulties with codling moth (*Cydia pomonella*) in Nebraska and San Jose scale in the eastern U.S. may have also been due to resistance (see Forgash, 1984). Resistance to hydrogen cyanide was observed in three additional scale species in 1912 and 1913, followed by codling moth resistance to lead arsenate in 1928. The number of cases of insecticide resistance gradually increased until 1947 when DDT resistance was first observed (Moberg, 1990), after which insecticide-resistant insects became more and more frequent (Forgash, 1984).

Herbicide and fungicide resistance arrived much later than for insecticides. The first cases of herbicide resistance were observed in 1957, when resistant wild carrot (*Daucus carota*) in Ontario (Whitehead and Switzer, 1963) and spreading dayflower (*Commelina diffusa*) in Hawaii were found to be resistant to 2,4-D (Hilton, 1957, cited by Schulz and Segobye, 2016). The significance of herbicide resistance in weeds increased with identification of common groundsel (*Senecio vulgaris*) resistant to some triazine herbicides, discovered in a Washington nursery in 1968 (Ryan, 1970).

Fungicide resistance was first noted in 1961, when Harding (Harding and Savage, 1961, cited by Harding, 1962) reported citrus green mold (*Penicillium digitatum*) resistant to biphenyl fungicide applied to California lemons (*Citrus limon*) treated for storage. Resistance of wheat bunt to hexachlorobenzene wheat seed treatments and resistance of leaf spot to organomercurial oat seed treatments were then reported in 1965 in Australia, and 1966 in the UK (Noble et al., 1966; Kuiper, 1965). While these cases appeared twenty to forty years after introduction of the corresponding fungicides, resistance to single site-of-action fungicides came more quickly (Brent and Hollomon, 2007; Smith, 1988). The first such occurrence was with the benzimidazole inhibitors of tubulin polymerization, where resistance was seen about two years after commercialization. The first case of benzimidazole resistance, confirmed in 1969 in New York, was resistance to benomyl fungicide in powdery mildew on greenhouse-grown curcurbits (Schroeder and Provvidenti, 1969).

Pesticide resistance first began with insecticides and by 2014, there were 586 species resistant to a total of 325 different insecticides (Sparks and Nauen, 2015). About 60 percent of these resistant species are agricultural pests (Whalon, Mota-Sanchez, and Hollingworth, 2008). For weeds, there were 255 weed species with documented herbicide resistance by 2018, covering 23 herbicide sites-of-action and at least 163 different herbicides (Heap, 2018). As of 2012, there were over 230 species by mode-of-action combinations among pesticide-resistant plant pathogen species of field or greenhouse origin (Anonymous, 2012).

Cost of Resistance

Poor pest control due to pesticide resistance can reduce crop yields and quality. Pimentel et al. (1993) estimated that 10 percent of the potential U.S. crop yield, valued at $1.4 billion, was lost due to pesticide resistance in 1984. Alternative pesticides or other control options, when available, are often more expensive and labor intensive. Resistance can have negative effects on land values as well.

Several studies have attempted to estimate costs associated with pesticide resistance. Among weeds, the cost of blackgrass (*Alopecurus myosuroides*) with metabolism-based herbicide resistance was estimated to be $154 per hectare in UK winter wheat in 1997 (Orson, 1999). Glyphosate-resistant horseweed (*Erigeron canadensis*) was estimated to cost western Tennessee farmers $30.46 per hectare per season when rotating cotton, soybean, and maize; while glyphosate-resistant waterhemp (*Amaranthus tuberculatus*) was estimated to have a cost of $44.25 per hectare per season in southern Illinois maize/soybean rotations (Mueller et al., 2005). Crop consultants in Arkansas estimated that managing a resistant weed in rice would add $65.60 per hectare to weed control costs (Norsworthy et al., 2007). Herbicide resistance tends to increase in weed populations over several seasons before poor weed control is first observed. The first season of noticeably poor weed control is often the most costly, not just due to yield losses, but also the costs of "trying again" by reapplying the same ineffective herbicide (Peterson, 1999). Not recognizing resistance until it is too late for alternative treatments is also most likely during the first season of poor weed control.

Palumbi (2001) estimated that the cost of insecticide resistance to the U.S. food industry totaled $2 to $7 billion. Grafius (1997) estimated that insecticide-resistant Colorado potato beetle cost Michigan potato growers $44 to $69 per hectare per year, totaling about $1 million per year in the state. The cost to comply with Bt resistance management for corn rootworm (*Diabrotica* spp.), requiring the planting of non-Bt corn refuges to reduce resistance selection in corn rootworm, was estimated at about $3.5 to $9 million (Hurley, Langrock, and Ostlie, 2006). However, benefits of the corn rootworm resistance management program were estimated at $14 to $33 million.

Biological Mechanisms of Pesticide Resistance

An impressive diversity of resistance mechanisms has been discovered among pests. This is in part due to the enormous population sizes of many pest species that inhabit arable cropland, especially when considered on a global scale. Large pest populations mean ample opportunities for novel resistance mutation events to occur. This is demonstrated by the existence of distinct, multiple resistance alleles that, either alone or in combination, can confer resistance to the same pesticide among individuals of many pest species (Hawkins et al., 2018; Powles and Yu, 2010). This results in diverse resistance mechanisms that can generally be categorized as alterations affecting a pesticide's target-site, metabolism, or localization.

Interaction between a pesticide, or more specifically a pesticide formulation's active ingredient, and its target site(s) is what leads to the death or suppression of pests. Target sites are usually enzymes, structural proteins, or other biochemical components to which pesticides bind or otherwise inhibit, resulting in reduction of the target site's normal function. Common pesticide resistance mutations are those that alter a target site's molecular structure, reducing the pesticide's ability to bind while still maintaining sufficient target-site function for pest survival and reproduction. While these alterations may directly affect the pesticide's binding site on the target protein, sometimes the target-site alteration is such that access to the binding site is blocked. Pests with target-site resistance sometimes but not always have reduced fitness in the absence of pesticide due to compromised target-site function (Hawkins et al., 2018).

The target-site alterations that give rise to pesticide resistance are often single amino acid substitutions. Such substitutions are very common among resistant pests and have been reviewed elsewhere (see Feyereisen et al., 2015; Ma and Michailides, 2005; Powles and Yu, 2010). Not all target-site resistance is due to substitution of a single amino acid, however. For example, more than one amino acid is replaced in some acetylcholinesterase target sites of organophosphate and carbamate insecticide-resistant *Drosophila melanogaster* fruit flies (Mutero et al., 1994). Deletion of an amino acid in protoporphyrinogen oxidase confers resistance to herbicidal inhibitors of this target site in waterhemp (Patzoldt et al., 2006).

Target-site pesticide resistance does not always involve alteration of the target protein. Rather, sometimes the target site is over-expressed (i.e., more target-site protein is produced than can be inhibited by the pesticide). The sterol 14α-demethylase enzyme CYP51 is over-expressed in the cherry leaf spot fungus *Blumeriella jaapii*, providing resistance to inhibitors of this enzyme (Ma et al., 2006). While over-expression in this case is likely due to insertion of an expression-boosting sequence near the single-copy *CYP51* gene, repetition of the target-site gene itself is sometimes responsible for over-expression. Glyphosate-resistant Palmer amaranth and kochia (*Bassia scoparia*) are both known to over-express 5-enolpyruvylshikimate-3-phosphate synthase (EPSPS), the target-site of glyphosate herbicide. But while Palmer amaranth has extra copies of EPSPS dispersed throughout its genome (Gaines et al., 2010), kochia has its extra copies together in tandem (Jugulam et al., 2014).

A second general mechanism of pesticide resistance is based on metabolic detoxification of the pesticide (Cobb and Reade, 2010; Ranson et al., 2002; Van Eerd et al., 2003; Yuan, Tranel, and Stewart, 2007). Such detoxification often occurs in two phases: (1) modification of the pesticide, often involving oxidation or hydrolysis, and (2) conjugation of the modified pesticide to glutathione, sugars, or other molecules. Modification and conjugation of pesticides and other xenobiotics is a normal process that occurs among most pests and other organisms. However, pests and other organisms, such as crop species, often have different metabolic detoxification rates for specific pesticides. This is a common but not the only explanation for normal pesticide selectivity, i.e., tolerance among crops and pests to some pesticides but not others. In pest species that are normally susceptible to a pesticide, mutations that establish or increase metabolic detoxification may result in pesticide resistance.

Phase I (modification) metabolic enzymes that may confer resistance to insecticides, fungicides, and/or herbicides include the cytochrome P450 monooxygenases and esterase families. Glutathione S-transferases are the major enzymatic family responsible for Phase II (conjugation) metabolic resistance. Pesticide resistance involving these metabolic enzymes may be the result of mutations that change their protein structure or increase their expression, either through an increase in the expression of a single gene or expansion of the number of gene copies.

A third general mechanism of pesticide resistance involves a variety of diverse processes related to the localization of pesticide molecules at the cellular or organismal level (FRAC, 2019; Sammons and Gaines, 2014; Sonderlund and Bloomquist, 1990). Resistant pests may display reduced penetration/absorption or increased excretion of pesticides. Movement of pesticide molecules within the organism may also be decreased, thus limiting exposure of target sites to the pesticide. At the cellular level, pesticide molecules may be pumped out of cells or sequestered into areas away from target sites.

Evolution of Resistant Pest Populations

Resistant pests usually originate from random mutations that are then selected for by pesticides. While resistant pests and the resistance genes they carry can spread from location to location, pest populations are often large enough at different locations to experience independent resistance mutation events among individuals, i.e., independent origins of resistance (ffrench-Constant, 2013). Sometimes a single mutation in a single gene is sufficient to provide the individual pest with sufficient resistance to allow reproduction. But sometimes additional resistance mutations/genes are necessary. The latter is especially the case for polygenic/quantitative resistance (Via, 1986), or where pests are resistant to more than one mode of action. Depending on the pest and resistance genes, hybridization may or may not play a significant role in resistance evolution.

Selection for resistance depends on the pesticide, rate, and timing of pesticide application, and any environmental influences on resistance expression. The specific pest and its genetics and expression of resistance are also important, along with fitness of the pest in the presence and absence of the pesticide. Some pesticide resistance traits have a fitness penalty where the resistant type does not thrive and reproduce as well as susceptible types. In such cases, resistant pest populations may slowly revert to susceptibility over time, if the pesticide is not present. However, equal fitness between resistant and susceptible types is common. In either case, the fitness of a resistant pest is, of course, much greater than that of susceptible pests when the pesticide is used, leading to resistance selection and an increase in resistance gene frequency.

Research into the parameters that most affect resistance selection is important to help design management strategies that are likely to minimize or prevent resistance evolution. Even with good information on the expression of resistance under different circumstances, predicting the outcomes of pesticide use can be difficult. Computer models continue to be developed for this purpose, but are often limited by lack of biological information on specific types of resistance.

Management of Pesticide Resistance

Goals of pesticide resistance management include preventing or delaying resistance, and reducing the severity and spread of resistance once realized. Strategies to manage pesticide resistance include both chemical and non-chemical strategies as well as integrated pest management (IPM), where diverse practices are adopted. Industry, academic, and government scientists have organized various "action committees" to compile, recommend, and disseminate information about pesticide resistance and recommend management strategies. Some of the largest and most globally focused include those organized by pesticide manufacturers, including the Fungicide Resistance Action Committee (www.frac.info), Insecticide Resistance Action Committee (www.irac-online.org), and the Herbicide Resistance Action Committee (www.hracglobal.com). Resistance management recommendations from these groups, as well as those from academic and government scientists, have been reviewed elsewhere (see Norsworthy et al., 2012; Sparks and Nauen, 2015; van den Bosch et al., 2014). The following is a generalized summary of resistance management practices for agricultural crop pesticides.

Avoiding overreliance on single pesticide modes-of-action is a highly recommended resistance management tool for insecticides, fungicides, and herbicides. Pesticides with different modes-of-action should be used within the same season, or season-to-season, depending on the pest. Mixtures of different modes-of-action are also generally effective in reducing selection for pesticide resistance, providing that the pesticides target the same pest. Pesticides with modes-of-action targeting multiple target sites and processes generally have lower resistance risk and are often desirable in a resistance management program.

The decision to use pesticides or not in a particular situation is not always easy. Withholding pesticide treatment until economic thresholds are met is sometimes recommended, especially for fungicides. Doing so reduces selection for resistance, but also may increase the chance that a rare resistant individual is present. This is also why prophylactic use of pesticides rather than eradicant use is sometimes recommended. Application of pesticides at recommended, labeled rates and at recommended pest sizes/stages is also critical for optimum resistance management. Within labeled rate ranges, whether high or low pesticide rates are best to reduce selection for resistance is often dependent on the genetics of resistance in the pest. Low rates can sometimes increase selection for minor resistance genes contributing to polygenic resistance. Higher rates, however, can also sometimes increase selection by reducing competition from susceptible pests.

Crop rotation is generally useful in reducing selection for pesticide resistance by not allowing individual pest species to have optimum growth conditions and hosts season after season. Crop rotation also encourages rotation of pesticide modes-of-action for different crops. Cultural pest management such as tillage is often helpful for pest and resistance management, but it is not always compatible with reduced- or no-tillage production systems. Alternative biological control methods are also useful, where available and appropriate.

Practices that reduce pest reproduction and spread can help reduce or delay pesticide resistance. Good pest control helps reduce the amount of disease inoculum, number of weed seeds, and insect populations overwintering to the next season. Planting pure and disease-free seed and other propagules, and cleaning equipment of pests between fields, not only helps prevent introduction of resistant pests into new locations, but also assists in keeping pest populations low to minimize the chance that a rare resistant individual will be present. It's also important to not neglect the control of pests in field borders. Routine scouting will help in the timely identification of new pest and resistance problems before major economic losses are experienced.

Integrated pest management (IPM), i.e., the combination of multiple pest control practices such as those above, is highly encouraged. Such practices help disrupt the continuous selection of resistant pests season after season.

Future Prospects

Discovery and commercialization of new pesticide modes-of-action have slowed dramatically, especially for herbicides (Duke, 2012). This limits the ability to manage pesticide resistance by switching or mixing different modes-of-action, as fewer options remain with increasing resistance. Rational design of pesticides, where molecules are specifically designed to inhibit specific target sites, might find increased success in the future with greater knowledge of plant biochemistry and increased ability to use computing power to predict molecular interactions (Cobb and Reade, 2010). While new commercially viable pesticide modes-of-action might be discovered in this way, success has been limited thus far.

Discovery of more biopesticides may help provide pest control alternatives to synthetic chemicals (Chandler et al., 2011). Research into insect baits and attractants may also help keep resistant insects from affecting crops. Increased use of biocontrol through the encouragement of pest competitors, predators, herbivores, or parasites may also be useful in combating pesticide resistance. Advances in organic agriculture may provide additional pest control methods.

Crop improvement and biotechnology are also tools that can be used to combat pests and pesticide resistance. The continued breeding of crops with improved inherent resistance to diseases and insect pests will help reduce reliance on synthetic pesticides. Crops with increased ability to compete with weeds would also be beneficial. Biotechnology and genetic engineering has assisted pest management by introducing pest resistance genes into crops, such as with insect-resistant *Bt* crops. However, such methods of pest control have their own resistance issues. Genetic engineering has also been used to produce crops resistant to herbicides, such as glyphosate, thus increasing the number of herbicides that can be used when resistance is a problem. However, overreliance on these genetically modified crops has also led to increased resistance problems.

Genomic research investigating pests holds promise for a greater understanding of pesticide resistance and potential new methods of control. One promising method is based on RNA interference (RNAi), where RNA either produced by crops or applied to pests could turn off genes necessary for pest survival or proliferation (see Huvenne and Smagghe, 2010). For example, RNAi could be used to inhibit the expression of resistance genes in the pest. Another emerging genetic technology for pest control, called a gene drive, is based on CRISPR-generated modifications to critical pest genes, or reversion of pesticide resistance genes back to susceptibility (Esvelt et al., 2014). In a gene drive, these modifications spread in pest populations generation-to-generation after initial introduction.

Monitoring pest populations and the efficacy of pesticide applications is necessary for optimal resistance management. New technologies such as drones for crop scouting or increased use of genetic tests for resistance confirmation are also very promising.

References

Anonymous (1895). *Fourth report of the state (massachusetts) board of agriculture on the work of extermination of the gypsy moth*. Boston: Wright and Potter Printing Co.

Anonymous (2011). *The contribution of crop protection products to the united states economy*. Washington, D.C.: CropLife America.

Anonymous (2012). *FRAC list of plant pathogenic organisms resistant to disease control agents*. Brussels: Fungicide Resistance Action Committee.

Anonymous (2015). *Response to the locust plague: programme - Campaign No. 2, 2014/2015*. Rome: FAO.

Bourke, P. M. A. (1964). Emergence of potato blight, 1843–46. *Nature, 203*, 805–808.

Bradshaw, L. D., Padgette, S. R., Kimball, S. L., and Wells, B. H. (1997). Perspectives on glyphosate resistance. *Weed Technology, 11*, 189–198.

Brent, K. J., and Hollomon., D. W. (2007). *Fungicide resistance in crop pathogens: how can it be managed?* 2nd Edition. Brussels: Fungicide Resistance Action Committee.

Chandler, D., Bailey, A. S., Tatchell, G. M., Davidson, G., Greaves, J., and Grant., W. P. (2011). The development, regulation and use of biopesticides for integrated pest management. *Philosophical Transactions of the Royal Society B-Biological Sciences, 366*, 1987–1998. doi: 10.1098/rstb.2010.0390.

Cobb, A., and Reade, J. P. H. (2010). *Herbicides and plant physiology*. 2nd edition. Chichester, West Sussex, UK: Wiley-Blackwell.

Dent, D. (2000). *Insect pest management*. 2nd edition. Wallingford, Oxon, UK: CABI Publishing.

Duke, S. O. (2012). Why have no new herbicide modes of action appeared in recent years? *Pest Management Science, 68*, 505–512.

Duke, S. O. (2018). The history and current status of glyphosate. *Pest Management Science, 74*, 1027–1034. doi: 10.1002/ps.4652.

Duke, S. O., and Powles, S. B. (2008). Glyphosate: a once-in-a-century herbicide. *Pest Management Science, 64*, 319–325. doi: 10.1002/ps.1518.

Esvelt, K. M., Smidler, A. L., Catteruccia, F., and Church, G. M. (2014). Concerning RNA-guided gene drives for the alteration of wild populations. *Elife, 17:3*, ii, e03401. doi: 10.7554/eLife.03401.

Ferguson, C. (1992). Bayer crop protection—the first 100 years: 1892–1992. *Pesticide Outlook, 3*, 32–34.

Feyereisen, R., Dermauw, W., and Van Leeuwen, T. (2015). Genotype to phenotype, the molecular and physiological dimensions of resistance in arthropods. *Pesticide Biochemistry and Physiology, 121*, 61–77. doi: 10.1016/j.pestbp.2015.01.004.

ffrench-Constant, R. H. (2013). The molecular genetics of insecticide resistance. *Genetics, 194*, 807–815.

Fleming, W. E., and Baker, F. E. (1935). *The use of carbon disulphide against the Japanese beetle.* Technical Bulletin No. 478. Washington, D.C.: United States Department of Agriculture.

Forgash, A. J. (1984). History, evolution, and consequences of insecticide resistance. *Pesticide Biochemistry and Physiology, 22*, 178–186. doi: 10.1016/0048-3575(84)90087-7.

FRAC. (2019). Mechanisms of fungicide resistance. Fungicide Resistance Action Committee. Retrieved from http://www.frac.info/resistance-overview/mechanisms-of-fungicide-resistance.

Gaines, T. A., Zhang, W., Wang, D., Bukun, B., Chisholm, S. T., Shaner, D. L., Nissen, S. J., Patzoldt, W. L., Tranel, P. J., Culpepper, A. S., Grey, T. L., Webster, T. M., Vencill, W. K., Sammons, R. D., Jiang, J., Preston, C., Leach, J. E., and Westra, P. (2010). Gene amplification confers glyphosate resistance in *Amaranthus palmeri. PNAS, 107*, 1029–1034. doi: 10.1073/pnas.0906649107.

Grafius, E. (1997). Economic impact of insecticide resistance in the Colorado potato beetle (Coleoptera: Chrysomelidae) on the Michigan potato industry. *Journal of Economic Entomology, 90*, 1144–1151. doi: 10.1093/jee/90.5.1144.

Gyrisko, G. G., Jodka, J. F. T., and Rawlins, W. A. (1945). DDT to control potato insects. *Journal of Economic Entomology, 38*, 169–173. doi: 10.1093/jee/38.2.169.

Harding, P. R. (1962). Differential sensitivity to sodium orthophenylphenate by biphenyl-sensitive and biphenyl-resistant strains of *Penicillium digitatum. Plant Disease Reporter, 46*, 100–104.

Harding, P. R., and Savage, D. C. (1961). Biphenyl-resistant strains of citrus green mold. *California Citrograph, 46*, 306–308.

Hartzler, B. (2017). Off-target dicamba injury: the Iowa experience. ISU Extension and Outreach. Retrieved from https://crops.extension.iastate.edu/cropnews/2017/07/target-dicamba-injury-iowa-experience.

Hawkins, N. J., Bass, C., Dixon, A., and Neve, P. (2018). The evolutionary origins of pesticide resistance. *Biological Reviews, 94*, 135–155. doi: 10.1111/brv.12440.

Heap, I. (2018). International survey of herbicide resistant weeds. Retrieved from www.weedscience.org.

Hilton, H. W. (1957). Herbicide tolerant strains of weeds. In *Hawaiian Sugar Planters' Association Annual Report*, 69–72.

Homer, Murray, A. T., and Dimock, G. E. (1995). *The odyssey.* The Loeb Classical Library No. 104–105. Cambridge, MA: Harvard University Press.

Hood, E. (2006). The apple bites back: claiming old orchards for residential development. *Environmental Health Perspectives, 114*, A470–A476. doi: 10.1289/ehp.114-a470.

Hurley, T. M., Langrock, I., and Ostlie, K. (2006). Estimating the benefits of Bt corn and cost of insect resistance management ex ante. *Journal of Agricultural and Resource Economics, 31*, 355–375.

Huvenne, H., and Smagghe, G. (2010). Mechanisms of dsRNA uptake in insects and potential of RNAi for pest control: a review. *Journal of Insect Physiology, 56*, 227–235.

Jarman, W. M. and Ballschmiter, K. (2012). From coal to DDT: the history of the development of the pesticide DDT from synthetic dyes till *Silent Spring. Endeavour, 36*, 131–142.

Johnson, G. F. (1935). The early history of copper fungicides. *Agricultural History, 9*, 67–79.

J. R. Simplot Co. (2017). Innate® second generation potatoes with late blight protection receive EPA and FDA clearances. Retrieved from http://www.innatepotatoes.com/newsroom/view-news/innate-second-generation-potatoes-with-late-blight-protection-receive-epa-a.

Jugulam, M., Niehues, K., Godar, A. S., Koo, D.-H., Danilova, T., Friebe, B., Sehgal, S., Varanasi, V. K., Wiersma, A., Westra, P., Stahlman, P. W., and Gill, B. S. (2014). Tandem amplification of a chromosomal segment harboring *5-enolpyruvylshikimate-3-phosphate synthase* locus confers glyphosate resistance in *Kochia scoparia. Plant Physiology, 166*, 1200–1207. doi: 10.1104/pp.114.242826.

Kataria, H. R., and Gisi, U. (1990). Interactions of fungicide-herbicide combinations against plant pathogens and weeds. *Crop Protection, 9*, 403–409. doi: 10.1016/0261-2194(90)90128-t.

Kenrick, W. (1844). *The new American orchardist*. Boston: Otis, Broaders, and Company.

Klittich, C. J. (2008). Milestones in fungicide discovery: chemistry that changed agriculture. *Plant Health Progress*. doi: 10.1094/PHP-2008-0418-01-RV.

Kuiper, J. (1965). Failure of hexachlorobenzene to control common bunt of wheat. *Nature, 206*, 1219–1220.

Kutney, G. (2013). *Sulfur: history, technology, applications and industry*. 2nd edition. Toronto: ChemTec Publishing.

LeBaron, H. M., McFarland, J., and Burnside, O., eds. (2008). *The Triazine herbicides*. San Diego, CA: Elsevier.

Ma, Z., and Michailides, T. J. (2005). Advances in understanding molecular mechanisms of fungicide resistance and molecular detection of resistant genotypes in phytopathogenic fungi. *Crop Protection, 24*, 853–863. doi: 10.1016/j.cropro.2005.01.011.

Ma, Z., Proffer, T. J., Jacobs, J. L., and Sundin, G. W. (2006). Overexpression of the 14α-demethylase target gene (*CYP51*) mediates fungicide resistance in *Blumeriella jaapii*. *Applied and Environmental Microbiology, 72*, 2581–2585.

McIndoo, N. E. (1943). *Insecticidal uses of nicotine and tobacco: a condensed summary of the literature, 1690–1934*. Washington, D.C.: United States Department of Agriculture .

Melander, A. L. (1914). Can insects become resistant to sprays? *Journal of Economic Entomology, 7*, 167–173.

Menudier (1879). The phylloxera in France translated. *Scientific American, 41*, 72–73.

Miller, J., Nolte, P., Olsen, N., Miller, T., Bohl, B., and Thornton, M. (2006). *Late blight management action plan for potatoes*. CIS 1132. Moscow, ID: University of Idaho Extension.

Moberg, W. K. (1990). Understanding and combating agrochemical resistance. In Green, M. B., LeBaron, H. M. and Moberg, W. K. (eds.) *Managing Resistance to agrochemicals: from fundamental research to practical strategies*, pp. 1–15. Washington, D.C.: American Chemical Society.

Morren, C. (1846). The potato disease: an authentic and valuable memoir on the potato disease. *Monthly Journal of Agriculture, 1*, 386–389.

Morton, V., and Staub, T. (2008). A short history of fungicides. *APSnet Features*. doi: 10.1094/APSnetFeature-2008-0308.

Mueller, T. C., Mitchell, P. D., Young, B. G., and Culpepper, A. S. (2005). Proactive versus reactive management of glyphosate-resistant or -tolerant weeds. *Weed Technology, 19*, 924–933. doi: 10.1614/WT-04-297R.1.

Mukula, J., and Ruuttunen, E. (1969). Chemical weed control in Finland in 1887–1965. *Annales Agriculturae Fenniae, 8*(suppl. 1), 1–45.

Mutero, A., Pralavorio, M., Bride, J. M., and Fournier, D. (1994). Resistance-associated point mutations in insecticide-insensitive acetylcholinesterase. *PNAS, 91*, 5922–5926. doi: 10.1073/pnas.91.13.5922.

Noble, M., Macgarvie, Q. D., Hams, A. F., and Leafe, E. L. (1966). Resistance to mercury of *Pyrenophora avenae* in Scottish seed oats. *Plant Pathology, 15*, 23–28.

Norsworthy, J. K., Burgos, N. R., Scott, R. C., and Smith, K. L. (2007). Consultant perspectives on weed management needs in Arkansas rice. *Weed Technology, 21*, 832–839. doi: 10.1614/WT-06-203.1.

Norsworthy, J. K., Ward, S. M., Shaw, D. R., Llewellyn, R. S., Nichols, R. L., Webster, T. M., Bradley, K. W., Frisvold, G., Powles, S. B., Burgos, N. R., Witt, W. W., and Barrett, M. (2012). Reducing the risks of herbicide resistance: best management practices and recommendations. *Weed Science, 60*(sp. 1), 31–62. doi: 10.1614/WS-D-11-00155.1.

Oerke, E.-C. (2006). Crop losses to pests. *Journal of Agricultural Science, 144*, 31–43. doi: 10.1017/s0021859605005708.

Oerke, E.-C., and Dehne, H.-W. (2004). Safeguarding production—losses in major crops and the role of crop protection. *Crop Protection, 23*, 275–285. doi: 10.1016/j.cropro.2003.10.001.

Oerke, E.-C., Weber, A., Dehne, H.-W., and Schöenbeck, F. (1994). Conclusion and perspectives. In Oerke, E.-C., Dehne, H.-W., Schöenbeck F., and Weber, A. (eds.) *Crop production and crop protection: estimated losses in major food and cash crops*, pp. 742–770. Amsterdam: Elsevier.

Orson, J. H. (1999). The cost to the farmer of herbicide resistance. *Weed Technology, 13*, 607–611.

Palumbi, S. R. (2001). Humans as the world's greatest evolutionary force. *Science, 293*, 1786–1790. doi: 10.1126/science.293.5536.1786.

Patzoldt, W. L., Hager, A. G., McCormick, J. S., and Tranel, P. J. (2006). A codon deletion confers resistance to herbicides inhibiting protoporphyrinogen oxidase. *PNAS, 103*, 12329–12334. doi: 10.1073/pnas.0603137103.

Pavela, R. (2014). Limitation of plant biopesticides. In Singh, D. (ed.) *Advances in plant biopesticides*, pp. 347–359. New Delhi: Springer.

Peterson, D. E. (1999). The impact of herbicide-resistant weeds on Kansas agriculture. *Weed Technology, 13*, 632–635.

Pimentel, D., Acquay, H., Biltonen, M., Rice, P., Silva, M., Nelson, J., Lipner, V., Giordano, S., Horowitz, A., and D'Amore, M. (1993). Assessment of environmental and economic impacts of pesticide use. In Pimentel, D., and Lehman, H. (eds.) *The pesticide question: environment, economics, and ethics*, pp. 47–84. New York: Chapman and Hall.

Pimentel, D. (2005). Environmental and economic costs of the application of pesticides primarily in the United States. *Environment, Development and Sustainability, 7*, 229–252. doi: 10.1007/s10668-005-7314-2.

Popp, J., Pető, K., and Nágy, J. (2013). Pesticide productivity and food security. a review. *Agronomy for Sustainable Development, 33*, 243–255. doi: 10.1007/s13593-012-0105-x.

Powles, S. B., Lorraine-Colwill, D. F., Dellow, J. J., and Preston, C. (1998). Evolved resistance to glyphosate in rigid ryegrass (*Lolium rigidum*) in Australia. *Weed Science, 46*, 604–607.

Powles, S. B., and Yu, Q. (2010). Evolution in action: plants resistant to herbicides. *Annual Review of Plant Biology, 61*, 317–347. doi: 10.1146/annurev-arplant-042809-112119.

Ranson, H., Claudianos, C., Ortelli, F., Abgrall, C., Hemingway, J., Sharakhova, M. V., Unger, M. F., Collins, F. H., and Feyereisen, R. (2002). Evolution of supergene families associated with insecticide resistance. *Science, 298*, 179–181.

Rashid, M. H., Alam, M. M., Rao, A. N., and Ladha, J. K. (2012). Comparative efficacy of pretilachlor and hand weeding in managing weeds and improving the productivity and net income of wet-seeded rice in Bangladesh. *Field Crops Research, 128*, 17–26. doi: 10.1016/j.fcr.2011.11.024.

Reynolds, E. (1869). Experiments with the potato bug. In *Report of the commissioner of agriculture for the year 1868*, p. 433. Washington, D.C.: Government Printing Office.

Ridler, K. (2017a). U.S. approves 3 more types of simplot GMO potatoes for sale this year. Retrieved from https://www.idahostatesman.com/news/business/article137891178.html.

Ridler, K. (2017b). Canada oks Idaho company's genetically engineered potatoes. Retrieved from https://www.usnews.com/news/news/articles/2017-08-03/canada-oks-idaho-companys-genetically-engineered-potatoes.

Ridley, R. T. (1986). To be taken with a pinch of salt: the destruction of Carthage. *Classical Philology, 81*, 140–146.

Robertson, J. (1821). On the mildew and some other diseases incident to fruit trees. *Transactions of the London Horticultural Society, 5*, 175–185.

Rosegrant, M. W., Koo, J., Cenacchi, N., Ringler, C., Robertson, R., Fisher, M., Cox, C., Garrett, K., Perez, N. D., and Sabbagh, P. (2014). *Food security in a world of natural resource scarcity: the role of agricultural technologies*. Washington, D.C.: International Food Policy Research Institute.

Russell III, E. P. (1999). The strange career of DDT: experts, federal capacity, and environmentalism in World War II. *Technology and Culture, 40*, 770–796.

Russell, P. E. (2005). A century of fungicide evolution. *Journal of Agricultural Science, 143*, 11–25. doi: 10.1017/s0021859605004971.

Ryan, G. F. (1970). Resistance of common groundsel to simazine and atrazine. *Weed Science, 18*, 614–616.

Sammons, R. D., and Gaines, T. A. (2014). Glyphosate resistance: state of knowledge. *Pest Management Science, 70*:1367–1377. doi: 10.1002/ps.3743.

Sanahuja, G., Banakar, R., Twyman, R. M., Capell, T., and Christou, P. (2011). *Bacillus thuringiensis*: a century of research, development and commercial applications. *Plant Biotechnology Journal*, *9*, 283–300.

Scherm, H., Christiano, R. S. C., Esker, P. D., Del Ponte, E. M., and Godoy, C. V. (2009). Quantitative review of fungicide efficacy trials for managing soybean rust in Brazil. *Crop Protection*, *28*, 774–782. doi: 10.1016/j.cropro.2009.05.006.

Schroeder, W. T., and Provvidenti, R. (1969). Resistance to benomyl in powdery mildew of cucurbits. *Plant Disease Reporter*, *53*, 271–275.

Schulz, B., and Segobye, K. (2016). 2,4-D transport and herbicide resistance in weeds. *Journal of Experimental Botany*, *67*, 3177–3179. doi: 10.1093/jxb/erw199.

Shaner, D. L., ed. (2014). *Herbicide handbook*. 10th Edition. Champaign, IL: Weed Science Society of America.

Shepard, H. H. (1951). *The chemistry and action of insecticides*. York, PA: The Maple Press Company.

Simms, C. (2017). The grape depression. *New Scientist*, *236*, 60–62.

Simpson, S. J., and Sword, G. A. (2008). Locusts. *Current Biology*, *18*, R364-R366. doi: 10.1016/j.cub.2008.02.029.

Smith, A. E., and Secoy, D. M. (1976). Early chemical control of weeds in Europe. *Weed Science*, *24*, 594–597.

Smith, C. M. (1988). History of benzimidazole use and resistance. In Delp, C. J. (ed.) *Fungicide resistance in North America*, pp. 23–24. St. Paul, MN: APS Press.

Smith, E. H. (1992). The grape phylloxera: a celebration of its own. *American Entomologist*, *38*, 212–221.

Sonderlund, D. M., and Bloomquist, J. R. (1990). Molecular mechanisms of insecticide resistance. In Roush, R. T. and Tabashnik, B. E. (eds.) *Pesticide resistance in arthropods*, pp. 58–96. Boston: Springer.

Sparks, T. C., and Nauen, R. (2015). IRAC: Mode of action classification and insecticide resistance management. *Pesticide Biochemistry and Physiology*, *121*, 122–128. doi: 10.1016/j.pestbp.2014.11.014.

Stokstad, E., and Grullón, G. (2013). Pesticide planet. *Science*, *341*, 730–731.

Templeman, W. G., and Marmoy, C. J. (1940). The effect upon the growth of plants of watering with solutions of plant-growth substances and of seed dressings containing these materials. *Annals of Applied Biology*, *27*, 453–471. doi: 10.1111/j.1744-7348.1940.tb07517.x.

Timmons, F. L. (2005). A history of weed control in the United States and Canada. *Weed Science*, *53*, 748–761. doi: 10.1614/0043-1745(2005)053[0748:AHOWCI]2.0.CO;2.

van den Bosch, F., Oliver, R., van den Berg, F., and Paveley, N. (2014). Governing principles can guide fungicide-resistance management tactics. *Annual Review of Phytopathology*, *52*, 175–195. doi: 10.1146/annurev-phyto-102313-050158.

Van Eerd, L. L., Hoagland, R. E., Zablotowicz, R. M., and Hall, J. C. (2003). Pesticide metabolism in plants and microorganisms. *Weed Science*, *51*, 472–495.

Via, S. (1986). Quantitative genetic models and the evolution of pesticide resistance. In Committee on Strategies for the Management of Pesticide Resistant Pest Populations (ed.) *Pesticide resistance: strategies and tactics for management*, pp. 222–235. Washington, D.C.: National Academy Press.

Walker, E. (1906). *Suggestions upon the care of apple orchards*. Bulletin No. 91. Fayetteville, AR: Arkansas Agricultural Experiment Station.

Walker, P. T. (1983). Crop losses: the need to quantify the effects of pests, diseases and weeds on agricultural production. *Agriculture, Ecosystems and Environment*, *9*, 119–158. doi: 10.1016/0167-8809(83)90038-5.

Ward, S. M., Webster, T. M., and Steckel, L. E. (2013). Palmer amaranth (*Amaranthus palmeri*): a review. *Weed Technology*, *27*, 12–27. doi: 10.1614/wt-d-12-00113.1.

Weller, S. C., Culbreath, A. K., Gianessi, L., and Godfrey, L. D. (2014). *The contributions of pesticides to pest management in meeting the global need for food production by 2050*. Issue Paper No. 55. Ames, IA: Council for Agricultural Science and Technology.

Whalon, M. E., Mota-Sanchez, D., and Hollingworth, R. M. (2008). *Global pesticide resistance in arthropods*. Wallingford, U.K.: CABI Publishing.
Whitehead, C. W., and Switzer, C. M. (1963). The differential response of strains of wild carrot to 2,4-D and related herbicides. *Canadian Journal of Plant Science, 43*, 255–262.
Yuan, J. S., Tranel, P. J., and Stewart Jr., C. N. (2007). Non-target-site herbicide resistance: a family business. *Trends in Plant Science, 12*, 6–13.
Zimdahl, R. L. (1993). *Fundamentals of weed science*. San Diego, CA: Academic Press.

RETHINKING WATER ON A GROWING, HUNGRY PLANET

Kate A. Brauman
University of Minnesota

THE WORLD IS NOT RUNNING out of water. Nearly all of the water on the planet has been here for a very long time, and it isn't going anywhere (Morbidelli et al., 2000). Water cycles, falling as rain or snow, moving across the planet's surface, continually evaporate then fall again, which means, for the most part, we get a fresh start with our water resources each year. Capricious weather varies the amount of water available from year to year. Climate change will make wet places wetter and dry places drier and will likely change the timing and intensity of storms (Donat et al., 2016). But water isn't going away.

We are, however, using water differently. As population grows and becomes more affluent, we demand more—more food, more hygiene, more stuff—and we use water to make most of it. As demand for water increases, we experience water stress because more of us are trying to share the same resource. And when we divvy up every last drop of available water, even small fluctuations in water supply can hit hard.

Water is surprisingly hard to conceptualize. You might have first thought about the liquid we drink and wash our hands with, or maybe regulations about where storm water can go, or endless irrigated green fields. Different people, professions, and users conceive of water in very different ways. Some measure biophysical properties. Some focus on the safety of tap water. Others are concerned with how much food or energy could be produced using water. Maybe you measure how far one has to walk to reach water. You might think about the laws that govern water. Or perhaps you worry about the people, states, or corporations who own water.

When we get into the details of evaluating water, most of what we evaluate isn't hydrology. We use water to make or clean or do things we value, but the things we value mostly aren't wet. We eat food grown with irrigation water, cleaned with water, cooked on a stove made in a factory cleaned with water and powered by electricity produced with water. We might drink a glass of water with dinner. But mostly we value the food, the energy, and the stove, not the water that went into making them. So water is instrumental, but it's frequently not the point. If we focus on managing and enhancing the things we value—making sure we have the food, the energy, the stoves we want—instead of fighting over the allocation of water among users, there is potential for managing water more effectively and getting more of what we want with the same water supply.

Water Use

What does it mean to use water? Sometimes we use water where it is—in rivers, lakes and streams—for activities like fishing, transporting goods on barges, and producing hydropower. These are in-stream uses. For all other water uses, we pull water out of surface water bodies or pump groundwater from wells that tap underground aquifers. We do lots of things with that water, which we can organize into five primary categories—household use, manufacturing, energy production, and two types of food production: crop irrigation and livestock watering. For most of these uses, the water is used for cleaning, powering, or growing—water is an instrument to produce something we value. Only occasionally do we explicitly value water itself.

What Do We Use Water For?

Household water use, sometimes called domestic water use, includes the water we drink, water for food preparation, water for cleaning ourselves and our belongings, water for disposing of waste, as well as the water we use for our household lawns and gardens. The basic physiological water requirements for human life are small, probably about three liters per person per day (Gleick, 1996), and though personal water requirements have been estimated at thirty times as much, it's still a manageably small amount of water. Some historical assessments have even suggested that personal water demand is so small that it has more or less never been denied to those in need (Salzman, 2012). However, subsistence water use means there is little available for washing and waste disposal, and many would consider it inadequate. The amount of water actually used in a household varies with custom and climate, as well as with ease of access to water and access to technology that can reduce water use. Differences in household property and the associated demand for household irrigation affect household water need as well; in the United States, outdoor water use constitutes about a third of total household use (Maupin et al., 2014). As a result of all this, there is dramatic variation in household water use around the world (Flörke et al., 2013).

At the global scale, household use accounts for about 4 percent of water consumption (Figure 1; Brauman et al., 2016). Though this number is small, household conservation is critical to good water management. This is because, in many cases, the specific water source that a household or municipality taps into may be limited, frequently because that source is groundwater. In other cases, it may be prohibitively expensive to tap an additional water source should demand exceed supply from an existing source. And even if a water source is not fully used up, building additional drinking water treatment and wastewater treatment infrastructure may be prohibitively costly. Finally, regardless of the abundance of supply, moving and treating water is energy intensive, and therefore costly, so reducing water use has both climate and financial benefits. And yet, the volume of water demanded for household use is small enough and valuable enough that it is plausible to consider energy intensive, expensive replacement sources such as desalination of seawater.

Manufacturing and industrial use of water is similar to household use, though on a larger scale. Most water in manufacturing is used for cleaning, cooling, lubricating, or as a solvent (Maupin et al., 2014). Industrial use accounts for about 5 percent of global water consumption (Figure 1; Brauman et al., 2016).

Energy and water are a different story. Perhaps when thinking of water for energy we first think of enormous dams harnessing rivers to run water through turbines and create hydroelectric power. This is a critical in-stream use of water. However, hydropower provides only about 16 percent of global electricity supply (IEA, 2012). When it comes to water for energy, thermoelectric energy generation is the big player. Whether burning fossil or nuclear fuels, power plants operate by generating steam to spin turbines. A tremendous amount of water is then used to cool that steam, so much water that it adds up to 15 percent of total global water withdrawals (Figure 1; Brauman et al., 2016). Most of that water, 97 percent, is put back into streams, though, so thermoelectric cooling accounts for only 1 percent of global water consumption. Water used for thermoelectric cooling that is put back into rivers and streams is warmer than it was when it was extracted, but in general it can be used again downstream. Newer technologies require much smaller water withdrawals, so water demand for energy is likely dropping (Maupin et al., 2014).

Household use, manufacturing, and thermoelectric cooling add up to just 11 percent of global water consumption. The rest of the water that people use—89 percent—goes to food production, so we'll focus on food for most of the rest of this chapter. Food production is the primary human use of water in part because we dedicate so much land to food: Close to 40 percent of the earth's ice-free surface is dedicated to growing crops or livestock for food, and about a quarter of global cropland is irrigated (Foley et al., 2011). On a global scale, about 70 percent of water withdrawals are for irrigation, and irrigation accounts for 87 percent of water consumption (Figure 1; Brauman et al., 2016). If a substantial amount of water is being used in a watershed, it is being used for irrigation. Though livestock accounts for just 2 percent of global water consumption, in sparsely populated arid regions it can amount to nearly all the water use.

FIGURE 1. Global water use, by sector. Globally, most water use by people is for irrigation. Water that is consumed, meaning it cannot be immediately reused, is a subset of the water that is withdrawn from surface and groundwater bodies. Data from Brauman et al. (2016).

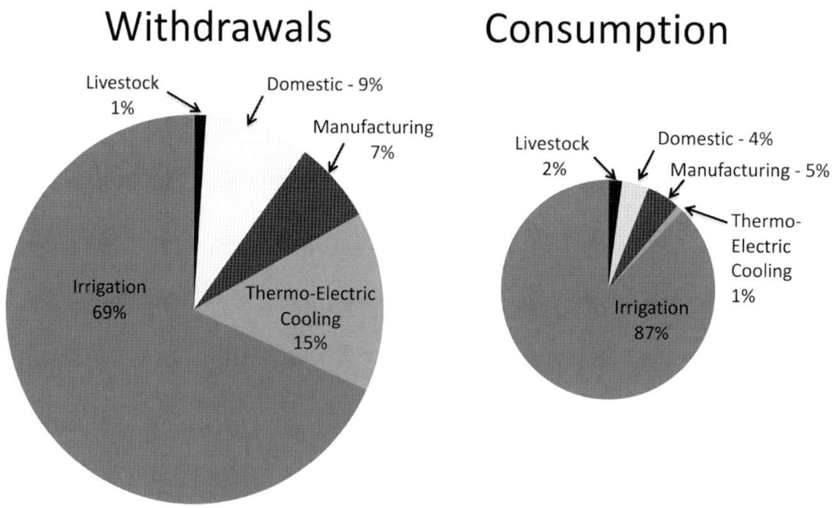

What Does It Mean to Use Water?

There are two sets of water-use numbers in the discussion above and in Figure 1. Both are important, and they have distinct management implications. Water withdrawals count for all the water removed from rivers, lakes, streams, and aquifers for use in our homes, businesses, and food production. This is usually the amount of water we think about when we figure out how much water we "need." Water withdrawals are the water we bring into our homes or onto our fields.

The thing is, most of this water isn't used up: It goes somewhere—down the toilet, off a farm field, out of the power plant and back into a river. Just a small part of the water we withdraw is used up when we use it—evaporated or moved through crops and returned to the atmosphere. We describe the water that is "used up" as consumed. Water consumption is a subset of water withdrawals. And though we can think of water consumption as the water we use up, it isn't gone—water will cycle and it will rain again—but consumed water doesn't flow downstream to be used again today.

What about the water that is withdrawn but not consumed? In our homes, nearly all the water we use goes down the drain. It moves through sewer lines or open channels, it might get cleaned up at a wastewater treatment plant, and it goes back into a river. This happens on farms, too: Water that isn't consumed may percolate down into an aquifer or run off a field. From the point of view of a homeowner or a farmer, that water is lost. But from the perspective of a whole watershed, that water can be (and probably is) used again (Perry, 2011). For individual users, managing water withdrawals is paramount. To make sure water is available downstream and to keep track of the available water in a watershed, managing consumption is critical.

Consumption may be critical at a watershed scale, but withdrawals are important, too. Moving and cleaning water is energy intensive, so reducing water withdrawals saves energy and money. Perhaps more importantly, water that is withdrawn but not consumed may not be returned to the source where it originated. Return flows could end up in a different lake, perhaps in a different watershed, and will almost certainly be released as surface water. If the original water source was groundwater, return flows may be effectively lost. Water infrastructure is generally built to get water from one place, so returning water that was pumped from groundwater into a river or lake may put it out of reach. Many water users may live in sight of a mighty river, but if their household taps are fed by groundwater wells, that water isn't accessible. So if a local aquifer is pumped dry, its users may be in in trouble, even if most of the groundwater was not consumed but transferred to a river as farm runoff and household wastewater.

Virtual Water Use

In contrast to keeping track of water withdrawals and consumption in a watershed, we can also account for the water used to make something—called virtual or embodied water (Allan, 2003). This is the water that was consumed—transferred to the atmosphere—as crops grew to maturity before the grain or fruit was harvested. It is called virtual because all that water isn't contained inside of a loaf of bread or within an orange. Instead, virtual water is a useful way of adding up and keeping track of all the water required to grow the wheat or the orange tree.

Identifying and accounting for the source of the water used to produce something is also useful. Rainfall on farm fields is a critical water source for food production—about three-quarters of global cropland is exclusively rainfed, and the average irrigated crop gets a little less than half of its water needs met by rainfall (Siebert and Döll, 2010). So in some sense, we use rainwater for food just as we use irrigation water for food. The rainwater is sometimes called "green" water and the irrigation by contrast called "blue" water.

However, rainwater for agriculture needs to be considered somewhat differently from irrigation water. Irrigation water comes from a limited source, and we must make a choice between water consumed for irrigation and water consumed in a household. For rainwater, if a particular farm field didn't exist, it's likely that rain would fall on that land anyway, and the same rainwater that is consumed by a crop would instead be consumed by the forest or meadow growing in the same spot. On average, crops actually consume less rainwater than non-agricultural vegetation, mostly because at the beginning and end of the season those crops aren't growing. Some studies have calculated that global runoff has increased because of the transition from native vegetation to crops (Scanlon et al., 2007). From that perspective, rainfed cropland is actually turning "green" water into "blue" water.

Can Technology Help?

Technology provides exciting opportunities for using water more effectively, but there are non-universal solutions. It is critical to evaluate the type of water savings desired and make sure that the technology can deliver.

For a household or business, technology can help reduce water demand. Low-flow toilets and high-efficiency machinery are examples. To do the same job with less water, these types of technology reduce the "excess" water that is returned to a water source (Batchelor et al., 2014). A low-flow toilet has less water in the tank, so less goes down the drain. However, there's usually very little impact on water consumption—about the same amount of water evaporates out of the toilet bowl. If you don't have enough water to meet your needs, or if water is expensive, these kinds of efficiency-improving technologies make it possible to operate with less water. However, reducing withdrawals and reducing runoff without changing water consumption means that there's not much impact on how much water is available downstream. So your water savings at home aren't much help to people who need water in the next town downstream (Ward and Pulido-Velazquez, 2008).

On farms, the basic math is the same, but the measurements are more complicated. Farmers, like households, tend to be interested in reducing water withdrawals. The water available to them may be limited, or they may have rights to withdraw only a certain amount of water. And it takes energy to pump water, so withdrawing less means saving money. Because reducing withdrawals is what farmers are interested in, that's what water-efficient irrigation technologies primarily focus on. Traditional forms of irrigation flood an entire field with water or run water through channels among crops. This requires moving a tremendous amount of water onto a field, and, though some of it soaks into the soil and is then consumed by growing plants, a large fraction is drained off. Pressurized irrigation, which usually means sprinklers, moves water through pipes so that a measured amount of water can sprayed on crops all across a field. Because it is piped under pressure, the whole field need not be flooded for water to reach the far corner. As a result, substantially less water is applied and there is much less runoff. Micro-irrigation, which includes drip irrigation, reduces water withdrawals even further. With this technology, water is still moved through pipes, but it is applied directly to the vicinity of plant roots at a slow rate.

Each of these irrigation technologies reduces water withdrawals by cutting down on the excess water applied—the water that would otherwise run off the field (Perry et al., 2009). But if on-farm technology that reduces withdrawals doesn't reduce consumption, it has little impact on cities, fish, or even farmers downstream who want to increase their available water (Frederiksen and Allen, 2011).

Making more water available downstream generally requires reducing water consumption upstream. The effect of different irrigation technologies on reducing water consumption is unclear, however, because on-farm water consumption research is limited. This is in part because farmers are more interested in withdrawals and in part because measuring water consumption—evapotranspiration—is complex (Allen et al., 1998).

In general, switching irrigation technologies has the potential to reduce on-farm water consumption not by altering plant transpiration—the water the plant needs to grow—but by reducing water "wasted" through evaporation from open water and bare soil (Burt et al., 2005). In flood-irrigated systems, open-water evaporation occurs while the field is flooded. Once the excess floodwater runs off, evaporation occurs from bare soil between plants. As a result, on a flooded day, evaporation with no benefit to the crop might be substantial. However, flooding saturates the soil, and the flood is repeated only when the soil dries out, perhaps every five to seven days; in between floods, once the top layer of soil dries, evaporation is limited. Pressurized spray irrigation applies water evenly and in smaller volumes. Under windy conditions, a significant fraction of airborne water droplets created by sprinklers evaporate, but when the air is still, evaporation of airborne droplets is minimal (Molle et al., 2012). Open-water evaporation is thus variable but likely small. Sprinkler irrigation is generally more frequent than flood irrigation, neither allowing the soil to dry out as much nor saturating it as completely as in a flood regime. Bare soil evaporation is thus likely higher under spray than flood irrigation. Micro or drip irrigation delivers a slow, steady stream of water directly to the vicinity of plant roots. Open water evaporation is negligible. Evaporation from bare soil between rows of plants is also negligible, as this area is not wet. For crops that are widely spaced, such as orchard trees, this could substantially reduce water consumption. However, because water is applied at a much lower rate, it is applied over a much longer period of time. As a result, the soil surface adjacent to the plant is nearly always wet and soil evaporation occurs from a small area but over a long period. So soil evaporation might be as high as with other types of irrigation. Drip irrigation that is buried below the soil surface means there is no soil evaporation, but it is relatively uncommon because it is expensive to install and maintain.

The overall impact of switching irrigation regimes on agricultural water consumption is unclear because of the lack of data, but given the processes outlined above, some researchers have estimated that switching from flood to drip irrigation would reduce water consumption by at most 10 percent and perhaps as little as 1 percent (Burt et al., 2005). Given the very large volume of water used for irrigation, a 10 percent reduction could make a big difference, but it is far less than the associated reduction in water withdrawals of 30 to 50 percent (Frederiksen and Allen, 2011).

Other approaches to reducing water consumption by irrigated crops include artificially changing the local climate or reducing transpiration. Net houses, which provide shade to crops, or glass houses, which increase humidity, reduce evaporative demand and thus reduce water consumption. Covering the soil surface between crop rows with mulch or plastic reduces evaporation from the soil. Changing crops, growing drought-tolerant or low-water-use varietals, or growing a crop under drought conditions will also reduce water consumption.

A final consideration about the potential for technology to reduce agricultural water use is related not to water consumption on field but in the process of storing water and moving it to farm fields. Evaporation from reservoirs and canals is much smaller than on-field water consumption but can still be substantial, and some efforts have been made to develop covers or coatings of reservoirs to reduce evaporation. Water can also leak out of canals into groundwater, and many canals have been lined with concrete as a water saving measure (Swamee, et al., 2002). As with farm runoff, however, water escaping into groundwater from a canal may be used downstream, and reducing this flow may have repercussions. Farmers in northern Mexico using groundwater to irrigate tomatoes (*Solanum lycopersicum*), for example, may find that the new concrete lining on the All American Canal across the border reduces their water availability, and thus American access to cheap tomatoes.

What Counts When We Add Up Water Use?

When adding up water use, deciding how long a window we keep count over affects what counts as high and low water use. With irrigation technologies, for example, flood irrigation has high withdrawal and consumption on the day it's applied, but it's put on the field much less frequently than spray or drip irrigation. So the math looks a little different when we compare a single application of water or the whole growing season.

When we look beyond the growing season, other factors come into play as well. All water contains salt, along with a whole host of other nutrients. Plants don't want the salt, so as they absorb water into their roots they leave the salt behind and it builds up in the soil around them (Rhoades and Loveday, 1990). With too much salt, the soil can become poisonous and the field has to be abandoned. More often, though, salt in the soil is diluted and washed away, but this only happens when there is enough water in the soil. Flood irrigation, and sometimes even spray irrigation, provides enough water to do this, and big rainstorms, either during or after the growing season, can do so, too. With drip irrigation, however, salt buildup doesn't get washed away. If it doesn't rain hard to clear salt from the soil after the growing season, farmers using drip irrigation have to add water to their fields to wash the salt away. The amount of water needed to control salt is called the leaching fraction, and leaching water is usually added after the crop is harvested (Ghassemi et al., 1995). If leaching water counts toward to the water needed to grow a crop with drip irrigation, the technology may require a lot more water.

Soil salinization is not a problem limited to drip irrigation systems (Ghassemi et al., 1995). If the water table below a field is brought close to the surface by excess water infiltrating from flood or spray irrigation systems, salt present in the soil or washed down from the root zone may work its way back up to the surface through capillary action to poison plants. So adding too much water can be a problem as well. If that water has to be pumped away, it might count as agricultural water use, too.

Even accounting for agricultural water use over the whole year may not really tell us how one type of irrigation technology compares to another. When farmers switch irrigation systems, they may also make changes in the extent of irrigated area, the crops that they grow, the seasons when they grow, and other farm management behaviors (Batchelor et al., 2014). The linked biophysical and behavioral changes that accompany changes in irrigation technology have not been studied broadly, but there is anecdotal and theoretical evidence from around the world that these switches often lead to increasing water use as irrigated area is expanded and farmers make changes in the crops they grow (Contor and Taylor, 2013).

EXAMPLES:

Pitfalls of Misguided Attempts at Water Savings

Managing withdrawals instead of consumption with a goal of increasing water available downstream has turned out badly more than once. Ward and Pulido-Velazquez (2008) document the plight of El Paso, which in search of more water convinced upstream farmers to switch to drip irrigation. Those farmers had no obligation to El Paso, however, and finding that with drip irrigation technology they were now able to irrigate their fields and still had a portion of their water rights unused, farmers expanded their irrigated cropland into the desert. Now withdrawing the same amount of water as at the start but irrigating more cropland and thus consuming more water, the total amount of runoff and therefore the water available downstream to El Paso has decreased.

In the Imperial Valley in inland southern California, farmers have traditionally used flood irrigation, sending their excess water into a low point in the valley called the Salton Sea. The Salton Sea, formed in 1905 from water pouring in from a break in diversion canals from the Colorado River, has since been supported by farm runoff and serves as a critical layover point on the Western Migratory Bird Flyway. In need of domestic water, the city of San Diego has purchased a portion of farmers' water rights, siphoning that water away from the Imperial Valley (Davids, Bliesner, and Eckhardt, 2008). At present, the water transfer is fulfilled by some farms that are choosing not to irrigate at all; some water is sent directly to refresh the Salton Sea. By 2017, however, water for San Diego is meant to be supplied without fallowing fields. "New" water is supposed to be made available by shifting to drip irrigation. With the decrease in runoff, however, it is not clear where water to sustain the Salton Sea will come from.

Water Productivity

How Well Are We Using Water?

One aspect of sustainable water use is to ensure that when we are using water, we're using it effectively. Because agriculture is the dominant way water is used, one way to quantify the effectiveness of water use is as crop water productivity, sometimes called "crop per drop," the calories produced for each liter of water consumed by crops. Crop water productivity is a useful metric because though crops need water to grow and water shortage can limit crop growth, crop yield doesn't increase linearly with water consumption (Hamilton et al., 2015). Crops may use water while yield is limited by factors such as nutrients or pests. Water also needs to be applied at the right time in order to effectively meet a crop's water need.

Evaluated worldwide, crop water productivity varies dramatically (Brauman, Siebert, and Foley, 2013). Some of this is due to climate: When the air is hot, dry, and windy, evaporative demand is higher, so crops require more water. However, even when controlling for climate, variability in crop water productivity persists. The range of crop water productivity varies among crops and for a given crop; it even varies within countries. For example, the water productivity of irrigated wheat is close to the world average in the United States but, as illustrated in Figure 2, water productivity of irrigated wheat is unevenly distributed, with some regions having very high water productivity and other regions having low water productivity. By comparison, water productivity of irrigated wheat in China is uniformly high and evenly distributed (Brauman et al., 2013).

FIGURE 2. Water productivity of irrigated wheat. The range of water productivity for irrigated wheat—the crop per drop—is shown for irrigated wheat producing areas of the United States. Water productivity is calculated as the total irrigation and rainwater consumed by the crop over the growing season divided by the caloric yield, following Brauman et al. (2013).

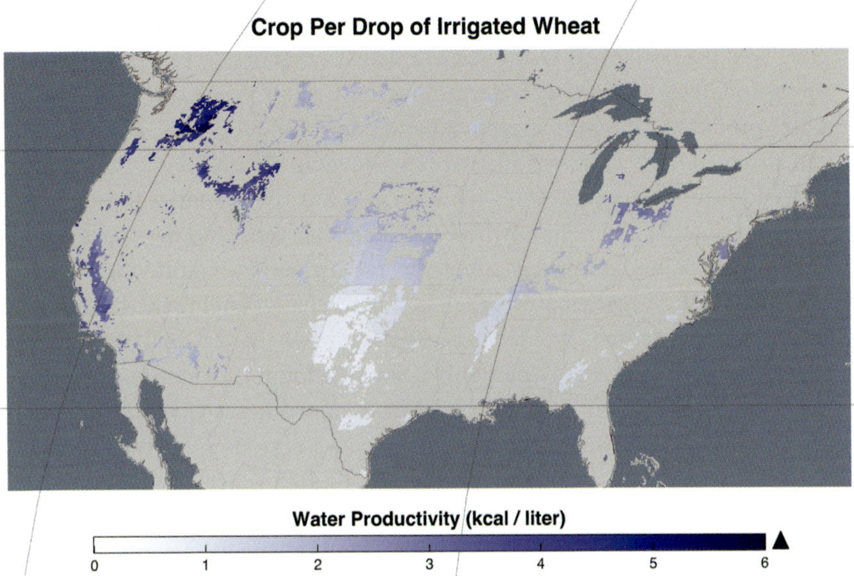

The good news about the existence of places with low crop water productivity is that it is possible to make substantial improvements to water utilization in agriculture. By raising crop water productivity, we could either produce more food with the same amount of water or consume less water while producing the same amount of food. The potential benefits are substantial. The impacts of improving very low crop water productivity—it's easier to raise low performance than to optimize performance that's already high—in arid regions could amount to food for an additional 110 million people from rainfed agriculture as well as water savings from irrigated agriculture big enough to meeting the annual domestic water demands of close to 1.4 billion people (Brauman et al., 2013).

• • •

Can We Use Water Better?

Despite the cases where attempts to save water in agriculture have gone awry (see box), there are many possibilities for using water in agriculture more effectively (Mueller et al., 2012; Foley et al., 2011). Most possibilities center around improving crop water productivity so that we can get more crop per drop. Low yields are often the result of inadequate plant nutrition, yet water consumption is frequently no lower on these fields than in more productive ones. Making both organic and mineral fertilizers more accessible in developing countries where crop water productivity is low is one way to increase water productivity. Soil management and control of crop pests and disease can also help raise water productivity (Brauman et al., 2013).

In addition, changing water management has the potential to increase water productivity. In regions where unpredictable rains and dry spells limit yields or kill plants completely, supplementary irrigation can provide protection against the vagaries of weather as well as changing climate. Irrigation schemes are sometimes avoided because they are costly relative to the return on investment. However, when only a small amount of water needs to be delivered, such as in an irrigation scheme designed to provide only a small fraction of a plant's water demand, inexpensive options such as rainwater harvesting are possible (Jägermeyr et al., 2016). Rainwater harvesting techniques, such as small dams built into dry channels on fields, can trap and store enough water to keep crops healthy. Timing irrigation application to meet crop needs effectively can also reduce water consumption while increasing yield.

Water Shortage

Are We Using Too Much Water?

Whether water use is a problem depends on availability. Thus, to evaluate water stress, we need to know about water use and water availability together. Water depletion, the ratio of water consumption to water availability assessed over a whole year, in dry months, and during dry years is a metric that does this (Brauman et al., 2016). The water depletion status of watersheds worldwide is illustrated in Figure 3.

FIGURE 3. Water depletion, a metric of water shortage. The fraction of renewable freshwater consumed by people and agriculture, called water depletion, for more than 15,000 watersheds worldwide. About 2 percent of watersheds are more than 75 percent depleted on an average annual basis, and an additional 30 percent of watersheds are depleted by more than 75 percent seasonally or in dry years. Analysis and map from Brauman et al. (2016).

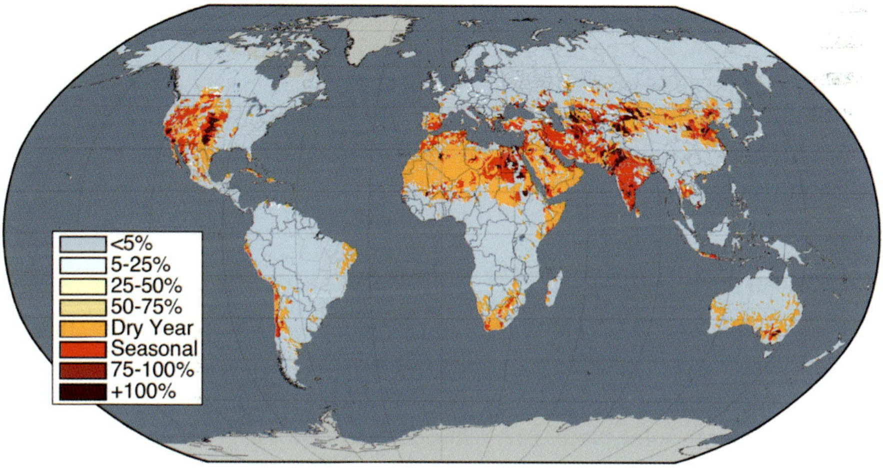

Water depletion is a biophysical indicator, so it's incomplete. Culture, governance, and existing infrastructure affect demand for and access to water. So the perception of water security is a function not just of biophysical availability but also of the context in which water scarcity occurs. Any evaluation of water security, stress, or crisis must account for the social structures that mediate whether water limitation is problematic (Srinivasan et al., 2012). That said, biophysical indicators are useful because they provide a baseline and allow us to compare different regions around the world. Water depletion tells us how much of the renewable water in a watershed, including both surface and groundwater, we're using up.

One major benefit of assessing water shortage based on the ratio of water use to water availability, as water depletion does, is that it doesn't conflate water shortage with water scarcity. A desert isn't water scarce if no one is using the water, it's just dry. Similarly, a humid place where people consume much of the available water might experience water scarcity.

Irrigation is the primary way people use water, and the majority of the world's irrigated agriculture occurs in places that are depleted (Brauman et al., 2016). While there are many *watersheds* where a tremendous amount of water is consumed for irrigation that are not depleted at all, water shortage poses a real risk to our food supply. Globally, just 2 percent of watersheds use more than 75 percent of their available water (Brauman et al., 2016). Nearly 15 percent of irrigated cereal grains are produced in these annually depleted watersheds, however. Perhaps more importantly, watersheds that are depleted seasonally and in dry years are the source of an additional 55 percent of irrigated grains. Irrigated crops grown in seasonally depleted watersheds may produce high yields, but conflict among water users is likely to occur in the dry season, making supply less reliable. Places that are depleted in dry years may be good producers most of the time, but they risk cutoffs in irrigation supply during drought. In total, 70 percent of irrigated cereals are produced in watersheds that are already facing some type of water shortage (Brauman et al., 2016).

Finding Solutions to Water Shortage

For the most part, it is not possible to use up our water. Capricious weather and changing climate affect how much water is available each year, and the way we use water on the landscape can even affect the climate (Mueller et al., 2016). Water in some places isn't renewed, like the fossil groundwater in the High Plains aquifer in the central United States. However, most of our water resources are renewable (Brauman et al., 2016). That means that if we stopped pulling water out, it would still be in streams and aquifers. If we decided that next year we wanted rivers to run full, we could stop taking out water to irrigate our crops and clean our homes. But we want food and clean homes, so we have to allocate and share the water we have at any given time. However, in most regions of the world we get to try again next year if what happens this year doesn't work out.

This is not to say that changing the way we manage water resources will be easy. Allocating resources is hard enough that, in the past, people have invested in enormous engineering schemes in order to make more water available instead of attempting to share or reallocate existing resources (Reisner, 1993). But opportunities for big water schemes are growing ever more limited, largely because there is probably already someone using that water. Cities in the United States' arid southwest, for example, sometimes make grandiose statements about bringing in water from the Great Lakes. But the Great Lakes States in the U.S. and Canada have an agreement not to export any water out of the watershed unless they all sign off (Olson, 2006). And perhaps more to the point, moving water is expensive. Agriculture is far and away the biggest water use in most regions of the world. For agriculture to be profitable and the food we eat to be affordable, water cannot be very expensive. Most of the large water infrastructure projects of the twentieth century, such as those built in the U.S. by the Bureau of Reclamation, were heavily subsidized and have still not been fully repaid (Reisner, 1993).

Water management of the future probably looks like some combination of reallocating water to uses we, as a society, deem critical, and using water in less critical applications in more flexible ways. Perhaps, for example, a farmer growing an annual crop could choose to take a payment instead of irrigating during a dry year. Or a power plant could be sited upstream from irrigation intakes so that the facility could withdraw, and return, as much water as it needs before farms take water out of the river and return far less.

Another approach is to use the available water we have more effectively. Doing so will require other inputs—energy, nutrients, labor—but in places where water is scarce, these investments likely make sense. In some places this is already happening. Corn (*Zea mays*) water productivity over the U.S. Great Plains aquifer is very high (Brauman et al., 2013). There, steadily dropping aquifer water levels in wells and the related increase in pumping expenses provide a price feedback that is likely spurring investment. In the many parts of India where water tables are dropping, water productivity is more varied. This may be in part because electricity is heavily subsidized, so there is no cost feedback encouraging farmers to invest in the inputs that will raise their crop per drop. The lesson, perhaps, is that costs of improving water productivity may not make it a good strategy in regions where water is abundant, but in regions where water is scarce the investment is worthwhile.

In our current system of trade, we don't necessarily move food products from water-rich regions to water-scarce regions (MacDonald et al., 2015). If water were a driving factor in crop value, however, we could certainly do this.

A Brief Aside about Water Quality

Water quality has largely been absent from this discussion, although of course the quality of water is integral to how it can be used and reused. This chapter does not explore water quality in depth, but because agriculture has such a major impact on water quality, we will at least touch on it here. Of note, first, is that water quality is simply a description of the constituents in water. Whether water quality is good or bad is entirely a function of what people want to use the water for (Keeler et al., 2012; Brauman et al., 2007).

For example, the fertilizers critical to attaining high agricultural yield also wash off farm fields and into rivers, lakes, and aquifers. The nitrogen and phosphorus in fertilizer are not inherently pollutants; they were put on farm fields to help plants grow, and they do the same with algae in surface water bodies. However, unlike acres of dense and thriving corn on land, an abundance of algae in a lake is generally considered at best a nuisance and at worst poisonous. As the algae die and begin to decompose, conditions are sometimes right for the process to remove all of the oxygen from the water, killing everything living there, which we generally perceive as negative (Bernhardt, 2013). Excess nutrients are also bad for human health, so we regulate the amount of nitrate in our drinking water (Keeler and Polasky, 2014).

Because water flows and cycles, and because evaporation is an effective way to separate water from many of the pollutants we mix into it, in many cases water quality can be improved simply by ceasing to add pollutants to water. Were all the farms in the Mississippi River watershed to stop applying fertilizer, presumably the hypoxic zone at its mouth in the Gulf of Mexico would rapidly disappear. The cost of this in diminished crop yield would be very high, however. At the moment, there is little incentive for farmers to spend money to reduce fertilizer running off farms into water bodies. As voluntary and regulatory incentives increase, however, so will management techniques ranging from keeping nutrients on farms to figuring out how to reuse nutrient-laden water on fields.

In other cases, changing water quality is harder and slower. Because groundwater moves slowly, impaired water quality in aquifers is particularly long-lived. Unwanted constituents in groundwater are hard to remove and are likely to remain a problem for a long time. Nitrogen fertilizer that has gotten into groundwater, for example, is harmful to health at high levels (Keeler and Polasky, 2014). Frequently, the only solution is to very carefully filter the water, or even to replace it with bottled water from some other source. The volume of water needed for drinking is small, so this type of replacement is feasible, but groundwater contamination that was poisonous to crops, for example, would effectively stop irrigation because the volume of replacement water would be so large.

For both surface and groundwater, the costs of changing water quality to meet the standards for current and future use are often high. Finding ways to reduce inputs so that cleaning is unnecessary, while also finding ways to reuse the constituents in water so they are valuable additions and not pollutants, will be critical to effective water reuse.

Conclusions

Whether there is a lot or a little water on earth or in your neighborhood is a question answerable only in the context of how much of it is or might be used. And when we use that water, whether we can use it again is critical, and thus we must better address where we move water and what we put in water. Though climate change will affect water availability and a growing and changing population will affect water demand, much of the water we use cycles annually, and so we get to try anew at managing our water resources each year. Reallocation and sharing requires innovative and flexible institutions, even more so than resilient infrastructure, but it is possible.

One innovative way to make the water we use go further is by focusing on maximizing the goods and services we create with water instead of on the water itself. We will need irrigation water to produce the food we want, but when we increase the amount of crop we get per drop we can use less water without decreasing food. Supporting farmers and food production through means other than irrigation creates the flexibility to ensure water is available when we need water itself—for drinking, for swimming, for fish.

References

Allan, J. A. (2003). Virtual water—the water, food, and trade nexus. useful concept or misleading metaphor? *Water International, 28*(1), 106–113. doi: 10.1080/02508060.2003.9724812.

Allen, R. G., Pereira, L. S., Raes, D., and Smith, M. (1998). Crop evapotranspiration: guidelines for computing crop water requirements. In *FAO irrigation and drainage paper*. Rome: Food and Agriculture Organization of the United Nations.

Batchelor, C., Reddy, V. R., Linstead, C., Dhar, M., Roy, S., and May, R. (2014). Do water-saving technologies improve environmental flows? *Journal of Hydrology, 518*(Part A), 140–149. doi: http://dx.doi.org/10.1016/j.jhydrol.2013.11.063.

Bernhardt, E. S. (2013). Cleaner lakes are dirtier lakes. *Science, 342*(6155), 205–206. doi: 10.1126/science.1245279.

Brauman, K. A., Richter, B. D., Postel, S., Malsy, M., and Flörke, M. (2016). Water depletion: An improved metric for incorporating seasonal and dry-year water scarcity into water risk assessments. *Elementa: Science of the Anthropocene, 4*, p.000083. doi: http://doi.org/10.12952/journal.elementa.000083

Brauman, K. A., Siebert, S., and Foley, J. A. (2013). STOImprovements in crop water productivity increase water sustainability and food security—a global analysis. *Environmental Research Letters, 8*(2), 02403024-30.

Brauman, K. A., Daily, G. C., Duarte, T. K., and Mooney, H. A. (2007). The nature and value of ecosystem services: an overview highlighting hydrologic services. *Annual Review of Environment and Resources, 32*, 67–98.

Burt, C., Mutziger, A., Allen, R., and Howell, T. (2005). Evaporation research: review and interpretation. *Journal of Irrigation and Drainage Engineering, 131*(1), 37–58. doi: 10.1061/(ASCE)0733-9437(2005)131:1(37).

Contor, B. A., and Taylor, R. G. (2013). Why improving irrigation efficiency increases total volume of consumptive use. *Irrigation and Drainage, 62*(3) , 273–280. doi: 10.1002/ird.1717.

Davids, G., Bliesner, R., and Eckhardt, J. R. (2008). Overview of the imperial irrigation district efficiency conservation definite plan. In *Urbanization of irrigated land and water transfers, A USCID Water Management Conference, Scottsdale, Arizona, May 28-31, 2008, pp. 273–280.*

Donat, M. G., Lowry, A. L., Alexander, L. V., Ogorman, P. A., and Maher, N. (2016). More extreme precipitation in the world's dry and wet regions. *Nature Clim. Change*, *6*(5), 508–513. doi: 10.1038/nclimate2941.

Flörke, M., Kynast, E., Barlund, I., Eisner, S., Wimmer, F., and Alcamo, J. (2013). Domestic and industrial water uses of the past 60 years as a mirror of socio-economic development: A global simulation study. *Global Environmental Change*, *23*(1), 144–156. doi: http://dx.doi.org/10.1016/j.gloenvcha.2012.10.018.

Foley, J. A., Ramankutty, N., Brauman, K. A., Cassidy, E. S., Gerber, J. S., Johnston, M., Mueller, N. D., O'Connell, C., Ray, D. K., West, P. C., Balzer, C., Bennett, E. M., Carpenter, S. R., Hill, J., Monfreda, C., Polasky, S., Rockstrom, J., Sheehan, J., Siebert, S., Tilman, D., and Zaks, D. P. M. (2011). Solutions for a cultivated planet. *Nature*, *478*(7369), 337–342. Retrieved from http://www.nature.com/nature/journal/v478/n7369/abs/nature10452.html- supplementary-information.

Frederiksen, H. D., and Allen, R. G. (2011). A common basis for analysis, evaluation and comparison of offstream water uses. *Water International*, *36*(3), 266–282. doi: 10.1080/02508060.2011.580449.

Ghassemi, Fereidoun, Anthony John Jakeman, and Henry Allan Nix. 1995. *Salinisation of land and water resources: human causes, extent, management and case studies*: CAB international.

Gleick, Peter H. 1996. Basic Water Requirements for Human Activities: Meeting Basic Needs. *Water International*, *21*(2), 83–92. doi: 10.1080/02508069608686494.

Hamilton, S. K., Hussain, M. Z., Bhardwaj, A. K., Basso, B., and Robertson, G. P. (2012) Comparative water use by maize, perennial crops, restored prairie, and poplar trees in the US Midwest. United Kingdom. doi:10.1088/1748-9326/10/6/064015.

IEA. Technology roadmap: hydropower. In *Energy Technology Perspectives*. Paris: International Energy Agency.

Jägermeyr, J., Gerten, D., Schaphoff, S., Heinke, J., Lucht, W., and Rockström, J. (2016). Integrated crop water management might sustainably halve the global food gap. *Environmental Research Letters*, *11*(2), 025002.

Keeler, B. L., and Polasky, S. (2014). Land-use change and costs to rural households: a case study in groundwater nitrate contamination. *Environmental Research Letters*, *9*(7), 074002.

Keeler, B. L., Polasky, S., Brauman, K. A., Johnson, K. A., Finlay, J. C., O'Neill, A., Kovacs, K., and Dalzell, B. (2012). Linking water quality and well-being for improved assessment and valuation of ecosystem services. *Proceedings of the National Academy of Sciences*, *109*(45), 18619–18624. doi: 10.1073/pnas.1215991109.

MacDonald, G. K., Brauman, K. A., Sun, S., Carlson, K. M., Cassidy, E. S., Gerber, J. S., and West, P. C. (2015). Rethinking agricultural trade relationships in an era of globalization. *BioScience*, *65*(3):275–289. doi: 10.1093/biosci/biu225.

Maupin, M. A., Kenny, J. F., Hutson, S. S., Lovelace, J. K., Barber, N. L., and Linsey, K. S. (2014). Estimated use of water in the United States in 2010. Reston, VA: U.S. Geological Survey.

Molle, B., Tomas, S., Hendawi, M., and Granier, J. (2012). Evaporation and wind drift losses during sprinkler irrigation influenced by droplet size distribution. *Irrigation and Drainage*, *61*(2), 240–250. doi: 10.1002/ird.648.

Morbidelli, A., Chambers, J., Lunine, J. I., Petit, J. M., Robert, F., Valsecchi, G. B., and Cyr, K. E. (2000). Source regions and timescales for the delivery of water to the Earth. *Meteoritics & Planetary Science*, *35*(6), 1309–1320. doi: 10.1111/j.1945-5100.2000.tb01518.x.

Mueller, N. D., Butler, E. E., McKinnon, K. A., Rhines, A., Tingley, M., Holbrook, N. M., and Huybers, P. (2016). Cooling of US Midwest summer temperature extremes from cropland intensification. *Nature Clim. Change*, *6*(3), 317–322. doi: 10.1038/nclimate2825.

Mueller, N. D., Gerber, J. S., Johnston, M., Ray, D. K., Ramankutty, N., and Foley, J. A. (2012). Closing yield gaps through nutrient and water management. *Nature*, *490*(7419), 254–257. Retrieved from http://www.nature.com/nature/journal/v490/n7419/abs/nature11420.html - supplementary-information.

Olson, J. M. (2006). Navigating the Great Lakes Compact: water, public trust, and international trade agreements. *Mich. St. L. Rev.*, 1103.

Perry, C. (2011). Accounting for water use: terminology and implications for saving water and increasing production. *Agricultural Water Management*, *98*(12), 1840–1846. doi: 10.1016/j.agwat.2010.10.002.

Perry, C., Steduto, P., Allen, R. G., and Burt, C. M. (2009). Increasing productivity in irrigated agriculture: agronomic constraints and hydrological realities. *Agricultural Water Management*, *96*(11), 1517–1524. doi: 10.1016/j.agwat.2009.05.005.

Reisner, M. (1993). *Cadillac desert: The American West and its disappearing water*. New York: Penguin.

Rhoades, J. D., and Loveday, J. (1990). Salinity in irrigated agriculture. In Stewart, B., and Nielsen, D. (eds.) *Irrigation of agricultural crops*, pp. 1089–1142. Madison, WI: American Society of Agronomy.

Salzman, J. (2012). *Drinking water: a history*. New York: Overlook Duckworth.

Scanlon, B.R., Jolly, I., Sophocleous, M., and Zhang, L. (2007). Global impacts of conversions from natural to agricultural ecosystems on water resources: Quantity versus quality. *Water Resources Research*, *43*(3).

Siebert, S., and Döll, P. (2010). Quantifying blue and green virtual water contents in global crop production as well as potential production losses without irrigation. *Journal Of Hydrology*, *384*(3–4), 198–217. doi: 10.1016/j.jhydrol.2009.07.031.

Srinivasan, V., Lambin, E. F., Gorelick, S. M., Thompson, B. H., and Rozelle, S. (2012). The nature and causes of the global water crisis: syndromes from a meta-analysis of coupled human-water studies. *Water Resources Research*, *48*(10), W10516. doi: 10.1029/2011WR011087.

Swamee, P. K., Mishra, G. C., and Chahar, B. R. (2002). Design of minimum water-loss canal sections. *Journal of Hydraulic Research*, *40*(2), 215–220. doi: 10.1080/00221680209499864.

Ward, F. A., and Pulido-Velazquez, M. (2008). Water conservation in irrigation can increase water use. *Proceedings of the National Academy of Sciences*, *105*(47), 18215–18220. doi: 10.1073/pnas.0805554105.

LIVESTOCK IMPACT ON BIODIVERSITY

Amy E. Young
University of California, Davis

Luke T. Macaulay
University of California, Berkeley

Stephanie R. Larson
University of California Cooperative Extension

Alison L. Van Eenennaam
University of California, Davis

Introduction

Biodiversity is a term that encompasses the widespread variability in form and function found in Earth's organisms, including differences in taxonomy, function, phylogeny, and genetics. The concept of biodiversity captures a wide array of complex, highly variable, independent components and is important to human well-being through the various ecosystem services it supports (MEA, 2005) (Fig. 1). As such, the role of biodiversity has been internationally accepted, not only because of its intrinsic value, but also because of the key role it plays in supporting ecosystem services that benefit human societies and economics.

• • •

FIGURE 1. Millennium Ecosystem Assessment conceptual framework of interactions between biodiversity, ecosystem services, human well-being and drivers of change (Reproduced with permission from MEA, 2005).

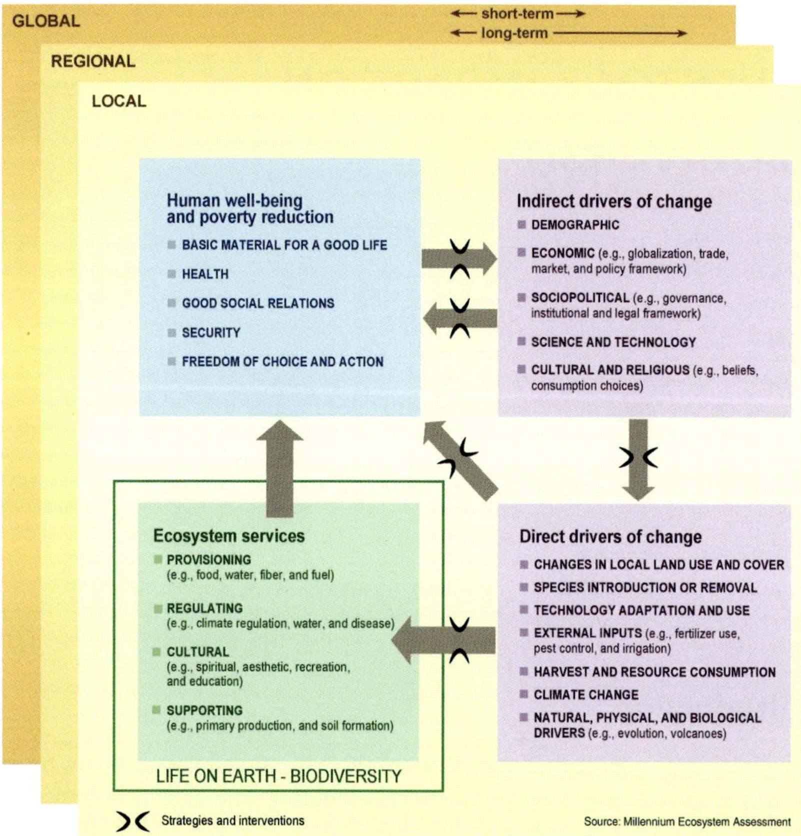

Biodiversity is at its lowest when an ecosystem is comprised of a single, genetically homogeneous population. Conversely, biodiversity is highest in genetically heterogeneous populations composed of many species (both recently evolved and ancient) that represent a broad taxonomic range and exhibit complex interactions with other populations through emigration or immigration (Naeem et al., 2016).

Livestock production occupies 30 percent of the planet's land surface: ~25 percent pastures and 5 percent feed crops (Monfreda et al., 2008; Ramankutty et al., 2008). Key ecosystem services that interact with or are supported by livestock production include biomass production (provisioning service); microorganism cycling of nutrients, soil formation, nitrogen fixation (supporting services); and pollination, pest control, climate regulation, and water purification (regulating services) (Garbach et al., 2014). As such, science-based rangeland grazing management strategies are important to the long-term sustainability of agriculture (Heitschmidt et al., 2004).

Animal products such as milk, meat, and eggs currently provide around 13 percent of the energy and 28 percent of the protein consumed globally; in developed countries, this rises to 20 percent and 48 percent, respectively (FAO, 2009). The production of meat from ruminants, hoofed animals that chew cud regurgitated from their rumen—e.g., cattle (*Bos taurus*), sheep (*Ovis aries*), and goats (*Capra aegagrus hircus*), but not horses (*Equus caballus*)—is associated with a number of polarizing topics. These include greenhouse gas emissions associated with the global ruminant population, deforestation, health effects of consuming red meat, impacts of grazing on ecosystems and biodiversity, concentrated animal feeding operations (CAFOs), religious and philosophical concerns about eating animals, and varied disparate opinions as to the best approach to alternatively provide for increased animal-source protein demand, or to reduce that demand.

Each of these topics is complex as it relates to ruminants, and not well suited to a simplistic dichotomous framing. Environmental and regional variation mean livestock production differs among geographies. For example, while most cattle are raised for beef consumption, they also produce a diverse range of products, such as dairy and leather products, which often involves a wide array of supply chains and production systems. This tremendous diversity in production systems, deliverable goods and services, environmental interactions, and options for improvement are often overlooked in the debate on beef production (Gerber et al., 2015). This makes it difficult to generalize about impacts and relative importance of cattle and other ruminants on biodiversity as they vary dramatically among different agro-ecologies.

In order to put ruminant impacts into perspective, it is important to recognize the larger context of biodiversity loss and the pressing threats, including those to the biodiversity of livestock themselves. Land use change and habitat loss associated with agriculture are arguably among the greatest threats to global biodiversity (Dorrough, et al., 2007). Invasive species also play a critical role in loss of biodiversity as non-natives can often outcompete and cause species losses in ecosystems (MEA, 2005). Ruminants play a role in these, and many other factors that affect biodiversity, but in most cases are a relatively small part of a much larger system that is undergoing significant reorganization and change due to human population growth and impacts not associated with livestock.

Global Ruminant Population and Production

Domesticated animals spread from their original habitats through thousands of years of trade and migration, which exposed them to a variety of new agro-ecological conditions. Natural selection and human-controlled breeding subsequently gave rise to a wide range of genetic diversity, which flourished in the mid-twentieth century when livestock were raised under varied conditions. Today, a number of factors, including developments in technology, the ability to transport genetic material over long distances, and the rise of large-scale commercial companies have reduced livestock diversity. Although these developments largely occurred in industrialized countries, rapidly rising demands for animal products have resulted in an increase in similar trends in the developing world (FAO, 2015).

The decline in the diversity of today's livestock and the extinction of many breeds has raised concerns about inbreeding depression, genetic defects, and the loss of locally adapted breeds, especially in developing regions. Livestock biodiversity will likely be essential in the face of emerging diseases, pressures on natural resources, and other environmental concerns resulting from a changing climate (FAO, 2015). In order to overcome identified shortcomings in countries' management of their animal genetic resources, the Global Plan of Action for Animal Genetic Resources was developed and adopted by 109 countries (Hoffmann, 2011; Hoffmann and Scherf, 2010). It includes 23 strategic priorities for action, including sustainable use, development and conservation of animal genetic resources to promote the wise management of these vital resources (FAO, 2007). While traditional breeds may provide cultural and aesthetic value, raising these rather than breed types selected for improved production performance can be associated with increases in methane emission intensities and net increased environmental burden. These trade-offs between productivity and local adaptation are an important consideration in efforts to preserve the biodiversity of local breeds as livestock keepers need a base level of production to be able to maintain their lifestyles and care for these traditional breeds.

Cattle

Cattle are found throughout the world, in almost all climatic zones, with the exception of high elevations. They have been bred for adaptations to heat, cold, humidity, extreme diet, water scarcity, mountainous terrain, dry environments, and for general hardiness. They produce meat, milk, fibers, hides, skins, fertilizer, and fuel, and are used for transportation and draft power

(FAO, 2015). As a result, they are found in more than two hundred countries around the world, with the highest densities in mixed-rainfed and mixed-irrigated systems, but also in grassland-based systems. In 2016, the world cattle inventory was reported to be approximately 1.5 billion head, the equivalent of one cow per five people, with the largest numbers found in Brazil, India, U.S.A., and China (FAO, 2018) (Table 1).

TABLE 1. Top 20 countries with the highest reported total cattle inventories in 2016 (data from FAO, 2018).

Rank	Country	2016 Stocks (Head)	% Global Inventory	2016 beef production (tonnes)	Beef production/ head (kg)	% of world production	Rank
1	Brazil	218225177	14.80	9284000	42.54	14.07	2
2	India	185987136	12.61	909027	4.89	1.38	15
3	United States	91918000	6.23	11470489	124.79	17.39	1
4	China	84523418	5.73	7011957	82.96	10.63	3
5	Ethiopia	59486667	4.03	389405	6.55	0.59	34
6	Argentina	52636778	3.57	2644000	50.23	4.01	4
7	Pakistan	42800000	2.90	922221	21.55	1.40	13
8	Mexico	33918906	2.30	1878705	55.39	2.85	6
9	Sudan	30559650	2.07	344756	11.28	0.52	37
10	Tanzania	27015712	1.83	322982	11.96	0.49	39
11	Australia	24971349	1.69	2360756	94.54	3.58	5
12	Bangladesh	23785000	1.61	193139	8.12	0.29	52
13	Colombia	22610101	1.53	828318	36.63	1.26	17
14	Nigeria	20560933	1.39	370324	18.01	0.56	36
15	Kenya	20529190	1.39	528990	25.77	0.80	24
16	France	19325515	1.31	1458284	75.46	2.21	8
17	Russian Federation	18991955	1.29	1618972	85.25	2.45	7
18	Venezuela	16574368	1.12	431716	26.05	0.65	30
19	Myanmar	16570928	1.12	402106	24.27	0.61	33
20	Indonesia	16092561	1.09	524109	32.57	0.79	25
	World Total	1474887717		65973820	44.73		

Globally, cattle are mainly raised for the production of meat or milk, but also provide draft power and manure that is used for fuel and fertilizer in many countries. Cattle are efficient at producing energy for human consumption from non-human edible feedstuffs. As such, they are important for global food security and providing protein, micronutrients, and energy to human populations (Gerber et al., 2015).

Cattle are particularly well adapted to diseases, harsh climatic conditions, and poor quality feed, which have historically enabled them to support human livelihoods in a variety of environments not suited for crop production (Gerber et al., 2015). Cattle and buffalo (*Bubalus bubalis*) breeds represent 25 percent of the world's 11,062 recorded mammalian livestock breeds (FAO, 2015). For comparison, only 3,807 avian breeds are reported, of which chickens represent more than 70 percent (FAO, 2015; The Poultry Site, 2014). Today's cattle breeds have been selected for high levels of productivity and quality. At the same time, the conservation of traditional cattle breed diversity has received increased attention in order to address the loss of livestock genetic resources (FAO, 2007).

Beef

Large numbers of animals do not necessarily correlate with high beef production on a per country basis. The U.S. is third globally in cattle inventory (6.23 percent) but is the world leader in beef production (17.39 percent), followed closely by Brazil (14.07 percent), which has the largest number of animals (14.80 percent). India, second in terms of global stocks (12.61 percent), accounts for only 1.38 percent of global beef production in part due to religious considerations associated with the consumption of beef (Table 1).

Land use per unit of beef also varies significantly by region. It has been estimated that 34 percent of total beef (35 percent cattle population) is produced on grassland-based grazing systems. In many rangeland systems, wild herbivores coevolved with vegetation, making them among the most species-rich in the world. Of these grassland-based systems, grazing in tropical and temperate zones accounts for 20 percent of global beef, with grazing systems in less temperate or arid environments accounting for the remaining 15 percent. Approximately 59 percent of beef (63 percent cattle population) is produced in mixed crop and livestock systems (defined as systems with more than 10 percent of feed coming from crops or crop byproducts), with 7 percent of beef cattle (2 percent cattle population) produced in intensive feedlot systems (Gerber et al., 2015).

Dairy

The global dairy cow population was reported at 270 million head in 2016, with India being the largest holder at almost 49 million head (FAO, 2018). In developed countries, dairy cattle production is extremely efficient compared with historical practices, having significantly reduced the number of animals, feed and water requirements, land use, and waste outputs per kilogram (kg) of milk since the mid 1940s (Capper et al., 2009). The adoption of artificial insemination (AI) increased the rate of genetic improvement, with a shift to reliance on the high milk volume-producing Holstein breed having driven much of this progress. Increased milk yield per cow is inversely proportional to the carbon footprint per kg milk, or carbon emission intensity, as illustrated in Figure 2.

FIGURE 2. Negative correlation between milk yield (green) and carbon footprint per kg milk (blue) (data from FAO, 2010).

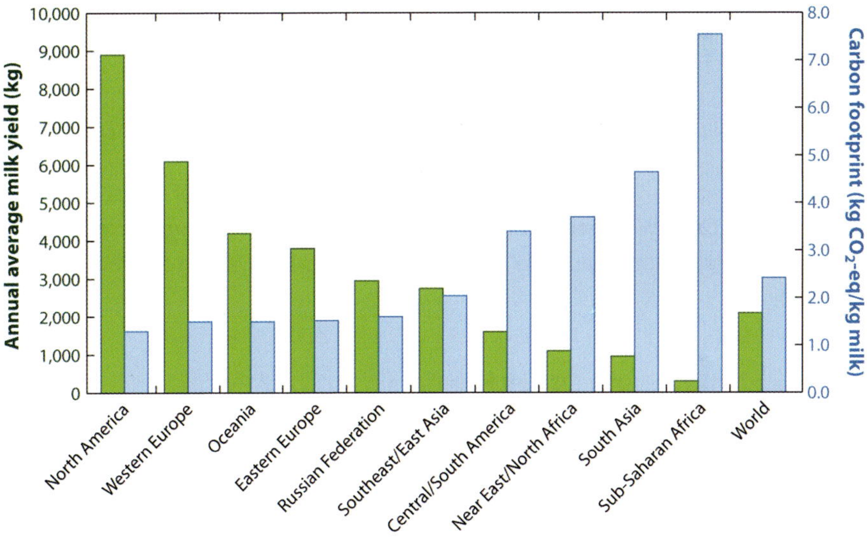

In 2016, 6.5 billion tons of cow milk, 15 million tons of goat milk, and 10 million tons of sheep milk were produced globally (FAO, 2018). For dairy cows, this represents more than twice the production reported in 1960, with only 1.5 times the number of animals, representing a 1.3 times increase in yield (Figure 3).

FIGURE 3. Dairy cattle production, number of dairy cows, and milk yield from 1961 to 2016 (data from FAO, 2018). Production in million tons and animals in million head are represented on the primary Y axis. Yield is represented by the dashed line on the secondary Y axis in thousand kg/Animal.

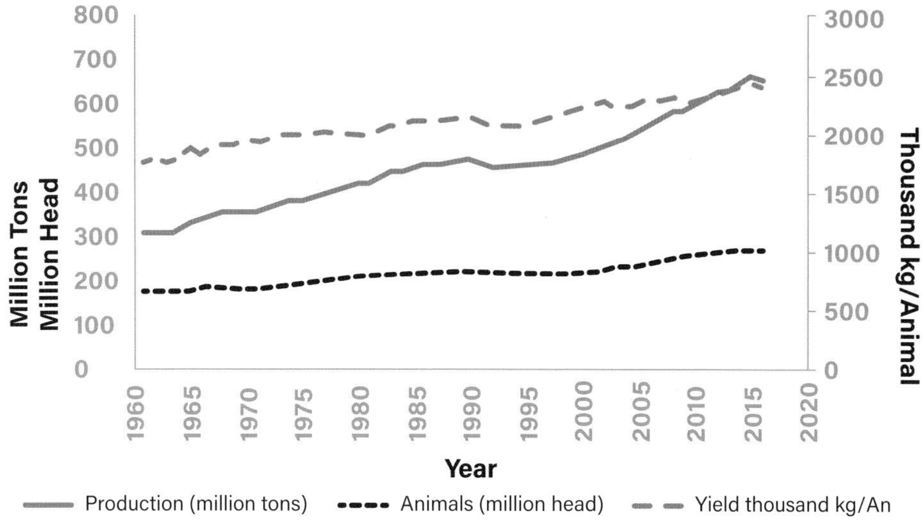

Approximately 88 percent of milk produced occurs within mixed crop and livestock systems and only 12 percent in intensive systems (Gerber et al., 2015).

Sheep, Goats, and Other Species

Globally, sheep and goats number about a billion head for each species. In 2016, world production was reported as 9.3 million tons of sheep meat and 5.6 million tons of goat meat (FAO, 2018). Many breeds of sheep and goats are well adapted to harsh conditions, including cold, hot, and arid environments, water scarcity and extreme diets. Sheep and goats are found in most geographic regions, with goats being common in the developing regions of Africa, Asia, and the Near and Middle East, but less so in developed regions. Sheep are more uniformly distributed throughout the world.

In terms of the global supply of animal protein, cattle and buffalo make up the largest contribution with 45 percent (including meat and milk), followed by chickens (*Gallus gallus domesticus*) (31 percent, including meat and eggs) and pigs (20 percent). Small ruminants—sheep and goats—produce only about 4 percent of global animal-source protein (Mottet et al., 2017); however, they are a very important source of such protein in the developing world as they are able to upcycle plants that are not edible for humans to high-quality protein (Figure 4).

FIGURE 4. Meat production by country in 2016. Countries listed include the top five producers of beef, chicken, pork, sheep, and goat meat. Scale is in thousand tons with the China pig total (pink bar) representing 54,400 thousand tons. Data from FAOSTAT (FAO, 2018).

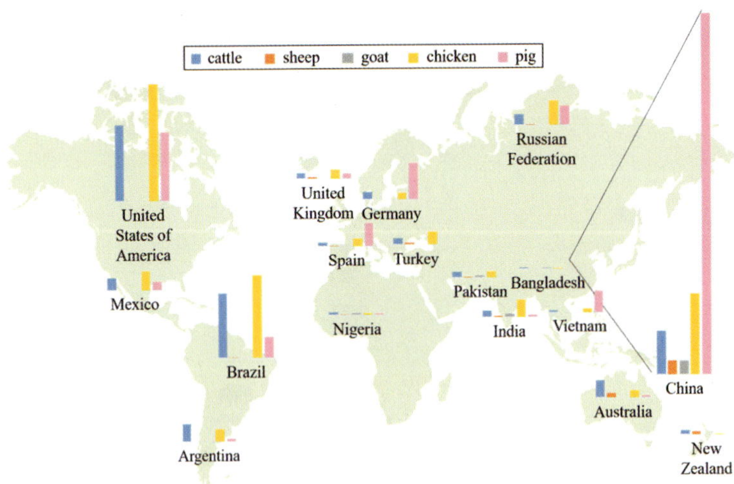

Pastoralists throughout the globe have traditionally herded large numbers of goats, camels (*Camelus*), yak (*Bos grunniens*), reindeer (*Rangifer tarandus*), llama (*Lama glama*), and alpaca (*Vicugna pacos*) to utilize land that is otherwise too steep, dry, cold, or hot for crop production. Many of these species can thrive and reproduce on exceptionally sparse vegetation and otherwise extreme environmental conditions (Figure 5). Although not currently part of global meat trades, these species are uniquely positioned to produce meat in extreme environments and may play an important role in food security in the face of climate change (Cawthorn and Hoffman, 2014).

• • •

FIGURE 5. Examples of ruminant production, by countries and main species. (Cawthorn and Hoffman, 2014; reproduced under a Creative Commons license).

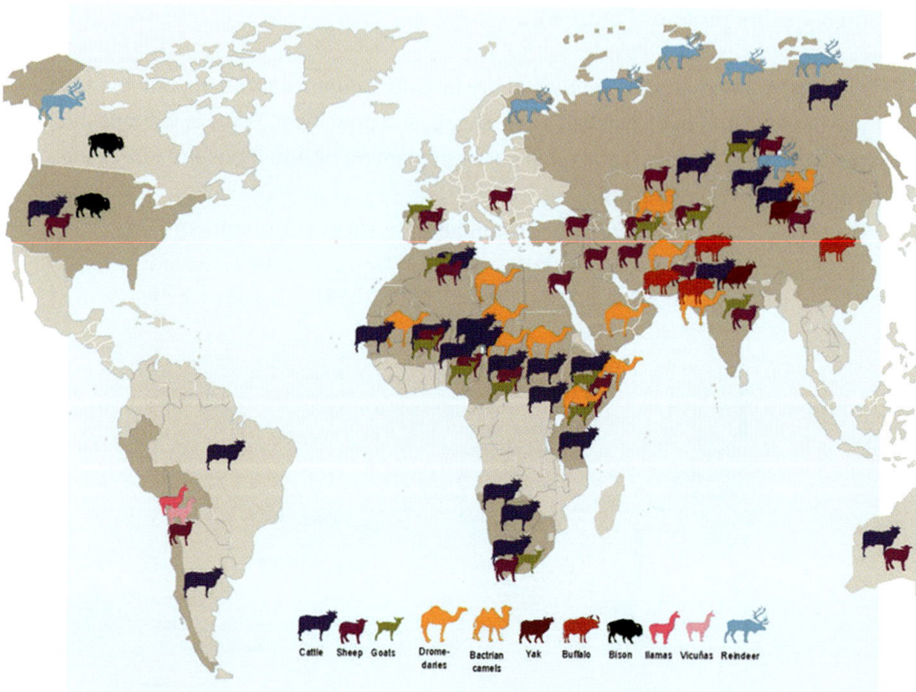

Feed and Land Use

Rangelands occur in all biomes and comprise between 18 percent and 80 percent of the world's land area, depending upon the definition used (Lund, 2007). They are generally considered to include grasslands, savannas, scrub-lands, many deserts, tundra, alpine communities, marshes, and meadows. The vegetation of rangelands is generally comprised of grasses, forbs and/or shrubs with various levels of tree canopy cover. The most widespread human activity and dominant land use in rangelands ecosystems is livestock grazing (Alkemade et al., 2013).

In grazing and mixed systems, approximately 90 percent of ruminant in-take is roughage: leaves, grass, silage, and crop residues. Livestock consume an estimated 6 billion tons of feed (dry matter or DM) annually, made up of 2.7 billion tons of grass and leaves, and 1.1 billion tons of crop residues (e.g., straws, stover, or sugar cane tops). Grains make up 13 percent of total live-stock feed (about a third of global cereal production), and oil seed cakes account for another 5 percent (with about 300 million tons DM) (Mottet et al., 2017).

At the global level, about 14 percent of the global livestock feed ration consists of materials edible to humans. On average, 2.8 kilograms of human-edible feed is required to produce 1 kilogram of boneless meat in ruminant systems; this increases to 3.2 kilograms in non-ruminants (Mottet et al., 2017). The type of production system greatly impacts the feed conversion ratios and competition for human food. Ruminants are able to digest roughage and leaves, while non-ruminants, such as pigs, can digest only simple carbohydrates, and consume very little roughage. As such, ruminants often do not compete directly for human food and are frequently raised on non-arable land where crops that produce food for humans cannot be grown (Figure 6).

FIGURE 6. Comparison of non-human edible dry matter (DM) and human edible dry matter (DM) and protein consumed per kilogram of protein produced for ruminants and non-ruminants (monogastrics) globally. Ruminants are able to digest roughage (non-human edible) whereas monogastrics consume small amounts of roughage and higher amounts of simple carbohydrates (human edible) (Mottet et al., 2017).

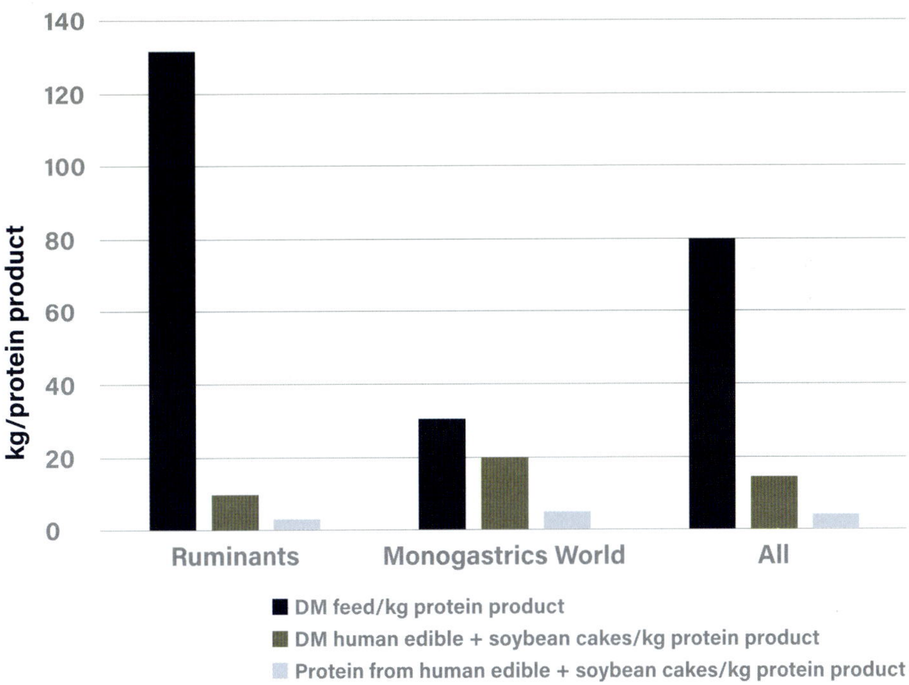

The Future of Animal Protein Production

Although animal protein production has made impressive advances in the past 50 years, it has been estimated that food production will need to grow by a further 50 percent to feed the projected global population of 9.8 billion by 2050 (Ingram et al., 2010). Predictions have suggested that the demand for animal-derived food in 2050 could be 70 percent higher than 2005 levels (Figure 7; Alexandratos and Bruinsma, 2012). The demand for beef and pork could grow by as much as 66 percent and 43 percent, respectively. The highest growth in demand (121 percent) is expected for poultry meat, especially in developing countries (United Nations, 2018), with the demand for eggs potentially increasing by 65 percent (Mottet and Tempio, 2017). As a result, livestock numbers are expected to continue to increase significantly, although at slower rates than in past years (Alexandratos and Bruinsma, 2012) (Figure 7).

FIGURE 7. Egg, beef, pork, chicken, fish, and milk production since 1970 and projected to 2050 (FAO 2018; Alexandratos and Bruinsma, 2012).

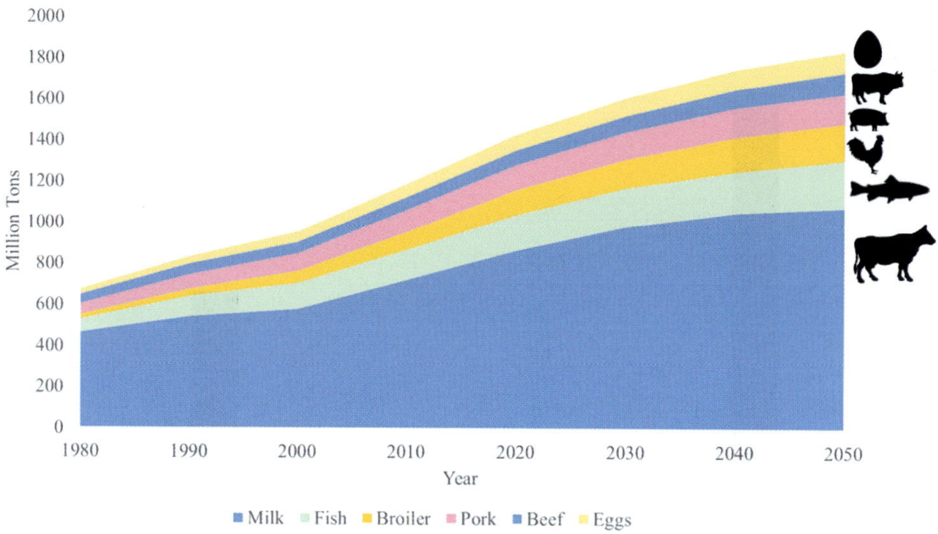

Livestock production currently represents more than 35 percent of the gross value of agricultural production globally, accounting for 50 percent of total production in developed countries and 33 percent in developing countries. Predictions suggest that developing countries' share of world meat production could account for 70 percent by 2050 (Alexandratos and Bruinsma, 2012). Growth in meat and dairy production is projected to be achieved through a combination of larger herds and higher output per animal, with global meat production increasing by almost 40 million tons (Mt) by 2026 (OECD/FAO, 2012).

The global demand for beef is expected to increase 1.2 percent per year toward 2050 (Alexandratos and Bruinsma, 2012). Bovine production exhibits great flexibility with respect to feeding regimes and extensive production. As a result, developing countries will likely continue to dominate beef production, representing 79 percent of additional bovine production, with less developed countries (United Nations, 2018) accounting for 13 percent of the additional 8 Mt of the beef produced by 2024. Developed countries are projected to see a 5 percent increase in beef production by 2026. This will be led by high growth in the U.S.A., Brazil, China and India, which combined are slated to account for 42 percent of the additional production supply. Sheep meat production is also projected to increase, registering 21 percent growth to 17.5 Mt by 2026 (OECD/FAO, 2012).

Although increases in ruminant production are likely, growth in world meat production is projected to be driven primarily by poultry production, which is expected to increase from 117 Mt to 132 Mt (13 percent) by 2026 (OECD/FAO, 2012). Poultry is the most efficient sub-sector within the livestock sector with respect to use of natural resources and protein production (Mottet and Tempio, 2017). This in turn has led to a rapidly increasing demand for poultry worldwide. At the global level, the annual growth in poultry meat production is expected to reach 1.8 percent, with growth in developing countries reaching 2.4 percent (Mottet and Tempio, 2017). Asia, China in particular, is projected to be a strong driver of growth in the global poultry market (OECD/FAO, 2012).

Pig meat production will expand by 12 percent by 2026, implying an additional supply of 13 Mt growth in this sector will be dominated by developing countries and will likely result in greater intensification. This increased production will depend on feed produced on croplands that are predicted to expand and be cultivated more intensively (Bouwman et al., 2011). The main environmental impacts of increased monogastric production will therefore be associated with cropland expansion and crop production intensification. Expansion in other, more developed regions is likely to be hindered by stricter environmental and animal welfare regulations, resulting in slower production growth.

With respect to the contribution of fish to 2050 animal protein demands, the current level of harvesting fish from natural sources is unlikely to increase significantly, since most wild fish populations are already being harvested at or beyond the level that they can support, although it may remain dominant for certain species and in some developing countries. For this reason, aquaculture is expected to continue to dominate growth in the production of fish in the future. Such production is expected to expand more than 30 Mt by 2030, 95 percent of which will come from developing countries. Aquaculture in China is likely to have a significant influence on global fish markets, accounting for 37 percent of total fish production in 2030. Steady improvements in feed and feeding efficiency in aquaculture will continue to contribute significantly to its growth (The World Bank, 2013).

Similar to the projected increases in meat production, milk production is expected to increase by an annual average of 1.8 percent, with world milk production projected to increase by 178 Mt (22 percent) over 2014–2016 levels by 2026. The bulk of this increase is projected to come from developing countries, particularly India. These increases in developing countries are likely to come about through improved feeding practices, higher milk production per animal, and herd expansion. In developed countries, however, where there are often severe land and water constraints, productivity gains for individual animals are likely to lead to smaller herds producing a given quantity of milk (OECD/FAO, 2012).

In order to feed 9.8 billion people by 2050, the livestock sector collectively will need to produce more with less while enhancing health and welfare for humans, animals, and the environment (Mottet and Tempio, 2017). The role of livestock in global food security today and in the future extends beyond animal-source proteins to include important contributions to livelihoods and economic opportunities for millions of people (Smith et al., 2017) as well as providing draft power, manure for crop production, a form of banking and savings, and many by-products (Mottet et al., 2017). It is clear that livestock will continue to play an important role in providing food, fiber, and agricultural by-products in the face of a rapidly changing world (Tolleson and Meiman, 2015).

A number of factors are expected to drive growth in global livestock production, including continued improvements in feed conversion ratios, animal genetics, health, feed additives, diversification, and overall herd management. In developing countries, increasing the proportion of dual-purpose animals, such as dairy cows or laying hens, which distribute maintenance energy over two products (Mottet et al., 2017) will result in more productivity from similar levels of inputs. Such an emphasis is also likely to increase the contribution of locally important, dual-purpose species such as guinea fowl (*Numida meleagris*), yaks, and camels (Smith et al., 2017). Studies have recognized that the livestock sector could have a significant role in "climate-smart agriculture." Some of the largest gains will likely be achieved by improved efficiency of production, especially in developing countries. Additionally, public policy needs to take into account the maintenance of food production in view of future climate outcomes and our rapidly growing population (Weindl et al., 2015).

Impacts of Ruminant Production Systems on Ecosystems and Biodiversity

The dynamics of global grazing systems are driven by a variety of complex interactions dependent upon demographic, socio-economic, agro-ecological, scientific, cultural, political, and institutional contexts. These vary by regions, villages, and households (Godde et al., 2017). The five principal direct drivers of biodiversity loss that apply to ruminants are habitat change, climate change, pollution, invasive species, and over-exploitation due to poor grazing management (MEA, 2005). Livestock are linked to these pressures, both positively and negatively, in a variety of ways (Figure 8).

• • •

FIGURE 8. Overview of the categories of influences that livestock have on biodiversity. The five main drivers of biodiversity loss, as recognized by the Millennium Ecosystem Assessment (2005) are shown in green circles. Livestock can put pressure on (black) or provide benefits to (gray) biodiversity for most of these drivers.

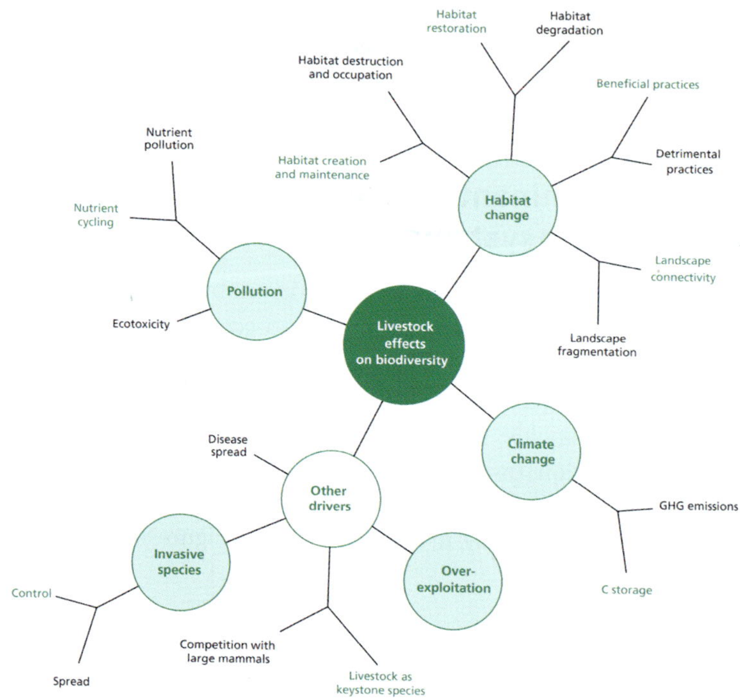

Reproduced with permission from Teillard et al., 2016.

Habitat Change

Deforestation

Biodiversity is often reduced or destroyed by changes in land use (Teillard et al., 2016). One of the most striking examples of this is the conversion of forests to pasture and croplands. The globe's forests shrank by an estimated 94,000 square kilometers per year in the 1990s and global deforestation continues at a high rate due to a variety of human actions including logging, mining, growing crops, and grazing livestock. Between 2001 and 2013, an estimated 57 percent of new grasslands replaced forests in Latin America

(Graesser et al., 2015). Deforestation has significant impacts on biodiversity, water cycling, greenhouse gas emissions, access to natural resources, and livelihoods of indigenous people (Fearnside, 2005; Agrawal and Gibson, 1999). Between 2 and 5 percent of the more than 13 million distinct rainforest species are estimated to become extinct in the course of a decade as a result of habitat loss caused by deforestation (FAO, 2006). With about a quarter of tropical forests having been cleared since the adoption of the Convention on Biological Diversity (CBD) twenty-five years ago, the problem is very acute.

Grazing can drive deforestation directly, through the need for increased areas for grazing activities, or indirectly, taking advantage of cleared areas after other activities have ceased, or clearing of land to plant crops for animal feed (Godde et al., 2017). Deforestation typically begins when roads are cut through forests, often for logging and mining activities. This in turn opens up the area for commercial and subsistence farmers to grow crops. Higher value soybean and oil palm cultivation in particular have been drivers of deforestation. Soybean seeds provide high protein feed for livestock; 80 percent of Amazon soy is destined for animal feed (FAO, 2016).

Although crops may initially thrive on cleared land, the forest soil is often not suitable for sustained crop growth, and nutrients are quickly depleted. Crops begin to fail as a result and it is at this stage that ranchers move in to graze cattle on the abandoned land. If grazing is not managed properly, it can lead to additional erosion and nutrient loss problems, further reducing the land's productive capacity. Many species of birds and invertebrates require diverse habitats, and while the initial clearing of forests provides the most substantial impact to biodiversity, intensive grazing without management can further contribute to those losses (FAO, 2006).

Land Degradation

Naturally occurring grasslands and rangelands are among the largest ecosystems in the world, covering a very large portion of the earth's ice-free land surface (Reynolds, 2005). These grasslands and their associated biodiversity in many cases are adapted to and maintained by herbivorous grazing by ruminants; however, overgrazing on these lands can drive grassland degradation (Stafford Smith et al., 2007). A 2002 study estimated that 7.7 percent of global grasslands were overgrazed, which equated to 260 million hectares.

Desertification refers to the loss of vegetative cover, which leads to erosion and soil losses, changes in water infiltration, and results in the permanent drying of a landscape. Loss of vegetation cover and the mismanagement of grazing by pastoralists is a frequently cited cause of desertification. While desertification is often cited as a cause of drought and dust storms, recent studies suggest that these outcomes can also be caused solely by climatic drivers and may not always be driven primarily by grazing (Vetter, 2005; Reid et al., 2014). Furthermore, Fairhead and Leach (1996) found that long-held desertification claims in Africa turned out to be based on incorrect assumptions, and noted that narratives and claims of desertification caused by grazing should be evaluated critically. Improving pastures by sowing, fertilizing, and irrigating in an effort to improve grazing systems can damage biodiversity, but this is also dependent on the agro-ecological context and farming practices that are involved (Godde et al., 2017).

The growing human population and increased incomes have driven the demand for livestock products. In order to meet this demand, grazing systems have expanded and intensified. However, clearly determining to what extent the expansion and intensification of grazing systems affect deforestation and land degradation is challenging due to the aforementioned complex interactions. Livestock are not the only contributors to land degradation and deforestation, and the main drivers of deforestation and land degradation vary by region (Teillard et al., 2016). Mechanisms to combat these outcomes include the enforcement of laws, supply chain interventions, expansion of indigenous reserves, and creation and protection of sustainable use areas. To date such efforts have led to a 70 percent decline in the annual deforestation rate (Nepstad et al., 2014).

Land Fragmentation

Another type of habitat change is land fragmentation, which can be driven economically as land ownership is split into smaller properties with increasing fencing and infrastructure, as well as ecologically as habitats are split into smaller pieces. This fragmentation can benefit some species and negatively impact others (Fahrig, 2003). Species that thrive in edge habitats are often beneficiaries of fragmentation, while those that require large expanses of habitat are harmed. Fragmentation can also impact hydrology, weed invasions, wildlife movement, and other ecological processes and features. Fragmentation and changes in resource allocation can affect neighboring landholders, compounding these issues. Although sometimes caused by actions intent on improving human livelihoods and well-being, the costs of fragmentation to ecosystems and human economies are often not thoroughly considered (Hobbs et al., 2008).

Other Land Use Change

Other mechanisms of biodiversity impacts associated with grazing systems include the loss of naturally occurring rangeland to more intensively used croplands or residential development. Similar to deforestation, this usually changes the prior ecosystem function completely. As crop varieties are developed that can grow in poorer soils and with less water, crops have become increasingly able to be grown on rangelands across vast areas, which results in significant losses of rangeland biodiversity.

The growth of human populations and their movement from rural to urban areas has also led to an expansion of urban and semi-developed areas onto rangelands, with accompanying impacts on the biodiversity supported by those rangelands. The interactions between biodiversity and urbanization are complex. Habitat loss and fragmentation caused by urbanization can physically and genetically isolate native species. This is problematic in both developed and developing countries where biodiversity hotspots are threatened by urban growth and few financial resources or political support is available to support land protection efforts (Elmqvist et al., 2016).

Climate Change

Climate change is projected to significantly affect a wide range of habitats and to be a major driver of biodiversity loss (Teillard et al., 2016). Livestock production will be influenced by a changing climate in a number of ways. These will include productivity changes in rangelands, pastures, and crop yields (Thornton and Gerber, 2010; Ghahramani and Moore, 2013), as well as increased heat stress which is known to negatively affect production (meat and milk yield and quality), reproductive performance, and animal health and welfare (Thornton et al., 2009; Nardone et al., 2010; Lara and Rostagno, 2013). As environments are impacted by climate change, the distribution of ruminants will likely also be affected.

Pollution

Nutrient pollution, mainly in the form of nitrogen or phosphorous, can occur at several livestock production stages, from feed crop fertilization to nutrient concentration in animal urine and manure (Teillard et al., 2016). It is of special concern for biodiversity in aquatic communities where excess nutrients in the water (eutrophication) exacerbate the growth of algae and weeds, causing oxygen shortages and the release of toxins that detrimentally affect aquatic organisms (Carpenter et al., 1998). Manure and fertilization are directly associated with volatilization of NH3 (Mosier et al., 1998) and thus contribute both to waste and to global warming. Studies have shown that the effect of nutrient pollution in soil or water is strongly influenced by farm-scale manure management.

Although most often associated with negative impacts, nutrient cycling can also be positively influenced by animal excreta. Nutrient loading, especially in grasslands, can benefit biodiversity. For example, fallow land is used by livestock for dry-season grazing and the resulting manure can provide fertilization and trigger the activity of beneficial arthropods such as dung beetles (Powell, 1986; Slade et al., 2016).

Invasive Species

Changes in land use have caused tremendous degradation to grasslands (Gornish and Ambrozio dos Santos, 2016). Many of these lands are now occupied with invasive alien species, defined by the Center for Biological Diversity as species whose introduction and/or spread threaten natural biological diversity. Invasion by alien species is increased through disturbances and degradation of natural systems; livestock production systems that cause such disturbances can contribute to the presence of invasive species. Intensive systems, which are more dependent on the trade of animal products, may have a more significant role in the dispersal of non-domesticated species, whereas extensive systems may be a source of feral organisms. It is often unclear whether invasive species are a cause or consequence of ecosystem degradation (MacDougall and Turkington, 2005). While it is known that livestock contribute to the seed dispersal of invasive plant species (Rejmanek et al., 2005), the issue is complex and there are no indicators to isolate the burden solely associated with livestock, as compared to other causes, in terms of invasive species (FAO, 2016).

In some cases, livestock can actually be used to exert positive pressure on invasive species. Establishment and spread of invasive species can be prevented through moderate grazing levels, which can also minimize soil disturbance and effects on the plant community. In cases where the invasive species is a plant that is edible by livestock, intensive grazing by multiple species can distribute the impact of grazing and control invasive species (FAO, 2016). In some cases, livestock grazing of non-native vegetation is useful in creating opportunities for many native grassland plants (Davis and Sherman, 1992; Hayes, 1998; USFWS, 2003; Marty, 2005).

Overexploitation

Overexploitation of rangelands is usually associated with the increase of livestock numbers beyond the land's ability to feed and maintain them. The clearest impacts are loss of vegetation, increases in bare soil, increased erosion, and cascading impacts to the ecosystem, which is often built upon on the primary productivity of vegetation. Increased erosion and formation of gullies can lead to changes in hydrology of an area and change the vegetation potential for an area (Belsky et al., 1999; Thurow et al., 1988).

In developed countries with profit-oriented production systems, animals will lose weight if forage supply is insufficient, and many times livestock producers will seek to sell or move animals to avoid losses as sales are often made based on animal weight. Furthermore, places with private property ownership or community-based norms to manage grazing often view their land as a long-term investment and seek to avoid long-term damage to their property by overexploitation. However, places with an absence of community norms and/or property ownership that may lead to incentives to maintain rangeland health, or in places where supplemental feeding is used to fatten livestock, livestock can denude areas of vegetation, leading to environmental impacts as described above. Furthermore, unexpected drought conditions may make stocking adjustments challenging in the short-term, leading to overstocked pastures and associated impacts.

Livestock Grazing Impact on Plant and Animal Diversity

The impact of grazing on biodiversity is complex and varies greatly with the species, location, timing, and intensity of grazing, the spatial and temporal scale under consideration, and factors such as the underlying climate, water resources, slope, ecosystem, and evolutionary history of grazing in the area. Highly differential results of the impacts of grazing on biodiversity from both an animal and a plant perspective can occur even between pastures that are only several hundred meters apart. As a result, addressing this issue on a global scale must be premised on the understanding that livestock grazing impacts vary even at small spatial scales.

Underlying Ecology of Grazed Area

An overarching component in assessing ruminant impacts on plant diversity around the world is the consideration of the underlying ecology of the area where grazing is taking place and the evolutionary pressures of grazing that in many cases have been present for millennia. In systems that are adapted to ruminant grazing, the removal of grazing can lead to significant changes to the ecosystem and plant communities over time. Evidence has shown highly site-specific increases and decreases in biodiversity with the removal of grazing in these systems.

Vegetation dynamics in different grazing systems has been described as a continuum between equilibrium and disequilibrium systems (Wiens, 1984). Equilibrium systems are those in which primary production is not limited by precipitation, but can undergo change according to biotic factors such as density dependence between herbivore and plant communities. Disequilibrium systems are described as those where primary production is driven by abiotic factors such as weather and soil. The theory posits that disequilibrium systems, which are characterized by high interannual variability, will experience forage limitations that decouple livestock numbers from forage production. In these systems, a dry year will result in such low forage production that there is little for livestock to persist on, while in wet years stocking rates cannot be increased sufficiently to utilize excess forage production. These systems tend to be characterized by pastoralists who migrate with their herds to areas of increased production in wet years, or who destock herds or use supplemental feeding in dry years (Sullivan and Rohde, 2002). Depending on the level in which a system is characterized by equilibrium or disequilibrium dynamics, grazing may have a greater or lesser impact on the plant communities and overall biodiversity.

It is important to consider the issue of temporal and spatial scale in understanding these systems. As the spatial scale increases, grazing heterogeneity also increases, leading to differential grazing impacts that may serve to enhance biodiversity over time given the ability of different plant communities to thrive under grazed and ungrazed regimes. Plant communities also change over time both in the presence and absence of grazing, and fluctuations can occur from changes at the seasonal, annual, and multi-year scales. In many disequilibrium systems such as California's Mediterranean climate, inter-annual variability in plant species based on weather factors often overrides the impact of grazing (Ratcliff et al., 2018).

Additional models evaluate grazing on gradients of aridity and evolutionary history of grazing as drivers of biodiversity change over time (Figure 9; Milchunas et al., 1988a). These models illustrate the complex interactions between grazing intensity, diversity, moisture, and location. These are useful concepts for discussing the highly differential impacts of grazing across the globe.

FIGURE 9. This conceptual model illustrates how grazing intensity may influence plant diversity in grassland communities, and how moisture and evolutionary history of grazing can influence the effects of grazing on biodiversity.

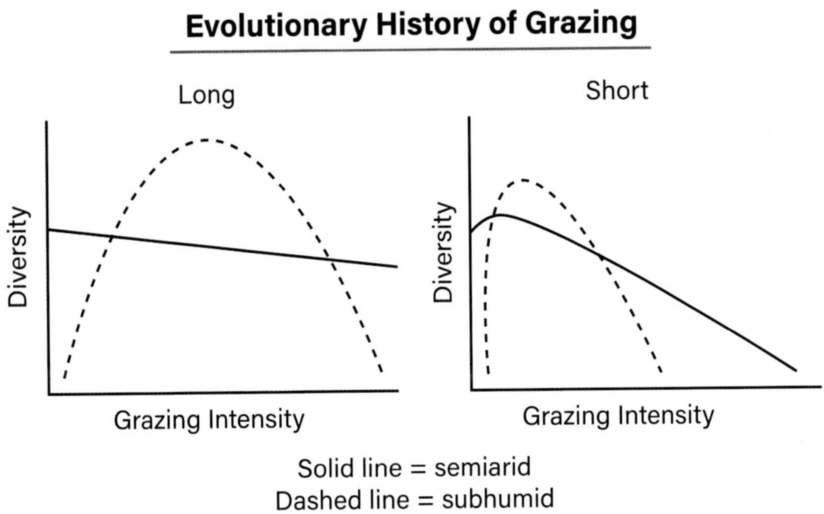

How Grazing Impacts Vegetation

Underlying these system components is the action of herbivory by ruminants, which affects vegetation and plant biodiversity in several interrelated ways, including removal of leaves, stems, and other plant parts; removal or redistribution of nutrients; and mechanical impacts on soil and plants through trampling (Vallentine, 1990). Herbivory by individual animals is mediated by four mechanisms: (1) the stocking rate, (2) the distribution of animals, (3) the type of grazing animal, and (4) the timing in which livestock are grazing an area. These four components make up the core principles of the discipline of range management, which seeks to manipulate rangeland components to obtain optimum combinations of goods and services for society on a sustained basis (Holechek et al., 2011). Biodiversity has increasingly become recognized as a management component related to the goods and services provided by rangelands.

Stocking Rate

The stocking rate of animals can have a significant impact on plant diversity as it directly affects grazing intensity. High stocking rates of grazing animals in confined areas can essentially eliminate plant cover from an area; conversely, removal of grazing has been documented to decrease plant diversity in some systems. This can occur through a variety of mechanisms such as successional change to less diverse shrublands or woodlands and loss of grassland habitat, or through the dominance of species that can crowd out other species (Watkinson and Ormerod, 2001). Residual dry matter, or the weight of remaining dry vegetation on the landscape at the end of the grazing season, is often used as a way to evaluate the grazing intensity of an area (Bartolome et al., 2002). It has been suggested that maximum local species diversity occurs when disturbance of the local ecology is neither too rare nor too frequent, a concept known as the intermediate disturbance hypothesis (Grime, 1973). With respect to grazing, this means that a light or moderate level of grazing may result in greater plant diversity than either grazing exclusion or heavy grazing, but these responses are habitat-specific (Schieltz and Rubenstein, 2016).

Animal Distribution

The distribution of animals within and between fenced pastures can also affect vegetation and lead to heterogeneity of grazing pressure across a landscape. This can serve to enhance diversity with plants that are more or less palatable or resistant to grazing succeeding in differential levels.

Type of Grazing Animal

Different livestock species have differing dietary preferences, digestive systems, mouth anatomy, hoof structure, and animal size and weight. Livestock behaviors such as herding ability, use of terrain, and willingness to travel also vary between species, breeds, and individual herds. Livestock species are described as grazers, browsers, or intermediate feeders according to the type of plants they eat (Hoffman, 1989). Grazers, including cattle and horses (although horses are not ruminants), eat mostly herbaceous plants such as grasses and forbs (herbs other than grasses), and have a digestive system that can handle large quantities of low-quality forage. Browsers, such as goats, select for the highest-quality leaves and stems from woody plants such as shrubs and trees. Intermediate feeders, such as sheep, graze selectively but can consume both herbaceous and woody vegetation. These categories are not absolute, as diet is also driven in part by the availability of different types of vegetation, nutritional needs, experiences, and inherited and learned behaviors, which can affect where and what a given animal eats.

Among ruminants, differences in dentition, digestive anatomy and body size can affect the impact of various species on grazed communities (Rook et al., 2004). For example, sheep have narrow mouths and highly curved incisors and require more energy relative to their gut capacity than larger herbivores such as cattle. As such, they preferentially select plants or plant parts, such as flowers, pods and young shoots, with high concentrations of nutrients. Cattle have a comparatively larger gut capacity in which they can digest food more thoroughly over a longer period of time, so they can consume plants with low nutrient contents. An animal's physiological state is also important; hungry animals are less selective about the food they consume (Liu et al., 2015).

Timing of Grazing

The seasonality of grazing should be taken into account, as some plant or animal species may be more or less vulnerable to grazing impacts in particular seasons. Livestock grazing can be used as a key management tool for maintaining healthy ecosystems; however, the effectiveness of using grazing to modify habitat for species of conservation concern depends on having a thorough understanding of how the grazing regime should be implemented (Davis et al., 2014).

Grazing Impacts on Plant Diversity

Grazing has been shown to have varied impacts on plant diversity. Most of the world's natural and managed grasslands are subject to grazing by a variety of herbivores. A two-year grazing experiment in the eastern Eurasian steppe showed that plant diversity was significantly increased for high diversity grassland under mixed grazing by cattle and sheep, but there was no significant effect on plant biomass, nor was there any effect of grazing by either species alone. For low diversity grassland, grazing by cattle alone and mixed grazing significantly increased plant diversity, but also significantly decreased plant biomass. No significant impact was observed on either plant diversity or biomass from sheep grazing only. This study underscores the complex and variable interactions between multiple herbivore species and plant communities (Liu et al., 2015).

Forage availability has a direct impact on the amount of time ruminants spend grazing. Dietary preferences and accessibility to preferred forage species have been shown to increase grazing time in particular areas. Diet consumed in a previous meal can also impact an animal's preferences and behaviors at subsequent meals, underlying the many factors that contribute to grazing interactions (Soder et al., 2007).

In the southeast Queensland Bioregion of Australia, grazing led to increased levels of some plants (known as increasers) while others decreased (decreasers). Plant diversity was reportedly maximized by grazing at different levels of intensity, including areas that have no grazing (Mcintyre et al., 2003). This concept has also been supported by additional research in the tallgrass prairie region in North America (Fuhlendorf and Engle, 2001). In the North American shortgrass prairie, slight increases in plant diversity were observed with light grazing. Decreasing diversity was shown as grazing pressure became heavier (Milchunas et al., 1998b).

Grazers that defoliate dominant plant species tend to increase diversity. However, it is important to recognize that not all grazing occurs in this manner, and that intensive grazing can lead to additional erosion problems, leaving very few grazing-resistant species. Grazing may lead to increases in both local extinction and colonization rates, leading to no significant differences between grazed and ungrazed areas in plant diversity (Olff and Ritchie, 1998). These effects on plant diversity vary across environmental gradients of soil quality and rainfall.

Grazing Impacts on Animal Diversity

The effects of grazing on vertebrate diversity vary greatly depending on the ecology and the animal species considered. For example, in the North American shortgrass steppe, rodent and bird diversity has been shown to increase under grazing; rabbit (*Oryctolagus cuniculus*) and hare (*Lepus*) diversity is highest under moderate grazing (Milchunas et al., 1998b). In Kenya, researchers found that competition between cattle and wild herbivores led to suppression of wild herbivores. However, they also found that practices implemented by pastoralists such as burning and construction of protective corral enclosures ("bomas") reduced woody biomass and improved grass quality, which benefitted many wild herbivores (Riginos et al., 2012). Managed livestock grazing is also thought to benefit the habitat for the California tiger salamander (*Ambystoma californiense*). Grazing keeps vegetation low, making the grasslands more sustainable for this species. Salamanders that inhabit vernal pools may also benefit from grazing. These ephemeral pools are wet only during the winter-spring rainy season, and too much vegetation in and around their edges can cause drying pools to lose depth too quickly, which reduces the vegetation. Grazing can keep the pools wet longer, giving salamander larvae more time to mature (USFWS, 2003).

Livestock production is generally thought to be detrimental to wildlife (Schieltz and Rubenstein, 2016), but as with all of the scenarios described, the situation is complex. Since many animals are dependent on the primary producers in the plant community, the impacts of grazing on plants and plant diversity is a relevant underlying base to describe how animals can be impacted by grazing.

Livestock can negatively affect wildlife species by competing directly for food or creating changes in vegetation structure, productivity, or composition that wildlife rely on for food, cover from predators, or nesting sites (Kauffman and Pyke, 2001). These alterations can further impact prey abundance in the food chain for higher trophic levels. Body size, lifestyle, seasonality, and diet are important factors in the response of wild species to grazing (Schieltz and Rubenstein, 2016).

Livestock can have positive effects on wildlife species as well. Grazing animals can be managed to produce differences in vegetation structure and composition for different animal species, birds in particular. For example, areas that are heavily grazed, resulting in higher amounts of bare ground, benefit species such as the mountain plover, *Charadrius montanus*. In other areas, reduced grazing pressure allows forage to grow taller, providing nesting habitat for grassland species such as the pintail, *Anas acuta,* or in sagebrush (*Artemisia tridentata*), the sage grouse (*Centrocercus, spp.)* Close, positive relationships have been well established between livestock and cowbirds (*Molothrus spp.*), cattle egrets (*Bubulcus ibis*) and barn swallows (*Hirundo rustica*) (Schieltz and Rubenstein, 2016). Livestock grazing is compatible with or supports grassland bird conservation in Mediterranean-type grasslands, including areas with high levels of exotic annual grass invasion, in part because grazing supports the persistence of native plants and heterogeneity in vegetation structure (Gennet et al., 2017) and can improve vegetation quality by removing old forage and stimulating new growth (Georgiadis et al., 1989). Approaches using livestock as a tool to alter vegetation structure provide a more flexible approach to complete cessation of livestock grazing on public lands, which may result in some unintended consequences that are negative for biodiversity goals (Fleischner, 1994).

In general, the response to grazing is species-specific, but positive effects are often seen for species that are adapted to more open habitats as opposed to those that need denser cover or specific vegetation, which are often negatively impacted (Schieltz and Rubenstein, 2016).

• • •

Conclusion

Given its wide prevalence, livestock production clearly affects biodiversity on a global scale. Whether the net effects are positive or negative is extremely context-dependent. The concept of biodiversity is itself complex and multifaceted as it is related to temporal and spatial scale and the life forms evaluated (plants, insects, birds, mammals, reptiles, amphibians, bacteria, etc.), so it is no surprise that assessing the factors that influence it is equally complicated. Furthermore, livestock grazing is only one of many biotic and abiotic factors that drive ecological communities. Many rangelands exhibit disequilibrium dynamics where livestock tend to be a minor influence on vegetation community. Current measurement frameworks and methodologies differ in their metrics and often evaluate different aspects of biodiversity, making a quantitative assessment of the effect of livestock on biodiversity difficult. While research is ongoing around the world to establish guidance with regard to the biodiversity and environmental consequences of different livestock systems (FAO, 2016), the literature suggests that site-specific analysis is crucial in making meaningful assessments of ruminant impact on biodiversity and recommendations for improvements. Broad generalizations about the impact of livestock on biodiversity are likely to lack nuance and be overly simplistic leading to conclusions that are inappropriate for many environments and situations.

References

Agrawal, A., and Gibson, C. C. (1999). Enchantment and disenchantment: the role of community in natural resource conservation. *World Development*, *27*(4), 629–49. doi: 10.1016/S0305-750X(98)00161-2.

Alexandratos, N., and Bruinsma, J. (2012). World agriculture towards 2030/2050: the 2012 revision. ESA Working Paper No. 12-03. Rome, FAO.

Alkemade, R., Reid, R. S., van den Berg, M., de Leeuw, J., and Jeuken, M. (2013). Assessing the impacts of livestock production on biodiversity in rangeland ecosystems. *Proceedings of the National Academy of Sciences*, *110*(52), 20900–20905. doi: 10.1073/pnas.1011013108.

Bartolome, J., Frost, W., and McDougald, N. (2002). California guidelines for residual dry matter (RDM) management on coastal and foothill annual rangelands. Division of Agriculture and Natural Resources, University of California. Rangeland Monitoring Series. Publication 8092.

Belsky, A. J., Matzke, A., and Uselman, S. (1999). Survey of livestock influences on stream and riparian ecosystems in the western United States. *Journal of Soil and Water Conservation*, *54*, 419–431.

Bouwman, L., Goldewijk, K. K., Van Der Hoek, K. W., Beusen, A. H. W., Van Vuuren, D. P., Willems, J., Rufino, M. C., and Stehfest, E. (2011). Exploring global changes in nitrogen and phosphorus cycles in agriculture induced by livestock production over the 1900–2050 period. *Proceedings of the National Academy of Sciences of the United States of America*, *110*(52), 20882–20887. doi: 10.1073/pnas.1012878108.

Capper J. L., Cady, R. A., and Bauman, D. E. (2009). The environmental impact of dairy production: 1944 compared with 2007. *Journal of Animal Science*, *87*, 2160–2167. doi: 10.2527/jas.2009-1781.

Capper, J. L., and Bauman, D. E. (2013). The role of productivity in improving the environmental sustainability of ruminant production systems. *Annual Review of Animal Biosciences, 1*, 469–489. doi: 10.1146/annurev-animal-031412-103727.

Carpenter, S., Caraco, N., Correll, R., Howarth, R., Sharpley, A., and Smith, V. (1998). Nonpoint pollution of surface waters with phosphorus and nitrogen. *Ecological Applications, 8*(3), 559–568. doi: 10.1890/1051-0761.

Cawthorn, D.-M., and Hoffman, L. C. (2014). The role of traditional and non-traditional meat animals in feeding a growing and evolving world. *Animal Frontiers, 4*(4), 6–12. doi: 10.2527/af.2014-0027.

Davis, L. H., and Sherman, R. J. (1992). Ecological study of the rare Chorizizanthe Valida (Polygonaceae) at Point Reyes National Seashore, California. *Madroño, 39*, 271–280.

Davis, S. C., Burkle, L. A., Cross, W.F., and Cutting, K. A. (2014). The effects of timing of grazing on plant and arthropod communities in high-elevation grasslands. *PLoS ONE, 9*(10), e110460. doi: 10.1371/journal.pone.0110460.

Dorrough, J., Moll, J., and Crosthwaite, J. (2007). Can intensification of temperate Australian livestock production system save land for native biodiversity? *Agriculture Ecosystems & Environment, 121*(3), 222–232. doi: 10.1016/j.agee.2006.12.012.

Elmqvist, T., Zipperer, W., and Güneralp, B. (2016). Urbanization, habitat loss, biodiversity decline: solution pathways to break the cycle. In Seta, K., Solecki, W. D., and Griffith, A. C. (eds.) *Routledge Handbook of Urbanization and Global Environmental Change*, pp. 139–151. London and New York: Routledge.

Fahrig, L.. (2003). Effects of habitat fragmentation on biodiversity. *Annual Review of Ecology, Evolution, and Systematics, 34*, 487–515. doi: 10.1146/annurev.ecolsys.34.011802.132419.

Fairhead, J., and Leach, M. (1996). *Misreading the African landscape: society and ecology in a forest-savanna mosaic.* Cambridge: Cambridge University Press.

Fearnside, P. M. (2005). Deforestation in Brazilian Amazonia: history, rates, and consequences. *Conservation Biology, 19*(3), 680–88. doi: 10.1111/j.1523-1739.2005.00697.x.

Fleischner, T. L. (1994). Ecological costs of livestock grazing in western North America. *Conservation Biology, 8*(3), 629–644. doi: 10.1046/j.1523-1739.1994.08030629.x.

Food and Agriculture Organization of the United Nations (FAO). (2006). Livestock policy brief 03: cattle ranching and deforestation. FAO Livestock Information, Sector Analysis and Policy Branch Animal Production and Health Division. FAO, Rome.

Food and Agriculture Organization of the United Nations (FAO). (2007). Global plan of action for animal genetic resources and the Interlaken Declaration. Commission on Genetic Resources for Food and Agriculture, Rome, Italy. Retrieved from http://www.fao.org/3/a-a1404e.pdf.

Food and Agriculture Organization of the United Nations (FAO) and World Bank (2009). *Awakening Africa's sleeping giant: prospects for commercial agriculture in the Guinea Savannah Zone and beyond*, Morris, Michael; Binswanger, Hans; Byerlee, Derek; Savanti, Paula; Staatz, John; Morris, Michael ; Binswanger, Hans ; Byerlee, Derek ; Savanti, Paula ; Staatz, John. 2009. *Awakening Africa's sleeping giant : prospects for commercial agriculture in the Guinea Savannah zone and beyond (English).* Directions in development ; agriculture and rural development. Washington, DC: World Bank. Retrieved from http://documents.worldbank.org/curated/en/312591468004457911/Awakening-Africas-sleeping-giant-prospects-for-commercial-agriculture-in-the-Guinea-Savannah-zone-and-beyond.

Food and Agriculture Organization of the United Nations (FAO). (2010). Greenhouse gas emissions from the dairy sector: a life cycle assessment. Retrieved from http://www.fao.org/docrep/012/k7930e/k7930e00.pdf.

Food and Agriculture Organization of the United Nations (FAO). (2015). *The second report on the state of the world's animal genetic resources for food and agriculture.* Scherf, B. D., and Pilling, D. (eds.) FAO Commission on Genetic Resources for Food and Agriculture Assessments. Retrieved from http://www.fao.org/3/a-i4784e/index.html.

Food and Agriculture Organization of the United Nations (FAO). (2016). *Principles for the assessment of livestock impacts on biodiversity.* Livestock Environmental Assessment and Performance (LEAP) Partnership. FAO, Rome, Italy.

Food and Agriculture Organization of the United Nations (FAO). (2018). *FAO statistical yearbook 2017*. Retrieved from http://www.fao.org/faostat/en/#data/QL.

Fuhlendorf, S. D., and Engle, D. M. (2001). Restoring heterogeneity on rangelands: ecosystem management based on evolutionary grazing patterns. *BioScience, 51*(8), 625–632. doi: 10.1641/0006-3568(2001)051[0625:RHOREM]2.0.CO;2.

Garbach, K., Milder, J. C., Montenegro, M., Karp, D. S., and DeClerck, F. A. J. (2014). Biodiversity and ecosystem services in agroecosystems. *Encyclopedia of Agriculture and Food Systems, 2*, 21–40. doi: 10.1016/B978-0-444-52512-3.00013-9.

Gennet, S., Spotswood, E., Hammond, M., and Bartolome, J. W. (2017). Livestock grazing supports native plants and songbirds in a California annual grassland. *PLoS ONE, 12*(6), e0176367. doi: 10.1371/journal.pone.0176367.

Georgiadis, N. J., Ruess, R. W., MaNaughton, S. J., and Western, D. (1989). Ecological conditions that determine when grazing stimulates grass production. *Oecologia, 81*(3), 316–22.

Gerber, P. J., Mottet, A., Opio, C. I., Falcucci, A., and Teillard, F. (2015). Environmental impacts of beef production: review of challenges and perspectives for durability. *Meat Science, 109*, 2–12. doi: 10.1016/j.meatsci.2015.05.013.

Ghahramani, A., and Moore, A. D. (2013). Climate change and broadacre livestock production across southern Australia. 2. Adaptation options via grassland management. *Crop and Pasture Science, 64*(6), 615–30. doi: 10.1071/CP13195.

Godde, C. M., Garnett, T., Thornton, P. K., Ash, A. J., and Herrero, M. (2017). Grazing systems expansion and intensification: drivers, dynamics, and trade-offs. *Global Food Security, 16*, 93–105. doi: 10.1016/j.gfs.2017.11.003.

Gornish, E. S., and Ambrozio dos Santos, P. (2016). Invasive species cover, soil type, and grain interact to predict long-term grassland restoration success. *Restoration Ecology, 24*(2), 222–229. doi: 10.1111/rec.12308.

Graesser, J., Aide, T. M., Grau, H. R., and Ramankutty, N. (2015). Cropland/pastureland dynamics and the slowdown of deforestation in Latin America. *Environmental Research Letters, 10*, 34017.

Grime, J. P. (1973). Competitive exclusion in herbaceous vegetation. *Nature, 242*, 344–47. doi: 10.1038/242344a0.

Hayes, G. (1988). The saga of the Santa Cruz tarplant. *Four Seasons, 10*(4), 18–21.

Heitschmidt, R. K., Vermeire, L. T., and Grings, E. E. (2004). Is rangeland agriculture sustainable? *Journal of Animal Science, 82*(suppl. 13), E138–E146. doi: 10.2527/2004.8213_supplE138x.

Hobbs, N. T., Galvin, K. A., Stokes, C. J., Lackett, J. M., Ash, A. J., Boone, R. B., Reid, R. S., Thornton, P. K. (2008). Fragmentation of rangelands: implications for humans, animals, and landscapes. *Global Environmental Change, 18*(4), 776–785. doi: 10.1016/j.gloenvcha.2008.07.011.

Hoffmann, I., and Scherf, B. (2010). Implementing the global plan of action for animal genetic resources. *Animal Genetic Resources, 47*, 1–10. doi: 10.1017/S2078633610001050.

Hoffmann, I. (2011). Livestock biodiversity and sustainability. *Livestock Science, 139*(1–2), 69–79. doi: 10.1016/j.livsci.2011.03.016.

Hoffmann, R. R. (1989). Evolutionary steps of ecophysiological adaptation and diversification of ruminants: A comparative view of their digestive systems. *Oecologia, 78*(4), 443–457.

Holechek, J. L., Pieper, R. D., and Herbel, C. H. (2011). *Range management: principles and practices*. 6th edition. London: Pearson.

Ingram, J., Ericksen, P., and Liverman, D. (2010). *Food security and global environmental change*. London: Earthscan.

Kauffman, J. B., and Pyke, D. A. (2001). Range ecology, global livestock influences. In Levin, S. A. (ed.) *Encyclopedia of Biodiversity*, volume 5, pp. 33–52. San Diego: Elsevier.

Lara, L. J., and Rostagno, M. H. (2013). Impact of heat stress on poultry production. *Animals, 3*(2), 356–69. doi: 10.3390/ani3020356.

Liu, J., Feng, C., Wang, D., Wang, L., Wilsey, B. J., Zong, Z. (2015). Impacts of grazing by different large herbivores in grassland depend on plant species diversity. *Journal of Applied Ecology, 52*(4), 1053–1062. doi: 10.1111/1365-2664.12456.

Lund, H. G. (2007). Accounting for the world's rangelands. *Rangeland, 29*(1), 3–10.

MacDougall, A. S., and Turkington, R. (2005). Are invasive species the drivers or passengers of change in degraded ecosystems? *Ecology*, *86*(1), 42–55. doi: 10.1890/04-0669.

Marty, J. T. (2005). Effects of cattle grazing on diversity in ephemeral wetlands. *Conservation Biology*, *19*(5), 1626–1632. doi: 10.1111/j.1523-1739.2005.00198.x

Mcintyre, S., Heard, K. M., and Martin, T. G. (2003). The relative importance of cattle grazing in subtropical grasslands: does it reduce or enhance plant biodiversity? *Journal of Applied Ecology*, *40*(3), 445–457. doi: 10.1046/j.1365-2664.2003.00823.x.

MEA (Millennium Ecosystem Assessment) (2005). *Ecosystems and human well-being: our human planet*. Summary for decision-makers. Washington, D.C.: Island Press.

Milchunas, D. G., Sala, O. E., and Lauenroth, W. K. (1988a). A generalized model of the effects of grazing by large herbivores on grassland community structure. *The American Naturalist*, *132*(1), 87–106.

Milchunas, D. G., Lauenroth, W. K., and Burke, I. C. (1998b). Livestock grazing: animal and plant biodiversity of shortgrass steppe and the relationship to ecosystem function. *Oikos*, *83*, 65–74.

Monfreda, C., Ramankutty, N., and Foley, J. A. (2008). Farming the planet: 2. Geographic distribution of crop areas, yields, physiological types, and net primary production in the year 2000. *Global Biogeochemical Cycles*, *22*(1), 1–19. doi: 10.1029/2007GB002947.

Mosier, A., Kroeze, C., Nevison, C., Oenema, O., Seitzinger, S., and van Cleemput, O. (1998). Closing the global N2O budget: nitrous oxide emissions through the agricultural nitrogen cycle. *Nutrient Cycling in Agroecosystems*, *52*, 225–248. doi: 10.1023/A:1009740530221.

Mottet, A., and Tempio, G. (2017). Global poultry production: current state and future outlook and challenges. *World's Poultry Science Journal*, *73*(2), 245–256. doi: 10.1017/S0043933917000071.

Mottet, A., de Haan, C., Falcucci, A., Tempio, G., Opio, C., and Gerber, P. (2017). Livestock: On our plates or eating at our table? A new analysis of the feed/food debate. *Global Food Security*, *14*, 1–8. doi: 10.1016/j.gfs.2017.01.001.

Naeem, S., Prager, C., Weeks, B., Varga, A., Flynn, D. F. B., Griffin, K., Muscarella, R., Palmer, M., Wood, S., and Schuster, W. (2016). Biodiversity as a multidimensional construct: a review, framework and case study of herbivory's impact on plant biodiversity. *Proceedings of the Royal Society B*, *283*(1844), 20153005. doi: 10.1098/rspb.2015.3005.

Nardone A., B. Ronchi, Lacetera, N., Ranieri, M. S., and Bernabucci, U. (2010). Effects of climate changes on animal production and sustainability of livestock systems. *Livestock Science*, *130*(1–3), 57–69. doi: 10.1016/j.livsci.2010.02.011.

Nepstad, D., McGrath, D., Stickler, C., Alencar, A., Azevedo, A., Swette, B., Bezerra, T., DiGiano, M., Shimada, J., Seroa da Motta, R., Armijo, E., Castello, L., Brando, P., Hansen, M. C., McGrath-Horn, M., Carvalho., O., and Hess, L. (2014). Slowing Amazon deforestation through public policy and interventions in beef and soy supply chains. *Science*, *344*, 1118–1123. doi: 10.1126/science.1248525.

OECD/FAO (2012). *OECD-FAO Agricultural outlook 2017–2026*. Paris: OECD Publishing.

Olff, H., and Ritchie, M. E. (1998). Effects of herbivores on grassland plant diversity. *Trends in Ecology & Evolution*, *13*(7), 261–265. doi: 10.1016/S0169-5347(98)01364-0.

Powell, J. M. (1986). Manure for cropping: a case study from central Nigeria. *Experimental Agriculture*, *22*(1), 15–24. doi: 10.1017/S0014479700014009.

Ramankutty, N., Evan, A. T., Monfreda, C., and Foley, J. A. (2008). Farming the planet: 1. geographic distribution of global agricultural lands in the year 2000. *Global Biogeochemical Cycles*, *22*(1), 1–19. doi: 10.1029/2007GB002952.

Ratcliff, F., Bartolome, J., Macaulay, L., Spiegal, S., and White, M. D. (2018). Applying ecological site concepts and state-and-transition models to a grazed riparian rangeland. *Ecology and Evolution*. doi: 10.1002/ece3.4057

Reid, R. S., Fernández-Giménez, M. E., and Galvin, K. A. (2014). Dynamics and resilience of rangelands and pastoral peoples around the globe. *Annual Review of Environmental Resources*, *39*, 217–242. doi: 10.1146/annurev-environ-020713-163329.

Rejmanek, M., Richardson, D. M., Higgins, S. I., Pitcairn, M. J., and Grotkopp, E. (2005). "Ecology of invasive plants: state of the art." In: Mooney, H. A. (ed.) *Invasive alien species: a new synthesis*. Washington, D.C.: Island Press.

Riginos, C., Porensky, L. M., Veblen, K. E., Odadi, W. O., Sensenig, R. L., Kimuyu, D., Keesing, F., Wilkerson, M. L., Young, T. P. (2012). Lessons on the relationship between livestock husbandry and biodiversity from the Kenya Long-term Exclosure Experiment (KLEE). *Pastoralism: Research, Policy and Practice, 2*, 10. doi: 10.1186/2041-7136-2-10.

Rook, A. J., Dumont, B., Isselstein, J., Osoro, K., WallisDeVries, M. F., Parente, G., and Mills, J. (2004). Matching type of livestock to desired biodiversity outcomes in pastures—a review. *Biological Conservation, 119*(2), 137–150. doi: 10.1016/j.biocon.2003.11.010.

Schieltz, J. M., and Rubenstein, D. I. (2016). Evidence based review: positive versus negative effects of livestock grazing on wildlife. What do we really know? *Environmental Research Letters, 11*, 113003. doi: 10.1088/1748-9326/11/11/113003.

Slade, E. M., Roslin, T., Santalahti, M., and Bell, T. (2016). Disentangling the "brown world" faecal–detritus interaction web: dung beetle effects on soil microbial properties. *Oikos, 125*(5), 629–635. doi: 10.1111/oik.02640.

Smith, J., Sones, K., Grace, D., MacMillan, S., Tarawali, S., and Herrero, M. (2017). Beyond milk, meat, and eggs: Role of livestock in food and nutrition security. *Animal Frontiers, 3*(1), 6–13. doi: 10.2527/af.2013-0002.

Soder, K. J., Rook, A. J., Sanderson, M. A., and Goslee, S. C. (2007). Interaction of plant species diversity on grazing behavior and performance of livestock grazing temperate region pastures. *Crop Science, 47*, 416–425. doi: 10.2135/cropsci2006.01.0061.

Stafford Smith, D. M., McKeon, G. M., Watson, I. W., Henry, B. K., Stone, G. S., Hall, W. B., and Howden, S. M. (2007). Learning from episodes of degradation and recovery in variable Australian rangelands. *Proceedings of the National Academy of Sciences of the United States of America, 104*(52), 20690–20695. doi: 10.1073/pnas.0704837104.

Sullivan, S. and Rohde, R. (2002). On non-equilibrium in arid and semi-arid grazing systems. *Journal of Biogeography, 29*, 1595–1618. doi: 10.1046/j.1365-2699.2002.00799.x.

Reynolds, S. G. (2005). Chapter 1: Introduction. In Suttie, J. M., Reynolds, S. G., and Batello, C. (eds.) *Grasslands of the world.* Rome: Food and Agriculture Organization of the United Nations. Retrieved from http://www.fao.org/docrep/008/y8344e/y8344e05.htm#TopOfPage.

Teillard, F., Anton, A., Duman, B., Finn, J., Henry, B., Souza, D. M., Manzano, P., Milá I Canals, L., Phelps, C., Said, M., Vijn, S., and White, S. (2016). A review of indicators and methods to assess biodiversity—application to livestock production at global scale. Rome, Italy: Livestock, Environmental Assessment and Performance Partnership (LEAP), FAO. Retrieved from http://www.fao.org/3/a-av151e.pdf

The Poultry Site. (2014). Breeds of poultry section. Retrieved from http://www.thepoultrysite.com/info/poultrybreeds.php.

The World Bank. (2013). Fish to 2030: Prospects for Fisheries and Aquaculture. World Bank Report Number 83177-GLB. Agriculture and Environmental Services Discussion Paper 03, Washington, D.C.

Thornton, P. K., Van de Steeg, J., Notenbaert, A., and Herrero, M. (2009). The impacts of climate change on livestock and livestock systems in developing countries: A review of what we know and what we need to know. *Agricultural Systems, 101*(3), 113–127. doi: 10.1016/j.agsy.2009.05.002.

Thornton, P. K., and Gerber, P. J. (2010). Climate change and the growth of the livestock sector in developing countries *Mitigation and Adaptation Strategies for Global Change, 15*(2), 169–84.

Thurow, T. L., Blackburn, W. H., and Taylor, C. A. (1988). Infiltration and interrill erosion responses to selected livestock grazing strategies, Edwards Plateau, Texas. *Journal of Range Management, 41*(4), 296. doi: 10.2307/3899382

Tolleson, D., and Meiman, P. (2015). Global effects of changing land-use on animal agriculture. *Animal Frontiers, 5*(4), 14–23. doi: 10.2527/af.2015-0042.

United Nations (2018). World economic situation and prospects 2018. Retrieved from https://www.un.org/development/desa/dpad/wp-content/uploads/sites/45/publication/WESP2018_Full_Web-1.pdf.

United State Fish and Wildlife Service (USFWS) (2003). Endangered and threatened wildlife and plants: Listing of the central California distinct population segment of the California tiger salamander; proposed rule. Federal Register 68:28648.

Vallentine, J. F. (1990). *Grazing management.* San Diego: Academic Press.
Vetter, S. (2005). Rangelands at equilibrium and non-equilibrium: recent developments in the debate. *Journal of Arid Environments, 62*(2), 321–341. doi: 10.1016/j.jaridenv.2004.11.015.
Wassenaar, T., Gerber, P., Verburg, P., Rosales, M., Ibrahim, M., Steinfeld, H. (2007). Projecting land use changes in the Neotropics: the geography of pasture expansion into forest. *Global Environmental Change, 17*(1), 86–104. doi: 10.1016/j.gloenvcha.2006.03.007.
Watkinson, A. R., and Ormerod, S. J. (2001). Grasslands, grazing and biodiversity: editors' introduction. *Journal of Applied Ecology, 38*(2), 233–37. doi: 10.1046/j.1365-2664.2001.00621.x.
Weindl, I., Lotze-Campen, H., Popp, A., Müller, C., Havlík, P., Herrero, M., Schmitz, C., and Rolinski, S. (2015). Livestock in a changing climate: production systems transitions as an adaptation strategy for agriculture. *Environmental Research Letters, 10,* 094021. doi: 10.1088/1748-9326-10-9-094021.
Wiens, J. A. (1984). On understanding a nonequilibrium world: myth and reality in community patterns and processes. In Strong, D. R., Simberloff, D., Abele, L. G., and Thistle, A. B. (eds.) *Ecological communities: conceptual issues and the evidence.* Princeton: Princeton University Press.

AGRICULTURE AND POLLUTION

Sharing the Crowded Two-Way Street

David I. Gustafson
Independent Scientist

Russell L. Jones
Bayer

Introduction

We may not live by bread alone, but we do need food. Modern-day civilization only became possible because of agriculture, by which we discovered ways to efficiently produce this needed food. And although agriculture is the world's oldest industry, it is now embroiled in controversies over its widespread use of technology to meet accelerating demands for food, feed, fiber, and fuel. The green revolution, powered mainly by the widespread adoption of inorganic nitrogen fertilizers, is claimed by some to be causing undesirable environmental effects as it enables rapid population growth and even more demand (IFPRI, 2002). This technology-driven cycle of accelerating crop production is feared by some to be unsustainable as the consumption of finite resources and degradative impacts of intensified crop production conspire to squeeze the ability of farmers to meet demand, in a kind of neo-Malthusian vortex (Dyson, 2001). For instance, the use of inorganic nitrogen fertilizers contributes to anthropogenic climate change, thereby worsening the very heat and moisture stresses that already limit the productivity of cropping systems (Stein and Yong, 2003; Mosier et al., 1998). Such concerns have caused some to express serious doubts about the long-term sustainability of modern agricultural techniques (Stewart and O'Malley, 2002).

Indeed, some experts have concluded that agriculture is the single largest threat of any human activity to the planet (NRC, 2012b). This assessment is based on a series of hard truths about agriculture. It is the leading cause of habitat conversion and deforestation, with growing demand for beef, palm oil, and soy driving much of this land use change. Agriculture uses twice as much water as all other uses combined, with the production of an average food calorie requiring about one liter of water. And agriculture is the largest source of global pollution, not only in lower income countries—where agriculture is the primary economic activity—but also in higher income nations, including the United States. It also uses more chemicals than any other activity, and has caused the world to lose more than half its topsoil—both of which enter surface water supplies, as unwanted chemical residues and sediment.

The loss of topsoil is a consequence of both tillage practices and—over the long-term—conversion of natural lands to crop and animal use. Over the past three hundred years, land planted with agricultural crops has increased nearly fivefold, to about 15 million km^2 (NRC, 1999b), which is roughly double the size of the lower forty-eight U.S. states. This net increase in cropland has been accompanied by a net decrease of forested land of similar magnitude. Another vast area has been degraded and lost to productive use since the emergence of agriculture about 10,000 years ago, as a result of erosion, salinization of irrigated lands, desertification, and permanent conversions of arable land to cities, roads, and reservoirs. During the twentieth century, this rate of loss of productive land accelerated sharply, to a rate of 70,000 km^2 (an area greater than the size of West Virginia) every year. Going forward, it is widely recognized that humanity can survive only by embracing truly sustainable pathways of agriculture development (NRC, 2010).

In this chapter, we do not attempt to provide ultimate answers to such existential questions, but we do seek to succinctly summarize current scientific understanding of how agriculture contributes to environmental pollution, and also how pollution in turn impacts agricultural systems. This bilateral interaction is unavoidable and has become an essential part of the challenge for humanity to create a sustainable world. As our title suggests, sustainably growing enough food for the 10 billion inhabitants expected by mid-century is indeed making the planet a very crowded two-way street: Food production and environmental considerations must each mutually yield to the other in order to avoid a catastrophic collision of interests.

The structure of our chapter is as follows. We begin by examining the primary way that agriculture is now believed to contribute to environmental pollution: via its impact on water quality, especially surface water. There are other agricultural processes that have polluting potential, such as in China, where heavy metals have been found to accumulate in agricultural soils under certain scenarios when copper has been used in animal feed, which then causes soil contamination when the animal manure is used as crop fertilizer (Xiong et al., 2010). If one takes the perspective that the generation of greenhouse gas (GHG) emissions is a form of air pollution, then agriculture is an important sector, serving as a major global source of both methane (primarily via rice farming) and nitrous oxide (primarily through the use of inorganic nitrogen-based fertilizers). However, most scientific and public policy attention has focused on the water quality impacts of agriculture, especially the loss of nutrients and soil (as sediment) into surface water. Accordingly, this is the topic that will be explored in greatest depth.

We then flip the perspective and ask how environmental pollution affects both the productivity of agricultural systems and the quality of the foods they produce. Although the pollution of soils by heavy metals is a cause of concern for any food crops grown in such areas (Möller et al., 2005), our examination of this literature found the most pressing concern to be the impact of air pollution on food crops, especially the highly variable but increasing effects of ozone. After further examining this issue, we close by turning our attention to several promising opportunities that are now being explored to lessen the pollution potential of current agricultural practices, as well as options to mitigate the effects of airborne ozone on food production systems. Before proceeding further, we acknowledge that much of the literature cited in this chapter is based on research conducted in the U.S. However, most of the mitigation options are at least somewhat relevant and in many cases—directly applicable to other parts of the world.

Agricultural Impacts on Water Quality

Until the mid-twentieth century, most farming was carried out on relatively small parcels of land and used organic fertilizers. Essentially all of the waste that was generated by these systems was easily assimilated by soils and any receiving water bodies (Novotny, 1999). However, beginning in the aftermath of World War II, there was a rapid global shift toward mechanization and intensification, primarily through the application of inorganic fertilizers and crop chemicals. These changes have provided many important societal benefits including greater crop yields, more efficient animal production enterprises, multiple new consumer products, and highly efficient global marketing systems, but they have also raised many concerns. From a water quality perspective, the most harmful alteration has been a rapid loss of water retention capacity, both within the soil profile and also at the landscape scale, through the installation of drainage systems and conversion of previous wetlands to agricultural uses. The intensification of both crop and animal operations increased overall agricultural impacts to water quality as a result of multiple factors: degradation due to land use conversion from native lands to agriculture; increased erosion and soil loss due to agricultural practices; chemical pollution by fertilizers and pesticides; and pollution from animal operations.

There are two different types of impacts agriculture has on the environment. One is the impact of devoting land (removing vast areas of natural habitat) and other resources such as water to agriculture. The second type is the impact of agriculture on non-agricultural areas. Examples are the effects of effluents such as soluble forms of nitrogen and soil eroded from fields. As noted above, the conversion of natural habitat to agricultural production is a very difficult issue to address, as our increasing population requires more food. The impacts can be minimized by adopting the techniques of modern agriculture to maximize yields and reduce losses, thus minimizing the amount of land required for food production (McConnell, Kelly, and Jones, 2017). Developments contributing to increased productivity and reduced water demand include improved properties of synthetic pesticides, use of emerging technologies such as genetic engineering and agricultural biologicals, improved application procedures including both spray drift and seed treatments, and the use of precision agriculture. Impacts of agricultural production can also be minimized by a variety of techniques (tillage practices, buffer strips, etc.), as discussed later in this chapter.

In most of the discussion that follows, agricultural impacts to surface water quality will be the major focus, as research has shown—with a few notable exceptions (Gustafson, 1993)—that groundwater quality is usually not impacted by agricultural activities. The presence of nutrients and pesticides in groundwater has not had major effects on organisms present in terrestrial or surface water environments and so have had minor environmental impacts, but there have been implications for human health when groundwater is used as a supply of drinking water. Groundwater resources are finite, and restoration of contaminated aquifers, even when feasible, is resource-intensive and time-consuming (NRC, 1997). Available aquifer restoration methods are also uncertain and unlikely to improve significantly in the future. Accordingly, it is almost always less expensive to prevent groundwater contamination rather than to clean it up, and so cautious preservation of groundwater resources is extremely prudent.

The fact that agriculture impacts surface water quality can be seen in the changes to the structure of invertebrate and fish communities in watersheds where agriculture dominates overall land use. Local habitat and biological diversity of streams and rivers are strongly influenced by landform and land use within the surrounding valley on multiple scales (Allan, 2004). The history of agriculture uses within the watershed is also important, as it may result in long-term modifications to and reductions in aquatic diversity, regardless of reforestation of riparian zones (Harding et al., 1998). However, empirical associations between land use and stream response only varyingly succeed in implicating pathways of influence. This is the case for a number of reasons, including covariation of anthropogenic and natural gradients in the landscape; the existence of multiple, scale-dependent mechanisms; non-linear responses; and the difficulties of separating present-day from historical influences. However, observed correlations between land use patterns and aquatic biodiversity are not always as expected, with a recent study in the central U.S. failing to find a correlation between agricultural land use and aquatic biodiversity (Sandefur et al., 2015). Despite the complexities associated with these relationships, a clear trend has emerged in which the presence of excess nutrients (nitrogen and phosphorous) in surface waters can cause excessive aquatic plant growth, depletion of oxygen, and deleterious changes in diversity of aquatic fauna, as in one detailed analysis of data from more than two hundred long-term monitoring stations (Chambers et al., 2012). This study showed that agricultural land cover was associated with increased nutrient concentrations in streams and this, in turn, was associated with increased algal abundance, loss of sensitive benthic macroinvertebrate taxa, and an increase in benthic diatom taxa indicative of eutrophication.

The U.S. Mississippi River basin is arguably the most productive agricultural region on the planet. But this productivity has come with an environmental cost, especially to the Gulf of Mexico. Over time, economic forces drove out most crops except for corn (*Zea mays*) and soybeans (*Glycine max*) in much of the watershed. This decline in crop diversity has resulted in longer periods in both the spring and autumn when no plants are growing and an alteration of hydrology processes (Gordon, et al., 2008). Cover crops can help mitigate that problem, but in their absence, precipitation during periods with minimal plant growth results in a gradual filling of available soil pore space and ultimately an increase in subsurface drainage loss of any soluble nutrients remaining in the soil profile. Nitrogen that mineralizes but is not captured by living plants becomes vulnerable to leaching loss.

In addition, to increase the area available for U.S. row crop production and to ease machinery operations, many wetlands have been tiled to improve drainage, streams have been straightened, and riparian vegetation has been removed. An unintended consequence of these actions is a reduced residence time for surface and near-surface water. This subsequently increased the speed of water movement from land to streams and resulted in "flashier" stream flows that subsequently led to more frequent and greater flood events. More rapid streamflow also increased transport of sediment, nutrients, and chemical contaminants to downstream water resources. This combination of higher nutrient loads and accelerated rate of water movement into major rivers, lakes, and coastal waters has resulted in accelerated eutrophication, wherein low dissolved oxygen concentrations makes water uninhabitable for aquatic organisms.

This is a real threat to the Gulf of Mexico, where the eutrophication-caused oxygen-poor waters ("Hypoxic Zone") on the inner continental shelf of the northern Gulf of Mexico can extend over an area as great as 20,000 km^2 (NRC, 2000a), nearly the size of New Jersey. Hypoxia has major impacts, from economic losses associated with reduced fisheries to potential human health impacts, and is likely to increase in severity as nutrient loading from upstream sources increases as a result of continuing urbanization, deforestation, agriculture, and atmospheric deposition. Given that over half the U.S. population lives in coastal communities, combating hypoxia has become an environmental priority for the United States, leading to the formation of the Gulf Hypoxia Task Force (EPA, 2017), as well as various other large national efforts, including the Mississippi River Basin Healthy Watersheds Initiative (MRBI), through which partners work to implement voluntary conservation practices (NRCS, 2017). Sustaining the quality of the nation's water resources and the health of our diverse ecosystems depends on the availability of sound water-resources data and information to develop effective, science-based policies. In 1991, Congress established the National Water-Quality Assessment (NAWQA) to address where, when, why, and how the nation's water quality has changed, or is likely to change in the future, in response to human activities and natural factors. Since then, NAWQA has been a leading source of scientific data and knowledge used by national, regional, state, and local agencies to develop science-based policies and management strategies to improve and protect water resources used for drinking water, recreation, irrigation, energy development, and ecosystem needs (USGS, 2017).

Of course, hypoxia is not unique to the U.S. and the Gulf of Mexico. Biological productivity in most of the world's oceans is controlled by the supply of nutrients to surface waters (Beman, Arrigo, and Matson, 2005). The relative balance between supply and removal of nutrients—including nitrogen, iron, and phosphorus—determines which nutrient limits phytoplankton growth. Although nitrogen limits productivity in much of the ocean, large areas are defined by extreme nitrogen depletion. In these regions, microbial denitrification removes biologically available forms of nitrogen from the water column, producing substantial deficits relative to other nutrients. The

Baltic Sea and waters off the coast of China are among the most impacted by this phenomenon (Diaz and Rosenberg, 2008). Hypoxic zones have spread exponentially since the mid-twentieth century and have serious consequences for ecosystem functioning. They now affect a total global area of more than 245,000 km^2 (the size of Michigan) and are probably a key stressor on marine ecosystems (NRC, 2000b). According to the "planetary boundary" assessment methodology put forth by the Stockholm Research Institute (Rockström et al., 2009), nitrogen and phosphorous emissions from the agricultural sector are now so large that they significantly perturb the global biogeochemical cycles of these two important elements (Foley et al., 2005; Gruber and Galloway, 2008). Human processes—primarily the manufacture of nitrogen fertilizer for food production and the cultivation of leguminous crops (such as soybeans)—convert around 120 million MT of N2 from the atmosphere per year into reactive forms—which is more than the combined effects from all of the other of Earth's terrestrial processes.

In addition to many efforts in understanding the linkages between agricultural intensification and surface water quality in North America, major studies have been reported in both the United Kingdom (Burt et al., 2011) and Spain (Monteagudo, Moreno, and Picazo, 2012). The study in Spain explored how spatial scale may affect the results when relating land use to nutrient enrichment of rivers and, second, to investigate which agricultural practices are more responsible for river eutrophication in the study area. Agriculture was split into three subclasses (irrigated, non-irrigated, and low-impact agriculture), which were correlated to stream nutrient concentration on four spatial scales. Analysis of environmental issues in the lower income countries of Sub-Saharan Africa and southern Asia suggests that agricultural activities are a major factor in both of their two surface water quality threats, namely: (1) increasing wastewater discharge due to urbanization and intensification of livestock farming, and (2) reductions in river dilution capacity due to climate change and water extractions (Wen, et al., 2017). Just as in higher income countries, organic pollutants in rivers stimulate microbial growth, leading to oxygen depletion and disturbance of the entire river ecosystem.

As noted above, the delivery of excess nutrients in surface water is the most important mechanism by which agriculture contributes to the global pollution burden. However, crop protection chemicals (commonly called pesticides and herbicides) are widely used in agricultural systems and can also reach water supplies (McConnell, et al., 2017; Bexfield, 2008; USGS, 2017). Pesticide concentrations in groundwater are typically quite low, and U.S. monitoring shows a downward trend (Bexfield, 2008). The potential for particular pesticides to reach groundwater supplies is related to their persistence and mobility in the soil profile (Gustafson, 1989). Pesticides are also able to reach surface water supplies, but commonly used drinking water treatment technologies are effective at removing residues before they enter tap water (Gustafson et al., 2003). The advent of herbicide-tolerant crops has been accompanied by some water quality benefits (Wauchope et al., 2002), albeit with some increasing concerns over weed resistance (Duke and Powles, 2008).

Air Pollution Impacts on Agriculture

Although progress has been made in combating air pollution in many countries, it remains a major threat to human health today and future food security. One out of eight deaths globally is caused by air pollution (NRC, 2016b), and such pollutants (especially ozone, O3) also harm food crops (NRC, 1989). The difficulties in addressing the effects of air pollution are made worse by the fact that effects of many types of disturbances are similar and that the responses of plants are responses to general stress and are not easily used to diagnose specific insults. Stress has many causes and collateral effects, such as diseases and pests that become important when plants are otherwise weakened. Sorting out cause and effect has frustrated pathologists and ecologists for years. Important physiological processes, photosynthesis, respiration, carbon allocation, and stomatal function are known to be affected by air pollutants (Darrall, 1989). A wide range in sensitivity of photosynthesis both within and between species is evident from the literature for the pollutants sulfur dioxide, ozone, nitrogen oxide, and hydrogen fluoride. Some of this variation is clearly due to genetic factors, but much is in response to differences in environmental conditions both prior to and during pollution. Exposure of plants to mixtures of pollutants generally reduced the threshold at which effects were first detected and increased the level of inhibitory responses. In the majority of studies on stomatal responses to air pollutants, opening occurs at low concentrations, below the threshold for effects on photosynthesis, and closure occurs at injurious concentrations; this latter response often following the inhibition of photosynthesis. Effects on carbon allocation have been reported in response to air pollutants. Changes usually favor leaf development over root growth, which can compensate for a decline in net assimilation rate up to a certain point but may limit water uptake from soils with low moisture content.

Future impacts of air pollution on agriculture have been estimated (Cofala et al. 2007). These researchers estimate that, with current expectations on future economic development and with the present air quality legislation, global anthropogenic emissions of sulfur dioxide (SO2) and nitrogen oxides (NOx) would slightly decrease between 2000 and 2030. For carbonaceous particles and carbon monoxide (CO), reductions between 20 percent and 35 percent are computed, while for methane (CH4) an increase of about 50 percent is calculated. Full application of currently available emission control technologies, however, could achieve substantially lower emissions levels, with decreases up to 30 percent for methane, 40 percent for carbon monoxide, and nearly 80 percent for sulfur dioxide.

Unlike soil and water pollution, air pollutants can be transported very large distances, even globally. Using global atmospheric modeling, intercontinental transboundary contributions to air pollution and subsequent regional crop losses have been estimated (Hollaway et al., 2012). The model calculations show that reductions in Southeast Asian anthropogenic NOx emissions tend to produce the greatest global reduction in production losses for wheat (*Triticum* spp.), rice (*Oryza sativa*), cotton (*Gossypium*), and potato (*Solanum tuberosum*), whereas reductions in North American NOx emissions results in the greatest global impact on crop production losses for maize (*Zea mays*) and soybean (*Glycine max*). Reduction of North American anthropogenic NOx emissions would produce the largest transboundary impact, resulting in large reductions in current European production loss reductions.

While many studies have estimated the impacts of air pollution on crop yields on experimental plots, few have estimated these impacts under actual farm production conditions. One attempt to do this used econometric modeling to estimate the impact of air pollution on corn and soybean yields, controlling for weather, soil quality, and management practices, using farm-level data for the eastern U.S. (Westenbarger and Frisvold, 1995). Ozone pollution was found to reduce yields for both crops. The mean elasticity of yield with respect to ozone exposure was 0.19 for corn and 0.54 for soybeans.

Peri-urban agriculture (agricultural production in or near cities and towns) is vital for the urban populations of many developing countries, such as India, where increasing industrialization and urbanization threaten urban food production and its quality (Agrawal et al., 2003). These researchers examined air pollution impacts in both rural and per-urban settings. Peri-urban concentrations of sulfur dioxide, NOx, and ozone were all found to impact yield and crop quality in beet (*Beta vulgaris*), mung bean (*Vigna radiata*), mustard (*Brassica*), and wheat (*Triticum* spp.). Ozone had the greatest impacts and its concentration in rural areas was also found to impact crops.

Among all air pollutants, ozone consistently shows the largest yield losses (on the order of 20 percent), and therefore warrants the greatest attention (Long et al., 2005). Mean surface ozone concentration is predicted to increase 23 percent by 2050 (Morgan et al., 2006). The majority of ozone formation in the lower atmosphere (or troposphere) occurs when NOx, carbon monoxide, and volatile organic compounds (VOCs), react in the atmosphere in the presence of sunlight (Cooper et al., 2010). Accordingly, these various pollutants are known as ozone precursors. Ozone itself is highly reactive and therefore quite ephemeral in the atmosphere, serving as an important source of the hydroxyl radical, an oxidant that breaks down most pollutants and some greenhouse gases. However, high concentrations of ozone are toxic, having a detrimental effect on humans and all life. Moreover, tropospheric ozone itself acts as an effective greenhouse gas. Much of the present tropospheric ozone burden is a consequence of anthropogenic emissions of ozone precursors resulting in widespread increases in ozone concentrations during the twentieth century. At present, east Asia has the fastest-growing ozone precursor emissions. Much of the springtime east Asian pollution is exported eastward toward western North America.

Elevated concentrations of ground-level ozone are frequently measured over farmland regions in many parts of the world (Fishman et al., 2010). These researchers collected soybean crop yield data during a five-year period over the U.S. Midwest and combined them with ground and satellite ozone measurements to provide evidence that yield losses on the order of 10 percent. Yield loss trends based on both conventional ground-based instrumentation and satellite-derived tropospheric ozone measurements were statistically significant and were consistent with results obtained from open-top chamber experiments and an open-air experimental facility (SoyFACE, Soybean Free Air Concentration Enrichment) in central Illinois (Long et al., 2005). Analyses suggest that such losses are a relatively new phenomenon due to the increase in background tropospheric ozone levels over recent decades. Extrapolation of these findings supports previous studies that estimate the global economic loss to the farming community of more than $10 billion annually.

But ozone affects other crops as well. Rice is arguably the most important food source on the planet and is consumed by over half of the world's population. A meta-analysis synthesized the research to date on rice responses to two elements of global change, rising atmospheric carbon dioxide and ozone (Ainsworth, 2008). On average, elevated carbon dioxide (627 ppm) increased rice yields by 23 percent. Modest increases in grain mass and larger increases in panicle and grain number contributed to this response.

Free air concentration enrichment (FACE) experiments showed only a 12 percent increase in rice yield. The rise in atmospheric carbon dioxide will be accompanied by increases in tropospheric ozone and temperature. When compared with rice grown in charcoal-filtered air, rice exposed to ozone (at 62 ppb) showed a 14 percent decrease in yield. Many determinants of yield, including photosynthesis, biomass, leaf area index, grain number, and grain mass, were reduced by elevated ozone. While there have been too few studies of the interaction of carbon dioxide and ozone for meta-analysis, the interaction of temperature and carbon dioxide has been studied more widely. Elevated temperature treatments negated any enhancement in rice yield at elevated carbon dioxide, which suggests that identifying high temperature tolerant germplasm will be key to realizing yield benefits in the future.

Wheat is also one of the world's most important staple crops. Field measurements and open-top chamber experiments using nine widely cultivated European winter wheat cultivars provided a data set for use in ozone risk assessments including effects on food security (Gruenhage et al., 2012). The results for Germany show negative effects of ozone on wheat grain yield and quality for most years since the mid 1980s. A subsequent effort examined the effects of ozone on wheat quality based on forty-two experiments performed in Asia, Europe, and North America (Broberg et al., 2015). Data were analyzed using meta-analysis and by deriving response functions between observed effects and daytime ozone concentration. There was a strong negative effect on grain size and weaker but significant negative effects on starch concentration and kernel weight. Ozone also impacted protein content and the levels of several micronutrients.

The effect of ozone on crops in China has begun to receive much more attention because of the country's very rapid economic development and increasing demand for food (Wang et al., 2007). It is evident that ozone concentrations in many parts of the country are high enough to cause significant reductions in crop yield for rice, wheat, and legumes, such as soybeans (Emberson et al., 2009). Analyses indicate that Asian grown wheat and rice cultivars are more sensitive to ozone than the North American varieties. For legumes, the scatter in the data makes it difficult to reach any equivalent conclusion in relative sensitivities. Even so, great care must be taken when attempting to extrapolate ozone crop sensitivity data from one region to another, and some of the early model-based risk assessments may have substantially underestimated the scale of the problem in Asia through use of North American derived dose-response relationships.

• • •

Modified In-Field Agricultural Practices

There are numerous opportunities to modify in-field agricultural practices in ways that reduce pollution potential and are also profitable for the producer. These include focusing on improved soil health; increasing nutrient, pesticide, herbicide, and irrigation use efficiencies; reducing erosion and runoff; and making greater use of buffer zones (NRC, 1993). Conservation tillage practices, especially continuous no-till, are a particularly effective way to enhance soil health and reduce the pollution potential of agricultural systems. Conservation tillage saves fuel, reduces soil erosion, retains soil tilth, prevents loss of soil organic matter, and avoids GHG emissions (CTIC, 2017a). Transgenic herbicide-tolerant crops facilitate the adoption of conservation tillage by enabling broad-spectrum herbicide mixtures, rather than mechanical tillage or hand-weeding, to be used for weed control (Wauchope et al., 2002). This increased reliance on herbicides raises some environmental questions, but the net impact on pesticide movement is generally in the desirable direction (Holland, 2004).

One part of the US where the benefits of conservation tillage practices have been extensively studied and documented is the Lake Erie basin, where agriculture impacts on surface water quality are particularly high (Gustafson et al., 2004). A major monitoring data analysis carried out by the Heidelberg College National Center for Water Quality Research (Richards et al., 2009) found that the Maumee and Sandusky Rivers in this region have shown continual decreases in sediment concentrations and loads throughout the period 1985–2004. The greatest decreases were observed in summer and fall and under low flow conditions, whereas the smallest decreases were observed in the spring and under high flow conditions. Analysis of concentration-flow relationships indicates that these changes are not due to weather, but reflect the successful use of conservation tillage to reduce erosion and prevent sediment loss. Similar positive trends were also seen for phosphorous (Baker and Richards, 2002), most of which enters surface water in a sediment-bound form. Opportunities for further reductions in suspended sediment and particulate phosphorus loads and concentrations lie in better management of sediment losses during winter and spring.

A companion practice that helps growers adopt conservation tillage is the incorporation of winter crops into the rotation (Snapp et al., 2005). Although the current extent of cover crop adoption in the U.S. remains rather low due to a variety of challenges and uncertainties (NASS, 2014), the growers who currently use cover crops are very knowledgeable, enthusiastic, and committed to continuing to include them in the rotation (CTIC, 2017b). Growers are experiencing soil health benefits and reduced nutrient losses from their fields. In addition, increases in the yields of succeeding commercial crops are common, while yield decreases remain very uncommon. Based on USDA data, the cover crop acres planted in the U.S. are rising steadily and have now likely surpassed 20 million acres, although this will need to be confirmed during the upcoming 2017 Ag Census. From an air pollution perspective (counting excess anthropogenic carbon dioxide as a pollutant), cover crops have a tremendous potential to capture atmospheric carbon and transform it into relatively stable forms in soil (so-called "carbon sequestration"). At the current adoption rate of 20 million acres, this corresponds to approximately 40 million metric tons of CO2 per year in the top six to ten inches of the soil. Additional sequestration can be expected at deeper soil depths, especially since many cover crops root to a depth of more than a meter. An additional benefit of cover crops is their climate adaptation potential, by moving more water deeper into the soil profile, thereby enhancing the resilience of agricultural fields to both drought and intense rainstorms.

Another conservation practice that lowers the pollution potential of agricultural systems is edge-of-field conservation buffers. They provide multiple ecosystem services, including wildlife cover, corridors for wildlife movement, below-ground carbon sequestration, slowing runoff, capturing nutrients from surface and subsurface flow, controlling erosion, protecting water quality, and mitigating pesticide spray drift. There is a substantial body of knowledge (Bentrup, 2008) available to help users understand the multiple functions served by conservation buffers and to provide basic guidelines for their design. Conservation buffers are often designed for only one or two specific objectives (e.g., reducing wind erosion or protecting water quality). Such buffers are typically not harvested, but one idea that has been suggested is to plant them with perennial grasses and woody species that could serve as bioenergy crops or "bioenergy buffers" (Gopalakrishnan et al., 2012). Given the continuing concerns about herbicide spray drift, the ability of buffers to mitigate such drift may become an increasingly valuable ecosystem service. Spray drift modeling and experimental data demonstrate that such buffers can reduce spray drift exposure to downwind areas, as long as the buffer provides some porosity (30 to 50 percent) to avoid acting like an impervious wall (QDNR, 1997). The use of buffers to separate areas with competing land uses and to thus ensure long-term protection of both the "conflict generating activity" and potential "impacted upon" areas is well-established throughout the U.S. and in other countries. Examples of such activities include sewage treatment works, composting sites, and intensive animal and plant production facilities. The principle of separating conflicting land uses also applies to the protection of natural resource areas (such as nature conservation reserves, streams, water supply storage areas and forest reserves). The "big data" approaches now available through the techniques of precision agriculture can be employed to determine which portions of field margin areas are least profitable, and might better be converted to such buffer areas (EFC, 2017).

Another approach for reducing the environmental impact of agriculture is "sustainable intensification" (Gustafson et al., 2013), a practice which has been shown to have resulted in the avoidance of significant GHG emissions (Burney, Davis, and Lobell, 2010). Advances in technology are improving nitrogen-use efficiency (NUE), one of the most effective means of increasing crop productivity while decreasing environmental degradation (Zhang et al., 2015). An industry-led initiative to better manage nutrient applications in the U.S. and Canada is known as "4Rs" (The Fertilizer Institute, 2017): the Right fertilizer source at the Right rate, at the Right time and in the Right place. The goal of fertilizer best management practices (BMPs) is to match nutrient supply with crop requirements and to minimize nutrient losses from fields (Snyder, 2016). Selection of BMPs varies by location, and those chosen for a given farm are dependent on local soil and climatic conditions, crop, management conditions, and other site-specific factors.

Additional resources for the management of nutrient applications include computer software applications (or "apps"), such as "Adapt-N" (Cornell University, 2017). The app uses models and biophysical data to simulate field conditions and derive an optimum nitrogen rate recommendation. The tool is currently calibrated for use on about 95 percent of the U.S. corn production area. Adapt-N focuses on an important component of the solution: getting farmers to apply the right N fertilizer rate for a particular production environment that optimizes crop uptake and minimizes environmental losses, while also facilitating the use of other beneficial technologies like better N application timing, cover cropping, and use of enhanced efficiency products.

Within the United States, agriculture is a significant source of nitrous oxide emissions, a GHG that is over 250 times more potent than carbon dioxide, on a per-molecule basis. Nitrous oxide emissions from synthetic nitrogen fertilizers and organic sources represent a major source of GHG emissions from U.S. agricultural production. Nitrous oxide emissions from synthetic fertilizer and organic amendments account for more than 50 percent of cropland agriculture nitrous oxide emissions. Management of nitrogen application can reduce these emissions. Nitrous oxide emissions from cropping practices depend on many factors, but several key factors are the timing, source, placement, and quantity of nitrogen (both from organic sources and synthetic sources) applied. Weather and soil health also influence nitrogen losses and the resulting efficiency with which it is taken up by crops. Taking these factors into account, GHG emissions can be reduced substantially through improved nitrogen management practices and increased nitrogen use efficiency on crop and pasturelands. In addition to reducing nitrous oxide emissions, these practices can also significantly reduce nitrate leaching, improving local and regional water quality, and reduce on-farm input costs.

Constructed and Restored Wetlands

Larger scale land management approaches, such as the construction or restoration of wetlands, provide another means of reducing the pollution potential of agricultural systems. A detailed study of their cost and effectiveness in the United Kingdom has been published (Ockenden et al., 2012). Ten wetlands were built on four farms in Cumbria and Leicestershire. The study focused on sediment retention within the wetlands. Estimates of annual sediment retention suggest higher trapping rates at sandy sites, compared to silty sites and clay sites. Establishment costs for the wetlands depended more on site-specific factors, such as fencing and gateways on livestock farms, rather than on wetland size or design. Wetlands with lower trapping rates would also have lower maintenance costs, as dredging would be required less frequently. The results indicate that field wetlands show promise for inclusion in agri-environment schemes, particularly if capital payments can be provided for establishment, to encourage uptake of these multi-functional features.

A comprehensive review found that constructed and restored wetlands remove or retain many water-quality contaminants in agricultural runoff if carefully designed and managed (O'Geen et al., 2010). Contaminant removal efficiency generally exceeded 50 percent for sediment, nitrate, microbial pathogens, particulate phosphorus, hydrophobic pesticides, and selected trace elements when wetlands were placed in the correct settings. There are some potentially adverse effects of constructed and restored wetlands that must be considered, including accumulation of mercury and selenium, increased salinity, mosquito habitat, and greenhouse gas emissions. Proper wetland management and design features can be incorporated in order to reduce these adverse effects, while optimizing contaminant removal. Phosphorous removal in wetlands has specific considerations that must be addressed as part of the design process (Roy-Poirier et al., 2010). Another review of wetland effectiveness was recently commissioned by the

United Kingdom's Department for Environment, Food and Rural Affairs (Palmer-Felgate et al., 2013). It examined the potential for wetlands to mitigate agricultural nitrogen, phosphorous, and sediment inputs to receiving surface water bodies in England. An even more recent systematic review specifically considered nitrogen and phosphorus removal rates in constructed or restored wetlands and related them to wetland characteristics, loading characteristics, and climate factors, with particular emphasis on Sweden (Land et al., 2016).

One of the ecological considerations associated with such constructed or restored wetlands is that they will also serve as habitat for organisms that could then be impacted by the various materials being trapped. The benefits of this increased wetland area, perhaps especially for amphibians, may be negated if habitat quality is insufficient to support persistent populations (Smalling et al., 2015). These researchers examined the presence of pesticides and nutrients in water and sediment as indicators of habitat quality and assessed the bioaccumulation of pesticides in the tissue of two native amphibian species at six wetlands (three restored and three reference) in Iowa. Concentrations of the pesticides frequently detected in water and sediment samples were not different between wetland types.

In another study, four restored wetlands were used to improve the quality of agricultural runoff in the Delta of the Ebro River in northeast Spain (Comín et al., 1997). The wetlands were continuously flooded with water from a rice-field irrigation network during the growing season and received water with biologically relevant levels of nitrogen. The nitrogen retention efficiency was close to 100 percent of the input, except for dissolved organic nitrogen at the end of the growing season. Emergent macrophytes accumulated most of the inflowing dissolved inorganic nitrogen. The wetlands also removed sediment. Plant uptake, detritus accumulation and decomposition, and nitrogen recycling in the sediment are major processes for nitrogen retention and recycling in the wetlands. This type of wetlands, restored from rice fields, act as highly efficient water polishing filters for agricultural runoff and, at the same time, can contribute to increase the habitat biodiversity of large areas where rice is cultivated extensively.

Policy and Research Options

Two kinds of options are available to reduce ozone damage to agricultural systems: either lowering emissions of ozone precursors or adapting crops to ozone exposure by selecting cultivars with demonstrated ozone resistance. These two options were studied in detail (Avnery, et al., 2013), and it was found that ozone precursor reductions would significantly increase global production of soybean, maize, and wheat—generating additional crops having a total value of $3.5 to $15 billion worldwide (USD, 2000) increasing the cost effectiveness of this ozone mitigation policy. Choosing crop varieties with ozone resistance (relative to median-sensitivity cultivars) could improve global agricultural production by up to $22 billion. Benefits are dominated by improvements for wheat in South Asia, where ozone-induced crop losses would otherwise be severe. Combining the two strategies generates benefits that are less than fully additive, given the nature of ozone effects on crops. Biotechnology approaches could be one approach for improving crop ozone tolerance (Ainsworth, Rogers, and Leakey, 2008).

As for impacts to water quality, combined economic and environmental modeling has demonstrated that a reduction in agriculture's impact on the environment is both possible and economically advantageous (NRC, 1999a). Alternative production practices that improve farm-level economics, even in the absence of policy incentives, are the ones that are most likely to be adopted (EFC, 2017). As noted above, an increasing number of U.S. farmers are now using winter cover crops, usually in combination with continuous no-till or at least some form of conservation tillage. This growing number of farmers has become convinced by the body of research and shared firsthand experiences that this renewed focus on soil health will pay dividends in the form of steadily increasing crop yields and overall financial returns. Many of these practices reduce production costs by improving input-use efficiency. Such alternative production practices could greatly help farmers conserve resources, improve productivity and profits, and reduce fiscal costs. Policy changes that remove the biases against such production practices would allow further improvement and save taxpayers money.

Decadal-scale economic modeling shows that if U.S. farmers fully accounted for the cost of long-term soil productivity changes and if alternative production practices were fully available, soil erosion and its off-site costs would go down significantly (NRC, 1999a). In the extended baseline, soil erosion is reduced nationally by 7 percent and damages by 11 percent. Soil depreciation cost estimates are relatively small, compared with other calculated production costs, and they are actually negative for production practices leading to yields estimated to increase over time. The larger effect in the baseline extension comes from including the alternative practices in the set of practices that the model can choose from. Many alternative practices turn out to be very competitive financially and come into the extended baseline model solution based solely on relative profits, even though they also provide environmental benefits. Policy analysis showed that green payment options to subsidize conservation also help to improve environmental performance, but not all green payment programs would work equally well, environmentally or fiscally. Targeted subsidies adjusted to account for the value of regional damages achieve lower program costs and have the greatest benefits. Environmental performance increases with increasing commodity program flexibility. Two changes account for the environmental results under increased flexibility: Both cultivated acreage and the use of mono-cultural practices decline as flexibility increases.

In assessing future challenges for U.S. agriculture, a group of experts concluded that although most farms have the potential and responsibility to contribute to different aspects of sustainability, U.S. agriculture needs both incremental and transformative changes to address the many challenges of the future (NRC, 2010). Incremental changes—such as pest-resistant varieties, conservation tillage, integrated pest management, and use of crop diversity including cover crops, crop rotations, and other biologically integrative technologies and practices—have been increasingly used in many regions, but have not yet been adapted to some fragile areas and to low-rainfall cropland. Transformative changes include the development of new farming systems that represent a dramatic departure from the dominant systems of present-day U.S. agriculture and capitalize on synergies and efficiencies associated with complex natural systems and broader social and economic forces using integrative approaches to research and extension at both the farm and landscape levels. Examples include development and broad adoption of water-conserving production systems in areas of water shortage and overdraft, landscape-scale reduction of nutrient and other materials runoff from agricultural lands that contributes to major hypoxic zones, and assessment of the potential and cost for broad adoption of alternative animal production systems that address many environmental and social concerns of some dominant production systems.

In addition to being driven by evolving water-quality problems, water-policy change is likely required to respond to tightened public budgets and increased concerns about efficiency in water-quality regulation (NRC, 2012c). To date, U.S. regulatory action on nonpoint source water-pollution control in agriculture has been pursued largely through voluntary compliance strategies supplemented by public assistance through the adoption of pollution control practices. Reduced federal and state budgets may require significant policy innovations if water-quality goals are to be achieved with reduced financial support.

It has long been realized that U.S. conservation efforts must be aggressively targeted in order to achieve water quality improvements in the Mississippi

River and its tributaries (NRC, 2008). Programs aimed at reducing nutrient and sediment inputs must include efforts at targeting areas of higher nutrient and sediment deliveries to surface water. Through the Hypoxia Task Force (HTF), the EPA, USDA, and USGS have been strengthening their level of cooperative activities designed to reduce impacts from agriculture on the water quality within this watershed and its receiving water body (the Gulf of Mexico). As new research and information have become available and systems of conservation practices are implemented on vulnerable lands across this large basin, the HTF has gained a better understanding of the complexities of hypoxia in the Gulf and the efforts and time that will be needed to achieve its goals (EPA, 2017). In February 2015, the HTF announced that it would retain its goal of reducing the areal extent of the Gulf of Mexico hypoxic zone to less than 5,000 km^2 by the year 2035. The HTF agreed on an interim target of a 20 percent nutrient load reduction by the year 2025 as a milestone toward achieving the final goal in 2035. The HTF also agreed to adopt quantitative measures to track progress in reducing point and non-point source inputs. To accelerate the reduction of nutrient pollution, the HTF will: (1) target vulnerable lands and quantify nutrient load reductions achieved through federal programs, subject to future appropriations; (2) implement state nutrient reduction strategies, including targeting vulnerable lands and quantifying nutrient reductions; (3) expand and build new partnerships and alliances with universities, the agricultural community, cities, and others; and (4) track progress toward the interim target and long-term goal, with intent to understand whether the current actions are appropriate to meet the goal.

The ultimate goals of the HTF will only be fully achieved when it has successfully engaged and involved all of the individuals within the Mississippi River basin. This point was highlighted in a recent workshop on Mississippi River water quality and interstate collaboration, sponsored by the National Research Council (NRC, 2016a). It is also critical to use structured decision making to develop and prioritize strategies, taking costs into account. Lastly, there needs to be a renewed recognition of the overall socio-economic and environmental value provided by healthy ecosystems, in addition to a more traditional focus on agricultural practices.

Conclusions

The projected doubling in global food demand projected over the next fifty years (Tilman et al., 2002) poses huge challenges for addressing the issues raised in this chapter. Better approaches to producing, processing, and consuming foods are clearly needed—both to avoid further pollution from agricultural systems and also to mitigate the increasing effects of pollution on food production. Agriculture will continue to dominate all other land uses and has already dramatically altered the appearance of the Earth and many of its natural processes. Some of these changes may be irreversible and so there is urgency around the growing efforts to transform our food systems (NRC, 2012a), such that human nutrition needs are met in truly sustainable ways.

Adding to the overall challenge, agriculture is just one of the industries that must be dramatically changed in order to achieve the "carbon-neutral world" envisioned in the 2015 Paris Agreement on climate change (Roe et al., 2017). For many countries, GHGs from the energy and transport sectors dominate overall emissions, but as those sectors continue to evolve toward carbon neutrality there will be increasing pressure for significant contributions from the food and agriculture sector, including both supply-side measures and demand-side measures related to healthier diets and reduced food waste. Land-sector emissions must peak by 2020 and become net-zero by 2040–2050 and net-negative thereafter, which implies a suite of new agricultural practices, including enhanced soil carbon sequestration and increased efficiency of synthetic fertilizer production and use.

Our overall goal in agriculture must be the production of high-quality food and other necessities at low cost and with minimal deleterious effects on humans or the environment (NRC, 2000c). As we have seen in this chapter, today's agricultural systems are currently having some unwanted effects, particularly through their contributions to GHG emissions and by adding to the overall pollution burden on receiving waters. In turn, the productivity of many cropping systems is being harmed by air pollution, especially ozone. The global scale of these problems is truly daunting. However, as we have seen above, there are new practices, policies, and technologies that can help bring the needed improvements in outcomes. It is now time to implement these improvements, such that humanity's oldest industry can be transformed into one that contributes to a sustainable future, free of pollution.

References

Agrawal, M., Singh, B., Rajput, M., Marshall, F., and Bell, J. N. B. (2003). Effect of air pollution on peri-urban agriculture: a case study. *Environmental Pollution*, *126*(3), 323–29. doi: 10.1016/S0269-7491(03)00245-8.

Ainsworth, E. A. (2008). Rice production in a changing climate: a meta-analysis of responses to elevated carbon dioxide and elevated ozone concentration. *Global Change Biology*, *14*, 1642–50.

Ainsworth, E. A., Rogers, A., and Leakey, A. D. B. (2008). Targets for crop biotechnology in a future high-CO2 and high-O3 world. *Plant Physiology*, *147*, 13–17.

Allan, J. D. (2004). Landscapes and riverscapes: the influence of land use on stream ecosystems. *Annu. Rev. Ecol. Evol. Syst., 35*, 257–84.

Avnery, S., Mauzerall, D. L., and Fiore, A. M. (2013). Increasing global agricultural production by reducing ozone damages via methane emission controls and ozone-resistant cultivar selection. *Global Change Biology*. doi: 10.1111/gcb.12118.

Baker, D. B., and Richards, R. P. (2002). Phosphorus budgets and riverine phosphorus export in northwestern Ohio watersheds. *Journal of Environmental Quality, 31*(1), 96–108.

Beman, J. M., Arrigo, K. R., and Matson, P. A. (2005). Agricultural runoff fuels large phytoplankton blooms in vulnerable areas of the ocean. *Nature, 434, 211-214*(March), 211. doi: 10.1038/nature 03370.

Bentrup, G. (2008). Conservation buffers: design guidelines for buffers, corridors, and greenways. Gen. Tech. Rep. SRS-109. Asheville, NC: U.S. Department of Agriculture, Forest Service, Southern Research Station. 110 p.

Bexfield, L. M. (2008). Decadal-scale changes of pesticides in ground water of the United States, 1993–2003. *Journal of Environmental Quality, 37*, 226–39.

Broberg, M. C., Feng, Z., Xin, Y., and Pleijel, H. (2015). Ozone effects on wheat grain quality—a summary. *Environmental Pollution*, 197, 203-213. doi: 10.1016/j.envpol.2014.12.009.

Burney, J. A., Davis, S. J., and Lobell, D. B. (2010). Greenhouse gas mitigation by agricultural intensification. *Proceedings of the National Academy of Sciences of the United States of America, 107*, 12052–57.

Burt, T. P., Howden, N. J. K., Worrall, F., Whelan, M. J., and Bieroza, M. (2011). Nitrate in United Kingdom rivers: policy and its outcomes since 1970. *Environmental Science and Technology*, 45, 1, 175-181. doi: 10.1021/es101395s.

Chambers, P., McGoldrick, D., Brua,R., Vis, C., Culp, J., and Benoy, G. (2012). Development of environmental thresholds for nitrogen and phosphorous in streams. *Journal of Environmental Quality, 41*, 7–20.

Cofala, J., Amann, M., Klimont, Z., Kupianinen, K., and Hoeglund-Isaksson, L. (2007). Scenarios of global anthropogenic emissions of air pollutants and methane until 2030. *Atmospheric Environment, 41*, 8486–99.

Comín, F. A., Romero, J. A., Astorga, V., and García, C. (1997). Nitrogen removal and cycling in restored wetlands used as filters of nutrients for agricultural runoff. *Water Science and Technology 35*(5), 255–61. doi: 10.1016/S0273-1223(97)00076-0.

Cooper, O. R., Parrish, D. D., Stohl, A., Trainer, M., Nédélec, P., Thouret, V., Cammas, J. P., et al. (2010). Increasing springtime ozone in the free troposphere over western North America. *Nature, 463*, 344–48.

Chang, A. (2017). Adapt-N for nitrogen management. Cornell Research, https://research.cornell.edu/news-features/adapt-n-nitrogen-management

CTIC (2017a). Conservation tillage. Retrieved from http://www.ctic.purdue.edu/Core4/Conservation Tillage/.

CTIC (2017b). *Report of the 2016–17 national cover crop survey*. West Lafayette, IN.

Darrall, N. M. (1989). The effect of air pollutants on physiological processes in plants. *Plant, Cell and Environment, 12*(1), 1–30. doi: 10.1111/j.1365-3040.1989.tb01913.x.

Diaz, R. J., and Rosenberg, R. (2008). Spreading dead zones and consequences for marine ecosystems. *Science, 321*(5891), 926–929. Retrieved fromhttp://science.sciencemag.org/content/321 /5891/926.abstract.

Duke, S. O., and Powles, S. B. (2008). Glyphosate-resistant weeds and crops. *Pest Management Science, 64*, 317–18.

Dyson, T. (2001). World food trends: a neo-Malthusian prospect? *Proceedings of the American Philosophical Society, 145*(4), 438–55.

EFC. (2017). Agronomic planning and sustainability. Retrieved from https://www.efcsystems.com/index.php/agronomicplanningandsustainability/.

Emberson, L.D., Büker, P., Ashmore, M. R., Mills, G., Jackson, L. S., Agrawal, M., Atikuzzaman, M. D., et al. (2009). A comparison of North American and Asian exposure-response data for ozone effects on crop yields. *Atmospheric Environment, 43*, 1945–53.

EPA. (2017). Mississippi River/Gulf of Mexico Watershed Nutrient Task Force, 2017 Report to Congress. Washington, D.C.

Fishman, J., Creilson, J. K., Parker, P. A., Ainsworth, E. A., Vining, G., Szarka, G., J., Booker, F. L., and Xu, X. (2010). An investigation of widespread ozone damage to the soybean crop in the upper Midwest determined from ground-based and satellite measurements. *Atmospheric Environment*. doi: 10.1016/j.atmosenv.2010.01.015.

Foley, J. A., DeFries, R., Asner, G. P., Barford, C., Bonan, G., Carpenter, S. R., Chapin, F. S., et al. (2005). Global consequences of land use. *Science, 309*(5734), 570–574. Retrieved from http://science.sciencemag.org/content/309/5734/570.abstract.

Gopalakrishnan, G., Negri, M. C., and Salas, W. (2012). Modeling biogeochemical impacts of bioenergy buffers with perennial grasses for a row-crop field in Illinois. *GCB Bioenergy, 4*(6), 739–50. doi: 10.1111/j.1757-1707.2011.01145.x.

Gordon, L. J., Peterson, G. D., and Bennett, E. M. (2008). Agricultural modifications of hydrological flows create ecological surprises. *Trends in Ecology and Evolution*. doi: 10.1016/j.tree.2007.11.011.

Gruber, N., and Galloway, J. N. (2008). An Earth-system perspective of the global nitrogen cycle. *Nature, 451*(January), 293. doi: 10.1038/nature06592.

Gruenhage, L., Pleijel, H., and Mills, G. (2012). Updated stomatal flux and flux-effect models for wheat for quantifying effects of ozone on grain yield, grain mass and protein yield. *Environmental Pollution, 165*, 147–57.

Gustafson, D. I. (1989). Groundwater ubiquity score: a simple method for assessing pesticide leachability. *J. Environ. Toxic. & Chem, 8*, 339–57.

Gustafson, D., Collins, M., Fry, J., Smith, S., Matlock, M., Zilberman, D., Shryock, J., Doane, M., and Ramsey, N. (2013). Climate adaptation imperatives: global sustainability trends and eco-efficiency metrics in four major crops—canola, cotton, maize, and soybeans. *International Journal of Agricultural Sustainability, 9*. doi: 10.1080/14735903.2013.846017.

Gustafson, D. I. (1993). *Pesticides in drinking water*. New York, NY: John Wiley & Sons.

Gustafson, D. I., Carr, K. H., Green, T. R., Gustin, C., Jones, R. L., and Richards, R. P. (2004). Fractal-based scaling and scale-invariant dispersion of peak concentrations of crop protection chemicals in rivers. *Environmental Science and Technology*. doi: 10.1021/es030522p.

Gustafson, D. I., Carr, K. H., Carson, D. B., Fuhrman, J. D., Hackett, A. G., Hoogheem, T. J., Snoeyink, V. L., et al. (2003). Activated carbon adsorption of Chloroacetanilide herbicides and their degradation products from surface water supplies. *AQUA: Journal of Water Supply: Research and Technology, 52*, 443–54.

Harding, J. S., Benfield, E. F., Bolstad, P. V., Helfman, G. S., and Jones, E. B. D. (1998). Stream biodiversity: the ghost of land use past. *Proceedings of the National Academy of Sciences of the United States of America, 95*, 14843–47.

Holland, J. M. (2004). The Environmental consequences of adopting conservation tillage practices in europe. *Agricultural Ecosystems & the Environment, 103*, 1–25.

Hollaway, M. J., Arnold, S. R., Challinor, A. J., and Emberson, L. D. (2012). Intercontinental trans-boundary contributions to ozone-induced crop yield losses in the northern hemisphere. *Biogeosciences, 9*, 271–92.

IFPRI. (2002). *Green revolution: curse or blessing?* Washington, D.C.

Land, M., Granéli, W., Grimvall, A., Hoffmann, C. C., Mitsch, W. J., Tonderski, K. S., and Verhoeven, J. T. A. (2016). How effective are created or restored freshwater wetlands for nitrogen and phosphorus removal? a systematic review. *Environmental Evidence*. 5, Article no. 9. doi: 10.1186/s13750-016-0060-0.

Long, S. P., Ainsworth, E. A., Leakey, A. D., and Morgan, P. B. (2005). Global food insecurity: treatment of major food crops with elevated carbon dioxide or ozone under large-scale fully open-air conditions suggests recent models may have overestimated future yields. *Philosophical Transactions of the Royal Society B, 360*, 2011–20.

McConnell, L. L., Kelly, I. D., and Jones, R. L. (2017). Integrating technologies to minimize environmental impacts. In Hester, R. E., and Harrison, R. M. (eds.) *Agricultural chemicals and the environment: issues and potential solutions*, 2nd edition, pp. 19. London, UK: Royal Society of Chemistry. doi: 10.1039/9781782626916-00001.

Möller, A., Müller, H. W., Abdullah, A., Abdelgawad, G., and Utermann, J. (2005). Urban soil pollution in Damascus, Syria: concentrations and patterns of heavy metals in the soils of the Damascus Ghouta. *Geoderma, 124*(1), 63–71. doi: 10.1016/j.geoderma.2004.04.003.

Monteagudo, L., Moreno, J. L., and Picazo, F. (2012). River eutrophication: irrigated vs. non-irrigated agriculture through different spatial scales. *Water Research.* doi: 10.1016/j.watres.2012.02.035.

Morgan, P. B., Mies, T. A., Bollero, G. A., Nelson, R. L., and Long, S. P. (2006). Season-long elevation of ozone concentration to projected 2050 levels under fully open-air conditions substantially decreases the growth and production of soybean. *New Phytol., 170*, 333–43.

Mosier, A., Kroeze, C., Nevison, C., Oenema, O., Seitzinger, S., and van Cleemput, O. (1998). Closing the global N2O budget: nitrous oxide emissions through the agricultural nitrogen cycle. In *Nutrient Cycling in Agroecosystems, 52*(2), 225–48. doi: 10.1023/A:1009740530221.

NASS (2014). *2012 census of agriculture highlights: conservation.* Washington, D.C.

Novotny, V. (1999). Diffuse pollution from agriculture—a worldwide outlook. *Water Science and Technology, 39*(3), 1–13. doi: 10.1016/S0273-1223(99)00027-X.

NRC (1989). *Biologic markers of air-pollution stress and damage in forests committee on biologic markers of air-pollution damage in trees, National Research Council.* Washington, D.C.: The National Academies Press. doi:10.17226/1414.

NRC (1993). *Soil and water quality: an agenda for agriculture.* Washington, D.C.: The National Academies Press. doi: 10.17226/2132.

NRC (1997). *Valuing ground water.* Washington, D.C.: The National Academies Press. doi: 10.17226/5498.

NRC (1999a). *Measures of environmental performance and ecosystem condition.* Washington, D.C.: The National Academies Press. doi: 10.17226/5147.

NRC (1999b). *Our common journey.* Washington, D.C.: The National Academies Press. doi: 10.17226/9690.

NRC (2000a). *Clean coastal waters.* Washington, D.C.: The National Academies Press. doi: 10.17226/9812.

NRC (2000b). *Global change ecosystems research.* Washington, D.C.: The National Academies Press. doi: 10.17226/9983.

NRC (2000c). *The future role of pesticides in U.S. agriculture.* Washington, D.C.: The National Academies Press. doi: 10.17226/9598.

NRC (2008). *Mississippi river water quality and the Clean Water Act: progress, challenges, and opportunities.* Washington, D.C.: The National Academies Press. doi: 10.17226/12051.

NRC (2010). *Toward sustainable agricultural systems in the 21st century.* Washington, D.C.: The National Academies Press. doi: 10.17226/12832.

NRC (2012a). *Exploring health and environmental costs of food: workshop summary.* Washington, D.C.: The National Academies Press. doi: 10.17226/13521.

NRC (2012b). *Exploring sustainable solutions for increasing global food supplies: report of a workshop.* Washington, D.C.: The National Academies Press. doi: 10.17226/13319.

NRC (2012c). *Science for environmental protection.* Washington, D.C.: The National Academies Press. doi: 10.17226/13510.

NRC (2016a). *A review of the landscape conservation cooperatives.* Washington, D.C.: The National Academies Press. doi: 10.17226/21829.

NRC (2016b). *The future of atmospheric chemistry research.* Washington, D.C.: The National Academies Press. doi: 10.17226/23573.

NRCS. (2017). Mississippi River Basin Health Watersheds Initiative. Retrieved from https://www.nrcs.usda.gov/wps/portal/nrcs/detailfull/national/home/?cid=stelprdb1048200.

O'Geen, A. T., Budd, R., Gan, J., Maynard, J. J., Parikh, S. J., and Dahlgren, R. A. (2010). *Mitigating nonpoint source pollution in agriculture with constructed and restored wetlands. Advances in*

Agronomy. In Advances in Agronomy, Vol. 10 8, Bulington Academic Press, pp. 1-76. doi: 10.1016/S0065-2113(10)08001-6.

Ockenden, M. C., Deasy, C., Quinton, J. N., Bailey, A. P., Surridge, B., and Stoate, C. (2012). Evaluation of field wetlands for mitigation of diffuse pollution from agriculture: sediment retention, cost and effectiveness. *Environmental Science and Policy*. doi: 10.1016/j.envsci.2012.06.003.

Palmer-Felgate, E. J., Acreman, M. C., Verhoeven, J. T. A., Scholz, M., Maltby, E., Stratford, C. J., Newman, J., Miller, J., and Coughlin, D. (2013). How effective are reedbeds, ponds, restored and constructed wetlands at retaining nitrogen, phosphorus and suspended sediment from agricultural pollution in England? *Environmental Evidence*. doi: 10.1186/2047-2382-2-1.

QDNR. (1997). *Planning guidelines separating agricultural and residential land uses*. Brisbane: Department of Natural Resources: Dept of Local Government and Planning.

Richards, P. R., Baker, D. B., and Crumrine, J. P. (2009). Improved water quality in Ohio tributaries to Lake Erie: A consequence of conservation practices. *Journal of Soil and Water Conservation*, *64*(3), 200–211.

Rockström, J., Steffen, W., Noone, K., Persson, Å., Chapin III, F. S., Lambin, E. F., Lenton, T. M., et al. (2009). A safe operating space for humanity. *Nature*, *461, 472-475*(September). doi: 10.1038/461472a.

Roe, S., Streck, C., Weiner, P. H., Obersteiner, M., and Frank, S. (2017). How improved land use can contribute to the 1.5°C Goal of the Paris Agreement.

Roy-Poirier, A., Champagne, P., and Filion, Y. (2010). Bioretention processes for phosphorus pollution control. *Environmental Reviews*. doi: 10.1139/A10-006.

Sandefur, H. N., Cummings, E., Johnston, R. Z., Matlock, M. D., and Gustafson, D. I. (2015). Analyzing correlations between stream benthic macroinvertebrate community structure, water quality, and watershed land use in the continental United States. *Biological Engineering Transactions*, *7*(4), 169–82.

Smalling, K. L., Reeves, R., Muths, E., Vandever, M., Battaglin, W. A., Hladik, M. L., and Pierce, C. L. (2015). Pesticide concentrations in frog tissue and wetland habitats in a landscape dominated by agriculture. *Science of the Total Environment*. doi: 10.1016/j.scitotenv.2014.08.114.

Snapp, S. S., Swinton, S. M., Labarta, R., Mutch, D., Black, J. R., Leep, R., Nyiraneza, J., and O'Neil, K. (2005). Evaluating cover crops for benefits, costs and performance within cropping system niches. *Agronomy Journal*, *97*, 322–32.

Snyder, C. (2016). Suites of 4R nitrogen management practices for sustainable crop production and environmental protection. International Plant Nutrition Institute, Issue Reivew, No. 1, Peachtree Corners, GA.

Stein, L. Y., and Yong, Y. L. (2003). Production, isotopic composition, and atmospheric fate of biologically produced nitrous oxide. *Annual Reviews of Earth & Planetary Science*, *31*, 329–56.

Stewart, B. A., and O'Malley, R. (2002). Croplands. In Lai, R. (ed.) *Encyclopedia of soil science*, pp. 369–71. New York, NY: Marcel Dekker.

The Fertilizer Institute. (2017). 4Rs nutrient stewardship. Retrieved from http://www.nutrientstewardship.com/4rs/4r-principles/.

Tilman, D., Cassman, K. G., Matson, P. A., Naylor, R., and Polasky, S. (2002). Agricultural sustainability and intensive production practices. *Nature*, *418, 671-677*(August), 671. doi: 10.1038/nature01014.

USGS. (2017). National water-quality assessment (NAWQA) program. Retrieved from https://water.usgs.gov/nawqa/.

Wang, X., Manning, W., Feng, Z., and Zhu, Y. (2007). Ground-level ozone in China: distribution and effects on crop yields. *Environmental Pollution*. 147(2), 394-400 doi: 10.1016/j.envpol.2006.05.006.

Wauchope, R. D., Estes, T. L., Allen,R., Baker, J. L., Gustafson, D. I., Hornsby, A. G., Jones, R. L., Richards, R. P., 9 Aventis Cropscience, and Alexander, T. W. (2002). Predicted impact of transgenic, herbicide-tolerant corn on drinking water quality in vulnerable watersheds of the Midwestern United States. *Pest Management Science*, *58*, 146–60.

Wen, Y., Schoups, G., and Van De Giesen, N. (2017). Organic pollution of rivers: combined threats of urbanization, livestock farming and global climate change. *Scientific Reports* 7 doi: 10.1038/srep43289.

Westenbarger, D. A., and Frisvold, G. B. (1995). Air pollution and farm-level crop yields: an empirical analysis of corn and soybeans. *Agricultural & Resource Economics Review, 24*, 156–65.
Xiong, X., Yanxia, L., Wei, L., Chunye, L., Wei, H., and Ming, Y. (2010). Copper content in animal manures and potential risk of soil copper pollution with animal manure use in agriculture. *Resources, Conservation and Recycling, 54*(11), 985–90. doi: 10.1016/j.resconrec.2010.02.005.
Zhang, X., Davidson, E. A., Mauzerall, D. L., Searchinger, T. D., Dumas, P., and Shen, Y. (2015). Managing nitrogen for sustainable development. *Nature, 528*(November), 51. doi: 10.1038/nature15743.

THE IMPACT OF AGRICULTURE ON GLOBAL BIODIVERSITY

Stuart L. Pimm
Duke University

Varsha Vijay
Duke University and National Socio-Environmental Synthesis Center

Introduction

In this chapter, we ask several simple questions in two sets. First, we review in broad outline what we know about biodiversity and how human actions are depleting it. We ask: how many species are there, both named and unnamed? How fast are species now going extinct? How fast did species go extinct before human actions inflated the extinction rate? And how fast do species diversify and thus possibly recover from the current massive losses? Finally, where are extinctions concentrated? The second set of questions considers how agriculture—and the demands of the growing human population—affect biodiversity. By "agriculture" we mean activities devoted to growing crops and rearing livestock. Within our classification, we include tree plantations, which produce food, like palm oil or cacao, and timber. We ask: where are these agricultural activities concentrated? We then address the potential impacts of these activities on terrestrial biodiversity and ask: Where does agriculture—now broadly defined—collide with the concentrations of species most at risk of extinction? Where are the frontiers where agricultural expansion most harms—or has the highest potential to harm—biodiversity?

• • •

How Many Species Are There?

"Biodiversity" includes all facets of the variety of life, but we will concentrate on species, since other measures are poorly mapped globally. This deceptively simple question of how many there are has a rich pedigree. Westwood (1833) speculated "On the probable number of species of insects in the Creation." In the last few decades, studies have drawn widely varying conclusions (Gaston, 1991; May and Beverton, 1990; Mora et al., 2011; Stork, 1993). A recent, thorough compilation of previous estimates by Chapman (2009), plus new studies that propose novel methods of estimation, motivated a review of progress (Scheffers et al., 2012).

That review dealt only with eukaryotic organisms. We are far from being able to make similar estimates for Eubacteria and Archaea, where the definition of species is problematic. The numbers of forms might even reach one trillion (Locey and Lennon, 2016).

The review counted the numbers of species that have been named within various taxonomic groupings. Deriving such estimates, even for well-studied groups like birds, is not straightforward. There are two kinds of problems. First, far more species exist than taxonomists have named. Second, taxonomists have complicated matters inadvertently by giving multiple names to many species. For large groups of poorly studied species, the determination of the number of such synonyms is difficult.

The completeness of global inventories varies greatly. Inventories are ~97 percent complete for mammals, 85 percent for flowering plants, 79 percent for fish, 67 percent for amphibians and no more than ~30 percent for arthropods (Hamilton et al., 2010; Joppa et al., 2010; Mora et al., 2008) (Table 1). Taxonomic effort is almost evenly distributed among vertebrates, plants, and invertebrates, yet plants have roughly ten times—and invertebrates at least a hundred times—more species than vertebrates.

The greatest uncertainties in species numbers involve "hyper-estimates"—by which we mean individual totals of 5 million species or more.

For example, Erwin (1982) estimated there might be 30 million species of arthropods. His calculation started with his assertion that 13.5 percent of the beetle species he found might live only in a single species of tree in Panama, then extrapolated to more tree species and other arthropod groups. His estimate generated much criticism, primarily from those concerned about the assumptions underlying such a "small to large" extrapolation. The method, however, spawned considerable interest and research.

Fundamental to the controversy was the degree of host specificity of herbivorous insects on their food plants—13.5 percent for the beetles he studied.

Novotny et al. (2007), ØDegaard (2000) and others found considerably lower host specificity than this, perhaps by a factor of four or five. Subsequent global estimates of insect species numbers have consequently been considerably lower. The subject requires much further study before we can attain a reasonable degree of precision.

Fungi are another poorly known group. Hawksworth (1991) started with the 6:1 ratio of fungi to flowering plant species found in Britain, where both groups are well known, and extrapolated this ratio to the global total for flowering plants, yielding an estimate of 1.62 million species of fungi globally. May (1991) was sharply critical of this estimate. A key concern, May argued, is that the species-rich tropics are unlikely to have the same ratio of fungi to plants as does Britain. Elsewhere, the ratio is even higher (Taylor et al., 2014). Even if the ratio were only 6:1, then collections of fungi from the tropics should find that >95 percent of the species encountered would be new, given that as of 2014 only about 70,000 fungi had been catalogued globally. May noticed that actual percentages of new species from tropical samples were much smaller. An alternative approach (Mora et al., 2011) estimates roughly 611,000 fungal species exist globally, which seems to support May's more conservative estimate of about 500,000 fungal species.

Quantitative Methods

Estimating the total number of species in a taxon would seem to be a straightforward task from the rates of their discovery. These rates should slow as the number of as-yet undescribed species falls and so the cumulative number of species should asymptote (Costello and Wilson, 2011; Ricotta et al., 2002; Solow and Smith, 2005). For birds globally, and for some taxa regionally, the rates of description are slowing over time, whence asymptotic approaches provide reasonable estimates. Mora et al. (2011) employ this approach to estimate the numbers of higher taxa as well—as described above.

Quite unexpectedly, for many taxa, not only are the rates of species description increasing, they are doing so exponentially. This precludes estimates of asymptotes. While there are widespread concerns that the numbers of taxonomists are decreasing, the numbers of individuals who describe species are, in fact, increasing and doing so exponentially. The pattern holds for mammals, amphibians, spiders, marine snails of the Conus (Joppa et al., 2011b), parasitic wasps (Bacher, 2012), and several other invertebrate taxa for which taxonomists have given Pimm access to their data. The discrepancy is surely that there are fewer dedicated taxonomists now than in the past for some groups, but that many more individuals likely describe species as part of their other research.

To account for this increase in the numbers of individuals describing species, a mechanistic model accounts for taxonomic effort and for changing taxonomic efficiency. Initially proposed by Joppa et al. (2011a, 2010) this model defines taxonomic effort as the number of people involved in describing species and taxonomic efficiency as an increase in the number of species described per person, adjusted for the continually diminishing pool of as-yet-unknown species. Rates of species description divided by taxonomic effort generally decline. This allows statistical estimation of the likely number of species in the taxon.

How well does this model perform? For plants, validating it by expert opinion revealed broad agreement with its predictions of how many species remained to be described (Joppa et al., 2010). In some plant taxa and for parasitic wasps, the numbers of species described per taxonomist remains roughly constant even as the pool of missing species inevitably declines. Taxonomists can describe only so many species in a year—regardless of how many missing species there are—and work through the backlog of undescribed species methodically, genus by genus.

Finally, the first description of a species is merely the first step. The number of recognized bird species has continued to rise, even though birds are the best-known taxon. Some of this increase represents genuinely new discoveries, but much is the recognition that what was once one species should now be best viewed as several. And, some of those changes stem from changing ideas of whether distinct allopatric populations should be considered the same or different species. Even for mammals, many species could not be considered as well-known, and for other taxa only a small fraction has been studied in detail.

Plants

Because plants are the base of the living food chain they demand special attention. For example, he expected large numbers of herbivorous insects likely depend on how many plant species there are. Collaboration between Kew and the Missouri Botanical Garden created the Global Plant List (www.plantlist.org), which lists 351,00 species of plants, comprised of 304,000 flowering plants, 1,100 species of gymnosperms, 10,600 species of ferns and lycopods, and 35,000 species of mosses and liverworts.

Plant taxonomists are working through the lists of plant species family-by-family to resolve synonyms. For the families assessed to date, there are about twice as many synonyms as valid names. For the species in the families not yet assessed, adding half the names to the already assessed total, estimates approximately 400,000 vascular plant species (Pimm et al., 2014; Pimm and Raven, 2017).

The rate of describing new plant species shows the taxonomic catalogue is far from complete. Both expert opinion and the Joppa et al. models suggest there will be 15 percent more plant species, resulting in an approximate total of 460,000 species (Pimm and Raven, 2017). If there are another 60,000 or so species to be named, perhaps half of them are already present as specimens in herbaria (Brummitt et al., 2015).

Summary

We likely have good estimates of how many vertebrate species there are and a plausible estimate of how many plant species there will be once the taxonomic catalogue is complete. Some other taxa are reasonably well known, but the estimates of the numbers of arthropods, especially insects, and fungi are widely uncertain. This makes any statements about how many species there are in total uncertain, and with that, we have no way to state how many species agricultural expansion exterminates per unit of time. To be able to quantify human impacts requires the different metric that we shall now introduce.

How Fast Are Species Going Extinct, and How Fast Should They Go Extinct?

> "...a mass extinction crisis, with a rate of extinction now 1,000 times higher than the normal background rate."—Al Gore, in the Academy Award–winning documentary *An Inconvenient Truth* (Gore, 2006, p. 152).

As Gore's remark testifies, it has become standard to quantify present extinctions as rates and then to compare them to some background that typifies geological history. Mass extinction events are understood to be exceptions. Gore's and similar statements emerged via secondary sources originally derived from Pimm et al. (1995) who estimated "recent extinction rates are 100 to 1,000 times their pre-human levels."

Before 1995, the literature quoted statistics on current extinctions in "species per day." Estimates ranged from a "minimum" of three (Myers, 1989) to "a hundred species per day" (Stork, 2010). More than uncertainties about the extinctions themselves, the numbers reflected the wide range of estimates for how many eukaryote species actually exist (see above).

The uncertainty in the "species per day" estimates also posed problems when dealing with critics of environmental concerns who demanded the scientific names of those recently extinct (Stork 2010). Of course, taxonomists have described only a small fraction of species, while the IUCN's Red List (www.iucnredlist.org) has assessed an even smaller fraction of those—about 53,000 species. To avoid the necessarily complex caveats for extinctions per day estimates, Pimm et al. (1995) deliberately chose to replace this metric with a proportional rate that they could calculate for a given taxon.

Estimating the current extinction rates is a straightforward process: one that follows a cohort of species for some specified interval and records what fraction succumbed (Pimm et al., 2006; Pimm et al., 1995; Pimm et al., 2014). For numerical convenience, we calculate extinctions per million species-years (E/MSY). This method does not count species, such as the dodo, that went extinct before the 1750s, when the current system of Latin names was first applied to organisms. For example, taxonomists have described 1,230 species of birds since 1900, of which 13 have become extinct (Table 2). This cohort accumulated 98,334 species years—meaning that an average number of species has been known for 80 years—and so the extinction rate is $13*10^6/98,334 = 132$ E/MSY.

Extinction rates from cohort analyses average about 100 E/MSY (Table 2). Given that most species are still undescribed and many species with small ranges are recent discoveries, these numbers are bound to be underestimates for global biodiversity. Many, if not most, species will have gone or be going extinct before taxonomists give them Latin names. This happens even for well-known birds (Lees and Pimm, 2015). The extinction rates of species described after 1900 are considerably higher than the rates for those described earlier, reflecting the fact that particularly in well-known groups, those species described more recently will on average be rare and restricted in range. (Pimm et al., 2014).

Rates of extinction increase with improved knowledge—since taxonomists describe widespread common species before local and rare ones, and the latter are more vulnerable to human actions. Thus, recent extinction rates based on poorly known taxa (such as insects) are bound to be substantial underestimates since so many species remain undescribed.

Finally, in addition to species deemed extinct, there are substantial fractions of species that are critically endangered (Pimm et al., 2014). Some of these may indeed be already extinct, while others survive only through energetic conservation efforts.

How Fast Should Species Be Going Extinct?

De Vos et al. (2015) explored three lines of evidence toward achieving the more difficult task of estimating this "background rate" of extinction. By this, we mean the geologically recent rates of extinction before human actions inflated them.

The fossil record provides broad information, but it is very limited and extremely uncertain in temporal resolution and taxonomic breadth (Purvis, 2008). Moreover, most paleontological studies assess genera, not species (Flessa and Jablonski, 1985). The one conclusion one can draw from (Barnosky et al., 2011) are the large uncertainties in the extinction rates of fossils provide. Making comparisons—such as the current rate of extinction (as estimated above) with (say) the rate of extinction of fossil taxa over hundreds of millions of years is meaningless. Claims of Earth being in the midst of a "sixth extinction" are best left to those seeking the attention of journalists. (That said, if current rates continue for decades or centuries to come, then Earth's biodiversity will take a very large hit.)

Molecular phylogenies provide an appealing alternative to the fossil record's many shortcomings, covering a large range of taxa, time periods, and environments. The simplest case expects a single lineage to grow to a clade size of N species, E(N), during time, t, according to

$$E(N) = \exp(r * t) \qquad (1)$$

The net diversification rate, r, is the speciation, λ, minus the extinction, μ, $(\lambda - \mu)$.

$$(\lambda - \mu) = r = \ln(N)/t \qquad (2)$$

Can one separately estimate the speciation and extinction rates, not just their difference? Equation (1) expects the plot of the logarithm of the number of lineages through time (LTT) to be linear, with slope $\lambda - \mu$ but with an important qualification. In the limit of the present day, there are no extinctions of the most recent taxa—they have not yet happened. Thus, near the present, the LTT slope should increase and approach λ, the speciation rate (Nee, 2006). This allows the separate calculation of the speciation and extinction rates.

Practice is considerably more complicated than this simple theory. To address them, De Vos et al. (2015) undertook an extensive set of simulations of known diversifications, with varying fractions of species removed from these modeled clades, and in some cases, different rates of diversification. At best, their simulations demonstrate that the underestimation of extinction due to complex diversification processes may be only slight: they always recovered the correct order of magnitude of absolute extinction across replicate phylogenies, even though individual estimates were associated with large uncertainties. They concluded that phylogenies of extant taxa should contain at least some information on extinction rates (Pyron and Burbrink, 2013).

Despite these difficulties, an important conclusion emerges. Of the 140 phylogenies, (De Vos et al., 2015) analyzed, all but four had median estimated extinction rates of < 0.4 E/MSY and only two (one arthropod and one mollusk) had rates > 1 and those were < 1.5. These estimates match those of Weir and Schluter (2007) who estimated bird and mammal extinctions that range from 0.08 E/MSY at the equator—where there are the most species—to ~0.4–0.6 E/MSY at 50° latitude—where they are far fewer. Thus, despite the methodological hurdles and the potentially confounding whims of taxonomists, there is consistent evidence that background extinction rates are below one extinction per million species-years and most likely much less than this.

How Fast Can Species Diversify?

In contrast to the methodological complexities of separating speciation and extinction rates, it is relatively easy to calculate their difference—the rates of diversification. Consequently, there are many estimates of diversification. Median diversification rates range from ~0.05 to ~0.2 new species per species per million years for disparate groups of animals and plants. Rates > 1.0 are exceptional.

Valente et al. (2010) explicitly addressed the issue of how fast taxa can diversify. They analyzed the genus *Dianthus* (carnations, Caryophyllaceae) and found net diversification rates of up to sixteen new species per species per million years. This puts them well above eleven other plant groups to which they compared them, the highest rate of which was Andean *Lupinus* (lupines, Fabaceae) at ~2 (Koenen et al., 2013). For birds, the record holder is the group of Southeast Asian *Zosterops* (white-eyes, Zosteropidae), at 2.6 new species per species per million years. Others classify some of these "species" as subspecies, which would reduce that rate; the distinction is quite arbitrary.

In contrast are lineages such as *Ginkgo,* which appears to have changed little since the Jurassic (Zhou and Zheng, 2003).

Summary

Combining the limited insights from the fossil record, from the separation of speciation and extinction rates from molecular phylogenies, and overall diversification rates (De Vos et al., 2015), concluded that background extinctions rates are typically ~0.1 extinctions per million species per year. This rate indicates that current extinction rates are a thousand times higher than the historical background.

Which Species Are at Risk of Extinction, and Where Are They?

The IUCN Red List (IUCN, 2016) assigns species to classes of Least Concern, through Near-Threatened to three progressively escalating categories of Threatened species (Vulnerable, Endangered, Critically Endangered) and Extinct. Other than the last category, these constitute an expert system of the relative likelihood of how soon species will become extinct in the future, given present conditions. The most important criteria in determining risk are the size of a species' geographic range and the degree to which the effects of human actions (including climate change) are shrinking its habitat. Other factors include whether humans hunt, collect, or kill the species for other reasons.

The Red List is incomplete. By January 2017, IUCN had considered 85,604 of >1.7 million named species. Groups of vertebrate animals get the most attention (~45,000), invertebrates constitute another 19,000, while plants ~22,000. Of the 63,303 total animals assessed, 11,923 were judged to be data deficient—meaning there is not enough information to know their status. This often means that they are rare, thus indicating that a large fraction of them may be threatened with extinction. Of the remaining 51,380, 1,285 were extinct or probably so, while 12,630 were threatened, with 2,696 of them deemed critically endangered. Birds, mammals, and amphibians constitute ~23,000 of the assessed species and are mostly terrestrial or partially terrestrial species. They are the only groups for which there are comprehensive species distribution maps (www.biodiversitymapping.org).

For plants, Kew (2016) tabulates of the various estimates are mostly in the 20 to 33 percent range, and highlights the complexities. First, the IUCN Red List has assessed a sample of only ~22,000 species, of which 7 percent had insufficient data, and 52 percent were deemed threatened. Kew (2016) argues that this is too high an estimate for all plants because those compiling the Red List had understandably emphasized taxa at particular risk of extinction.

An estimate of 20 percent is broadly comparable with the percentage determined for well-known vertebrate taxa, for which one can separately calculate extinction rates directly by following species fates from their year of scientific description (see above). For these vertebrates, it has been estimated that human actions are now driving species extinct a thousand times faster than their normal background rates of extinction. If plants have similar rates of extinction, these would also be one thousand times faster than the rates of plant speciation derived from molecular phylogenies (De Vos et al., 2015).

Kew's (2016) tabulations show higher percentages of species at risk when one sensibly adds in the predicted numbers of species as-yet unknown to the total of threatened species. They are likely to be rare, which is why we have not yet found them. Moreover, they are likely to be where recent discoveries are being made which, unfortunately, are places where habitat loss is proceeding rapidly (Joppa et al., 2011a)

Where Are Species at Risk of Extinction?

These concerns motivate the next question of asking where species are at risk of extinction. There are at several "laws" to describe the geographical patterns of where species occur. By "law," we mean a general, widespread pattern, that is, one found across many groups of species and many regions of the world. Wallace (1855) famously provides one, in describing the general patterns of evolution in his famous "Sarawak Law" paper:

- *LAW 1. The following law may be deduced from these [preceding] facts:—Every species has come into existence coincident both in space and time with a pre-existing closely allied species. He would uncover natural selection, as the mechanism behind this law, a few years later, independently of Darwin. There are other laws too.*
- *LAW 2. Most species' ranges are very small; few are very large.* Pimm et al. (2014) show that the range size of flowering plants, amphibians, mammals, birds, and the marine gastropod genus Conus are highly skewed. There are species with very large ranges—some >10 million km^2, for example. However, for example, over half of all amphibian species have ranges smaller than 4,300 km^2. The comparable medians for the other taxa range from ~115,602 km^2 (mammals) to ~729,770 km^2 (plants). For plants, the number is certainly a substantial overestimate, since the range data come from coarse regions

(e.g., single countries) and a plant species will not likely occur everywhere within a given region.

- *LAW 3. Species with small ranges are locally scarce.* There is a well-established relationship across many geographical scales and groups of species that link a species' range to its local abundance. The largest-scale study is that of Manne and Pimm (2001) who used data on bird species across South America. A species is "common" if one is nearly guaranteed to see it in a day's fieldwork, then "fairly common," "uncommon" down to "rare"—meaning it likely takes several days of fieldwork to find one even in the appropriate habitat. Almost all bird species with ranges greater than 10 million km^2 are "common," while nearly a third of species with ranges of less than 10,000 km^2 are "rare" and very few are "common."
- *LAW 4. Species with small ranges are geographically concentrated and...*
- *LAW 5. ...those concentrations are generally not where the greatest numbers of species are found.* The simplest expectation is that where there are more species, there will be more large-ranged species, more small-ranged species (and indeed everything in between)—and, of course, more threatened species. Reality is strikingly different.

Figure 1 shows the patterns for all species and small-ranged species of birds. Broadly comparable maps for mammals and amphibians are at www.biodiversitymapping.org. At a much coarser spatial resolution, they mirror the patterns for plants (Jenkins et al., 2013).

The concentrations of small-ranged species are often not where the greatest numbers of species are. Several interesting consequences emerge. Most of the species at greatest risk of extinction concentrate geographically and, broadly, such species in different taxa concentrate into the same places—the hotspots. This is of huge practical significance for it means that conservation efforts can be and should be concentrated in these special places. Moreover, priorities set for one taxonomic group may be sensible for others, at least at this geographical scale. These ideas were first developed by Norman Myers (Myers, 1988).

Other things being equal, extinctions will concentrate where there are many species with small ranges. Other things are not equal, of course, and the other important driver is human impact. Myers added the second—and vital criterion—that these regions have less than 30 percent of their natural vegetation remaining. Myers's idea is a very powerful one. It creates the "number of small-ranged species times habitat loss equals extinction" idea

FIGURE 1. The distribution of all bird species (top) and those with smaller than the median geographical range size (bottom).

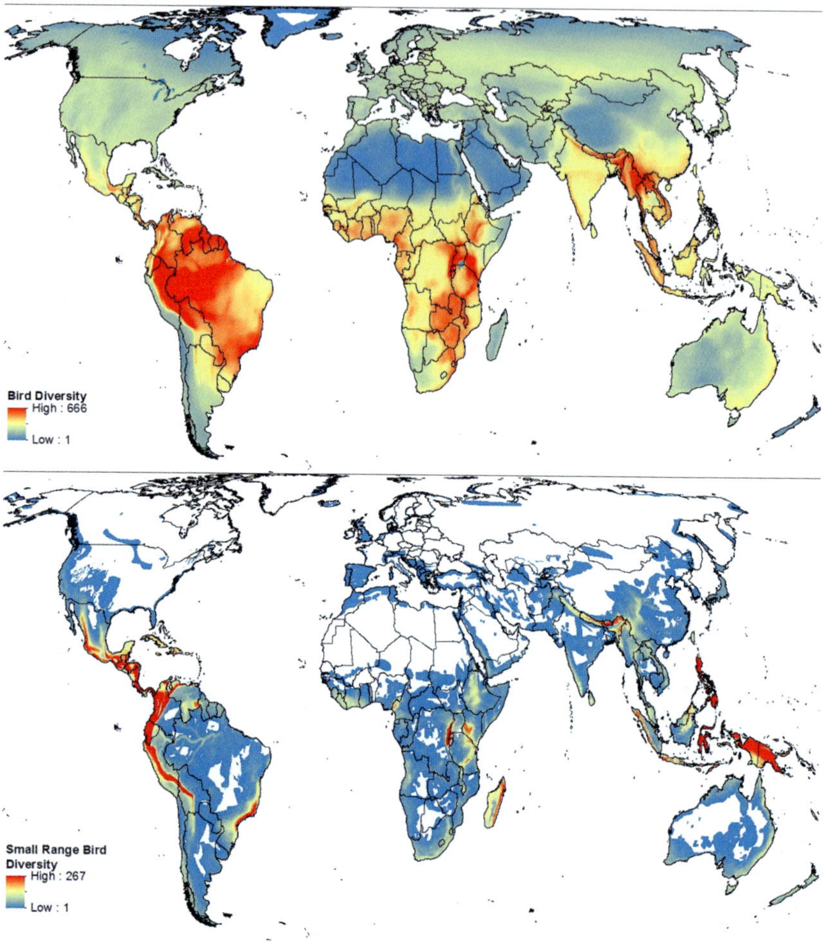

Data from www.biodiversitymapping.org.

with another key and surprising insight. What is surprising is that there are few examples of concentrations of small-ranged species that do not also meet the criterion of having lost 70 percent or more of their natural habitat. The island of New Guinea is an exception, though habitat losses there are quickly depriving it of that distinction. Hotspots that have a disproportionate human impact measured in other ways besides their habitat loss. Cincotta et al. (2000) show that hotspots generally have higher human population densities and that almost all of them have annual population growth rates that are higher (average = 1.6 percent per annum) than the global average (1.3 percent per annum).

We considered each of the twenty-five hotspots using the statistics on endemic bird species, original area, and the current area of remaining natural vegetation (Pimm and Raven, 2000). This provides a best-case scenario of what habitat might remain. We predicted that ~1,700 species of birds should be eventually lost just from the loss of habitat to date—a number similar to current estimates of the number of threatened species.

Species can obviously linger in small habitat fragments for decades before they expire (Brooks et al., 1999; Ferraz et al., 2003; Pimm and Brooks, 2013). Based on these rates of species loss, perhaps three-quarters of these species—1,250—will likely go extinct this century. If so, Pimm and Raven (2000) estimated future extinction rates of ~1,000 E/MSY from human actions to date—about ten times larger than the ones we currently observe. Put another way, extinction rates should accelerate.

Two extrapolations are possible. The worst-case scenario for the hotspots assumes that the only habitats that will remain intact will be the areas currently protected. This assumption increases the prediction of the number of extinctions to 2,200 (Pimm and Raven, 2000). The second adds in species from areas not already extensively deforested. These changes would even further accelerate the rates of extinction.

Summary

Very roughly, a quarter of well-known species are threatened with extinction. Terrestrial plant and vertebrate species are concentrated in the tropical moist forests of the world. There are far fewer species in more temperate climes. Importantly, the distribution of geographical range sizes is such that there are many species with small ranges. Moreover, such species tend to be locally rare, while species with large ranges tend to be locally common. Such small-ranged species account for about 90 percent of threatened species; the remainder are large-ranged species that conflict with human activities—such as large predators. The distributions of small-ranged species (and so threatened species) are geographically concentrated.

Where Is Agriculture?

We define agricultural areas as both cropland (areas used for crop cultivation) and pastureland (non-cultivated land used for livestock grazing). Our estimates of agricultural land are from Teleguntla et. al (2016) for cropland (Figure

2 top) and Ramankutty et al. (2008) for pasture (Figure 2 bottom), re-scaled to 10 km to match the spatial resolution of our biodiversity datasets. Based on these estimates, agricultural areas cover approximately 30 percent of the terrestrial land area. Comparing the current extent of agriculture to the maps of biodiversity (Figure 1) shows that only a minority of croplands and pasturelands overlap with either the areas with the greatest numbers of species, or the numbers of small-ranged species. The extensive areas of croplands—and certainly intensive cropping—are in midwestern North America, northern Argentina, Europe, West Africa, India, and eastern China. Only West Africa and India have exceptional numbers of species. There is even less overlap with pasturelands, though of course many croplands also have some mixed use.

FIGURE 2. The distribution of croplands (top) and grazing lands (bottom).

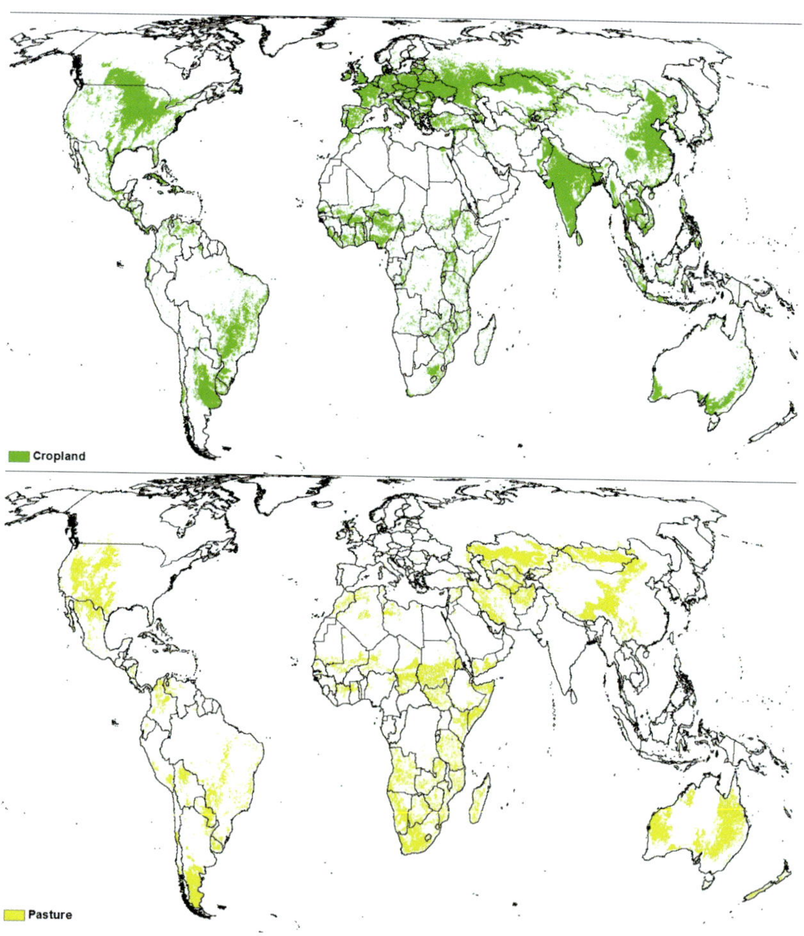

Broadly, most species live in the topical moist forests of the world—two-thirds of all plants, for example (Pimm and Joppa, 2015)—and most agricultural land is elsewhere. One important caveat to this assessment is that agricultural data in the tropics is more incomplete than in the temperate regions of Europe, North America, and East Asia. Where hotspots and other areas of high biodiversity overlap with agricultural production, especially those where rapid expansion has occurred, available natural habitat areas may be even smaller than previously understood. This means that species may be more threatened than their current IUCN status indicates, an extension of ideas presented in (Ocampo-Peñuela et al., 2016).

While recent increases in food production have reduced food scarcity (Godfray et al., 2010), growing human populations will increase demand for agricultural production (for food, feed, fiber, and bioenergy) through the first half of this century (Tilman et al., 2011). There are two primary approaches to addressing the need for additional agricultural production. The first approach is the expansion of agriculture into areas that are not currently cropland. The second is the intensification of production in existing croplands. Both approaches increase production but may pose potential issues for biodiversity directly through habitat loss and indirectly through effects like increased human wildlife conflict and reduced connectivity between intact areas. Both current and future croplands remain constrained by land suitability, dictated by climatic, edaphic, and topographic conditions.

Are Species Where They Are Because Agriculture Isn't?

The obvious question is whether past agricultural expansion has already eliminated species so their concentrations (Figure 1) reflect what remains. We think this is unlikely. There is nothing to suggest that Europe, for example, ever had large numbers of plants and animals comparable to tropical forests, but lost to agricultural expansion over the last millennia.

In eastern North America, the more recent clearing of forests following settlement by Europeans is easier to document. Pimm and Askins (1995) document that the minimum forest cover was in about 1870, but then the region partially recovered as settlers moved into the prairies. The destruction of about half the forest resulted in the predicted number of bird extinctions—four—but that was because so few species were endemic to the area.

The extensive conversion of the American prairies in the late 1800s and the "Virgin Lands" policies of the former USSR in today's Kazakhstan in the 1950s, certainly harmed grassland species. Again, bird examples are well-documented: The former exterminated the Eskimo curlew (*Numenius borealis)*, the latter critically endangered the sociable plover *(Vanellus gregarius)*. Surely many more species other than birds were lost, but there is no evidence for massive losses: These areas had relatively few endemic species.

In short, the clearing of temperate forests and prairies for agricultural cultivation, likely did not eliminate many species. This, perhaps surprising, conclusion does not preclude agriculture and its expansion as being a major threat to species.

In fact, there is reason to believe that agricultural expansion may affect areas more vulnerable to high numbers of species extinctions. Given current climate conditions, analyses show that broad areas of the tropics are suitable for rainfed and irrigated agriculture (Zabel et al., 2014). Other studies suggest that agricultural production in Africa, Southeast Asia, and South America could be increased by increasing the amount of land used for multiple harvests per year (Ray and Foley, 2013). By 2100, with the effects of changing climate, many of these areas may see reductions in suitable area (Zabel et al., 2014), thus suggesting that short-term expansion or intensification in the tropics or other areas of high biodiversity may occur at a high cost to species without long-term gains in productivity.

Where Agriculture and Biodiversity Collide

As we address the challenge of meeting the goal of global food security, through either agricultural intensification or expansion, it is clear that both strategies pose some risk for biodiversity (Foley et al., 2011). Possible harmful effects from agricultural activities include pollution from agrochemicals, unsustainable reductions in water availability, increased human wildlife conflict, and habitat loss. Habitat fragmentation and loss is a global driver of species extinctions, particularly in highly biodiverse areas (Hosonuma et al., 2012). These effects have consequences for both the species within the terrestrial systems, discussed here, as well as species found within the aquatic biome.

Therefore, a close examination of these risks is needed, especially in cases where food insecurity coincides with high biodiversity. Risk to these species from agriculture is occurring in the context of other threats. A recent assessment of projected land uses suggested that, by 2040, more than a thousand threatened species may lose a majority of their habitat (Pouzols et al., 2014). A global analysis of conflict zones between biodiversity and food security goals identified several key areas like Madagascar and the Peruvian Andes (Molotoks et al., 2017). These areas and others emphasize the need for interdisciplinary approaches by a diversity of disciplines including agro-ecologists, conservation scientists, and social scientists focused on understanding food demand and cultivation practices to minimize, mitigate, and predict possible adverse effects of agricultural production on biodiversity and the environment.

These studies and many others also emphasize the need to understand these risks at a local level. Here, we discuss three case studies exemplifying harmful effects of current and expanding agriculture. We discuss both indirect impacts of agricultural activity, human wildlife conflict, and the direct impact of habitat loss. The first case is a discussion of human wildlife conflict in savanna Africa. The second and third cases, focusing on habitat loss, discuss the Mata Atlântica of Brazil and the expansion of oil palm cultivation across the global tropics, respectively.

Crucially, even in areas where agricultural expansion does not result in species extinctions it may affect local populations in a way that alters overall community structure. One classic example of this is the increase in white-tailed deer populations (*Odocoileus virginianus*) throughout the eastern United States. Deer are habitat generalists and easily adapt to vegetation changes. They benefit from local population declines and extirpations of predators including puma (*Puma concolor*) and gray wolves (*Canis lupus*). In the state of Missouri, the population is so large that the state has proposed to extend the hunting season to reflect the fact that less than 5 percent of the population is killed during the current hunting season. These imbalances affect both plant and animal species diversity, vegetation structure, and may play a role in propagation of invasive species. It is important to note that not all effects of agriculture on biodiversity are negative. These need to be evaluated as well. For example, some methods of crop cultivation may provide important pollen sources for a diversity of pollinator species.

Human Wildlife Conflict in African Savannas

The public's perception of threatened species is surely the charismatic large-bodied species. For instance, the World Wildlife Fund lists giant pandas, tigers, elephants, gorillas, polar bears, and rhinos as its terrestrial species of primary concern. In particular, African savannas and their wildlife readily capture the public's imagination.

The largest extent of savanna is in Africa: 13.5 million km^2—if we define it as areas having between 300 and 1,500 mm of rain annually (Riggio et al., 2013). Most of these areas show as grazing lands in figure 2. Sub-Saharan Africa has had the fastest human population growth and agriculture has expanded into savannas. Livestock cultivation produces three main consequences for large mammals: loss of habitat, reduction in wild prey densities for predators, and increases in human-wildlife conflicts. This and other anthropogenic drivers of habitat loss have resulted in dramatic population and range changes for these species. The top predator—the lion—now occupies only 25 percent of its original range, barely survives in West Africa, and its range is everywhere massively fragmented (Riggio et al., 2013). Another of Africa's iconic species, the elephant, occurs at only a quarter of its estimated equilibrium densities even within the protected areas designed to conserve it (Robson et al., 2017). Another threatened big cat species native to this region, the cheetah, has strongholds in areas with high densities of livestock cultivation. Where the two overlap, a minority of farmers who persecute the cheetah can result in significant reductions in localized population densities (Weise et. al., 2017). Large mammal species are particularly vulnerable to these conflicts since they are very wide-ranging and move through both intact and human-impacted landscapes.

Viewed together, these three species typify large-bodied species that cause conflict with humans through crops damage, livestock predation and, occasionally, through direct attacks on humans. Not surprisingly, species, including tigers in Asia and wolves in temperate regions, have also disappeared from large fractions of their original ranges and faced dramatic reductions in population size and density. Wherever these conflicts occur, the survival of these charismatic mammal species depends on creative solutions to minimize conflict. Two potential solutions are the creation of more effective livestock enclosures and compensation schemes for victims that mitigate damages caused by wildlife (Karanth et al., 2012; Lichtenfeld et al., 2015).

The Mata Atlântica of Brazil—A Biodiversity Hotspot

Myers's (1988) key insight was not just that the most vulnerable species—those with small geographical ranges—were concentrated, but that these concentrations had exceptional levels of habitat loss. These areas are the principal foci for biodiversity loss. Nonetheless, how agriculture drives species extinction in these places is more complicated than a simple conversion of croplands from once-forested areas with a concomitant loss of species. The roughly 1.2 million km^2 of the coastal moist forests of Brazil provides an example. This includes ecoregions of the Alto Paraná Atlantic forests, Araucaria moist forests, Atlantic Coast restingas, Bahia coastal forests, Bahia interior forests, Caatinga Enclaves moist forests, Campos Rupestres montane savanna, Pernambuco coastal forests, Pernambuco interior forests, Serra do Mar coastal forests, Southern Atlantic Brazilian mangroves, and Brazilian Atlantic dry forests. Figure 1b shows the concentration of small-ranged birds, while www.biodiversitymapping.org provides comparable maps for mammals and amphibians. These forests are vitally important for known plant species (Joppa et al., 2013) and models predict them to contain relatively large numbers of the species that taxonomists have not yet described (Joppa et al., 2011a; Pimm et al., 2010).

FIGURE 3. (Top) flat, lowlands of Rio de Janeiro State, Brazil, are mostly cropland, while upland areas (bottom) are a mosaic of forest remnants and cattle pasture.

Clearing the forests here caused some bird extinctions—again birds are the best-known taxa—and very probably some extinctions were missed, since some species were extinct when taxonomists first described them (Lees and Pimm, 2015).

As Dean (1997) documented, the clearing of all but about 5 percent of the forests started soon after European colonization. In brief, the flat lowlands went first, cleared for sugar cane (*Saccharum officinarum*) (figure 3, top). Later, the upland forests were felled for coffee (*Coffea*) that, after several decades, depleted the soil. What remains are isolated forest patches interspersed between extensive areas of land that provides very poor pasture for cattle (Figure 3, bottom).

The land use classifications of this area show it at as neither being extensive croplands nor grazing lands. Today cropland and pastureland occupy approximately 480,000 km^2 and 193,000 km^2 respectively—i.e., about 55 percent of the original mostly forested land in the Mata Atlântica. About 40 percent of the land is neither natural vegetation nor agricultural land: So what is it? On-the-ground inspection of this cleared land shows that much of it is just bad land—land that was once-extensive forest is now unproductive (Jenkins and Pimm, 2003).

Production data confirm this. As a country, it has the most cattle—over twice as many as the United States or the European Union. Its best cattle-producing areas are in the southwest, while those in former coastal forests are often <0.3 head per hectare, about a quarter of the national average (ABIEC, 2016).

The pattern is global and typifies many of the world's biodiversity hotspots. Human actions have shrunk the world's tropical moist forests from an original area of from 14 to 18 million km^2 to about 7 million km^2 at present (Pimm, 2001). Yet only about 2 million km^2 of croplands are in what were formerly humid forests. As in Brazil, these can include very productive sugar cane fields, rice (*Oryza sativa*) paddies, or other crops. That said, 5 to 9 million km^2 of humid forests have become, at best, nominally grazing land, again much of it with very low stocking rates. Globally, *Lantana camara*, *Opuntia* species, *Ulex europaeus*, *Mimosa pudica*, and other introduced, unpalatable plant species dominate much of what are currently putatively pasture lands. Stands of *Eucalyptus*, once likely planted for some reforestation scheme, now appear abandoned.

The Mata Atlântica is typical in another way. Most of the biodiversity hotspots are in mountains—for tropical mountains generate large numbers of small-ranged species. This has protected their unique diversity to some extent, but increasing human population has pushed agriculture into these relatively inaccessible areas.

Not all the biodiversity hotspots are tropical moist forests: Some of them are Mediterranean ecosystems, such as Southwestern Australia and the South African Cape. Many of these areas are now planted in grain and support productive agriculture.

The Expansion of Oil Palm

The last large extents of tropical forests and the large numbers of species that they contain straddle the equator in South America, Africa, and Asia. These areas have been too remote from human population centers and, rice excepted, not been suitable for most commodity crop production. This situation is now changing rapidly with the expansion of oil palm.

Palm oil, derived from the fruit and seeds of the African oil palm (*Elaeis guineensis*), is the most widely consumed vegetable oil in the world. Crop production has doubled in recent years (FAO, 2017), and further expansions are predicted to meet global demand (Corley, 2009; FAPRI, 2012). Oil palm could affect more than half of all threatened birds and mammals globally, if it expands into all biophysically suitable areas (Meijaard et al., 2018). Thus, oil palm is a commodity crop of singular concern to conservation, as it grows almost exclusively in areas that were once tropical moist forests. Because of this, the question is not whether or not the areas that produce oil palm were once tropical moist forests, but how recently such land was deforested and if oil palm will drive further deforestation.

To date, over 80 percent of oil palm production occurs in Southeast Asia, primarily Malaysia and Indonesia. As of 2005, the majority of land in oil palm production had been deforested since 1990 (Koh and Wilcove, 2008). However, at least forty-three countries have some land in oil palm production and in its native range in West Africa, it has been used for as long as 5,000 years (Kiple and Ornelas, 2000). Vijay et al. (2016) examined deforestation patterns in twenty countries in four major oil palm–producing regions. They found regional trends suggesting that deforestation since 1989 was much higher for palm oil grown in Southeast Asia and South America than that grown in Mesoamerica and Africa. However, not all countries conformed to regional trends and future expansion of industrial plantations could differ. This study identified large areas of forest in South America and Africa that are climatically and edaphically suited for oil palm, making them vulnerable to such expansions. All regions of oil palm production were also found to contain areas of the greatest diversity of bird and mammal species at high risk of extinction.

As oil palm plantations continue to expand, so does the need to assess its impact on biodiversity and to develop conservation management plans to prevent further species extinctions. As part of that goal, there is a need to understand differences in patterns of development for oil palm. For example, in contrast to many parts of Southeast Asia, palm plantation establishment in the Peruvian Amazon occurred far from population centers in areas made accessible through the establishment of road networks (Vijay, 2018). This is particularly important in Peru, where almost 50 percent of oil palm plantations were established in areas deforested since 2000. As added emphasis for the need to understand variability in these processes, another study of oil palm plantation expansion in Latin America showed the pattern of deforestation in Peru to be an outlier, with lower levels of recent deforestation observed in other study countries (Furumo and Aide, 2017). Palm oil is but one of several plantation crops expanding into tropical forests, along with rubber and teak, in particular. Rubber plantations in the Amazon devasted native peoples when rubber was first commercially exploited. In the late 1800s, rubber production moved to Southeast Asia, where its production was more efficient. Regional studies—such as Abreds et al. (2015)—make similar points to ones already noted. Much of the production involves conversion of marginal areas including areas of high biodiversity. Perhaps 57 percent of the plantations may not be sustainable.

An inescapable observation is the inadequacy of the data on tree plantations worldwide. The Food and Agriculture Organization reports on the total area of tree plantations, claiming that they increased from about 2 million km^2 in 1990, to 2.8 million km^2 in 2010, the last year for which they provide data. We discuss the many inadequacies of these data elsewhere (Pimm, 2001). Advances in remote sensing may address difficulties in distinguishing natural forest from oil palm, teak (*Tectona grandis*), rubber (*Hevea brasiliensis*), or indeed other tree crops and allow for more comprehensive assessments.

Summary

The areas of extensive agriculture do not overlap extensively with the areas where the most vulnerable species are concentrated. It is unlikely this is a consequence of agriculture eliminating so many species that few remain. Western Europe and eastern North America surely never had the concentrations of species still found in tropical moist forests. Nonetheless the expansion of agriculture has massively reduced biodiversity, both locally and globally. We present three examples to illustrate the complexities.

First, agriculture has eliminated particularly large mammals and predators across large swaths of the planet. Second, agriculture in biodiversity hotspots is of critical concern. In these areas, most of the original habitat has been lost due to anthropogenic activities, agriculture being a main driver. The biodiversity hotspots are also where most species are at risk of extinction. They have lost most of their original habitats—predominantly tropical moist forests and Mediterranean ecosystems—but have not provided extensive, sustainable, and productive agriculture. In the biodiversity hotspots with high food insecurity, increased food production may be a sufficient motivation for land clearing, even when the benefits are transient. While large areas of tropical lowland forest still remain, the rapid expansion of palm oil and other tree plantations are a major threat to the species that live there.

Finally, the data to understand the impacts of agriculture on biodiversity are too few. To understand these impacts we need to understand not only current agricultural extent but patterns of past conversion. Counterintuitively, data on current LULC often fail to accurately identify certain crops such as tree plantations that occupy key areas for biodiversity and which may not be sustainable.

References

ABIEC/Brazilian Beef/Apex Brasil. (2016). *Brazilian livestock profile: annual report*. São Paulo, Brazil.

Bacher, S. (2012). Still not enough taxonomists: reply to Joppa et al. *Trends in Ecology and Evolution, 27*, 65–66.

Barnosky, A. D., Matzke, N., Tomiya, S., Wogan, G. O. U., Swartz, B., Quental, T. B., Marshall, C., McGuire, J. L., Lindsey, E. L., and Maguire, K. C. (2011). Has the Earth's sixth mass extinction already arrived? *Nature, 471*, 51–57.

Brooks, T. M., Pimm, S. L., and Oyugi, J. O. (1999). Time lag between deforestation and bird extinction in tropical forest fragments. *Conservation Biology, 13*, 1140–1150.

Brummitt, N. A., Bachman, S. P., Griffiths-Lee, J., Lutz, M., Moat, J. F., Farjon, A., Donaldson, J. S., Hilton-Taylor, C., Meagher, T. R., and Albuquerque, S. (2015). Green plants in the red: A baseline global assessment for the IUCN sampled Red List Index for plants. *PLoS ONE 10*, e0135152.

Chapman, A. D. (2009). *Numbers of living species in Australia and the world*, Toowoomba, Australia.

Cincotta, R. P., Wisnewski, J., and Engelman, R. (2000). Human population in the biodiversity hotspots. *Nature, 404*, 990–992.

Corley, R. (2009). How much palm oil do we need? *Environmental Science & Policy, 12*, 134–139.

Costello, M. J., and Wilson, S. P. (2011). Predicting the number of known and unknown species in European seas using rates of description. *Global Ecology and Biogeography, 20*, 319–330.

De Vos, J. M., Joppa, L. N., Gittleman, J. L., Stephens, P. R., and Pimm, S. L. (2015). Estimating the normal background rate of species extinction. *Conservation Biology, 29*, 452–462.

Dean, W. (1997). *With broadax and firebrand: the destruction of the Brazilian Atlantic Forest*. University of California Press.

Erwin, T. L. (1982). Tropical forests: their richness in coleoptera and other arthropod species. *The Coleopterists Bulletin, 36*, 74–75.

Ferraz, G., Russell, G. J., Stouffer, P. C., Bierregaard, R. O., Pimm, S. L., and Lovejoy, T. E. (2003). Rates of species loss from Amazonian forest fragments. *Proceedings of the National Academy of Sciences, 100*, 14069–14073.

Flessa, K. W., and Jablonski, D. (1985). Declining Phanerozoic background extinction rates: effect of taxonomic structure? *Nature, 313*, 216–218.

Furumo, P. R., and Aide, T. M. (2017). Characterizing commercial oil palm expansion in Latin America: land use change and trade. *Environmental Research Letters, 12*, 024008.

Gaston, K. J. (1991). The magnitude of global insect species richness. *Conservation Biology, 5*, 283–296.

Godfray, H. C. J., Beddington, J. R., Crute, I. R., Haddad, L., Lawrence, D., Muir, J. F., Pretty, J., Robinson, S., Thomas, S. M., and Toulmin, C. (2010). Food security: the challenge of feeding 9 billion people. *Science*, 1185383.

Gore, A. (2006). *An inconvenient truth: The planetary emergency of global warming and what we can do about it*. Rodale Books.

Hamilton, A. J., Basset, Y., Benke, K. K., Grimbacher, P. S., Miller, S. E., Novotný, V., Samuelson, G. A., Stork, N. E., Weiblen, G. D., and Yen, J. D. L. (2010). Quantifying uncertainty in estimation of tropical arthropod species richness. *The American Naturalist, 176*, 90–95.

Hawksworth, D. L. (1991). The fungal dimension of biodiversity: magnitude, significance, and conservation. *Mycological Research, 95*, 641–655.

Hosonuma, N., Herold, M., De Sy, V., De Fries, R. S., Brockhaus, M., Verchot, L., Angelsen, A., and Romijn, E. (2012). An assessment of deforestation and forest degradation drivers in developing countries. *Environmental Research Letters, 7*, 044009.

IUCN. (2016). Retrieved from http://www.iucnredlist.org.

Jenkins, C. N., and Pimm, S. L. (2003). How big is the global weed patch? *Annals of the Missouri Botanical Garden*, 172–178.

Jenkins, C. N., Pimm, S. L., and Joppa, L. N. (2013). PNAS Plus: Global patterns of terrestrial vertebrate diversity and conservation. *PNAS, 110*(28), E2602–E2610; published ahead of print June 26 (2013). doi: 10.1073/pnas. 1302251110.

Joppa, L., Visconti, P., Jenkins, C., and Pimm, S. (2013). Achieving the convention on biological diversity's goals for plant conservation. *Science, 341*, 1100–1103.

Joppa, L. N., Roberts, D. L., Myers, N., and Pimm, S. L. (2011a). Biodiversity hotspots house most undiscovered plant species. *Proceedings of the National Academy of Sciences, 108*, 13171–13176.

Joppa, L. N., Roberts, D. L., and Pimm, S. L. (2010). How many species of flowering plants are there? *Proceedings of the Royal Society B: Biological Sciences, 278*, 554.

Joppa, L. N., Roberts, D. L., and Pimm, S. L. (2011b). The population ecology and social behaviour of taxonomists. *Trends in Ecology & Evolution (Personal edition), 26*, 551–553.

Karanth, K. K., Gopalaswamy, A. M., DeFries, R., and Ballal, N. (2012). Assessing patterns of human-wildlife conflicts and compensation around a central Indian protected area. *PLoS ONE, 7*, e50433.

Kew, R. (2016). *The state of the world's plants report–2016*. Kew, UK: The Board of Trustees of the Royal Botanic Gardens.

Kiple, K. F., and Ornelas, K. C. (2000). *The Cambridge world history of food*. Cambridge University Press.

Koenen, E., de Vos, J., Atchison, G., Simon, M., Schrire, B., de Souza, E., de Queiroz, L., and Hughes, C. (2013). Exploring the tempo of species diversification in legumes. *South African Journal of Botany*.

Koh, L. P., and Wilcove, D. S. (2008). Is oil palm agriculture really destroying tropical biodiversity? *Conservation Letters, 1*, 60–64.

Lees, A. C., and Pimm, S. L. (2015). Species, extinct before we know them? *Current Biology, 25*, R177–R180.

Lichtenfeld, L. L., Trout, C., and Kisimir, E. L. (2015). Evidence-based conservation: predator-proof bomas protect livestock and lions. *Biodiversity and Conservation, 24*, 483–491.

Locey, K. J., and Lennon, J. T. (2016). Scaling laws predict global microbial diversity. *Proceedings of the National Academy of Sciences*, 201521291.

Manne, L. L., and Pimm, S. L. (2001). Beyond eight forms of rarity: which species are threatened and which will be next? *Animal Conservation*, *4*, 221–229.

May, R. M. (1991). A fondness for fungi. *Nature*, *352*, 475–476.

May, R. M., and Beverton, R. J. H. (1990). How many species? [and Discussion]. *Philosophical Transactions of the Royal Society of London. Series B: Biological Sciences*, *330*, 293–304.

Meijaard, E., Garcia-Ulloa, J., Sheil, D., Wich, S., Carlson, K., Juffe-Bignoli, D., and Brooks, T. (2018). Oil palm and biodiversity: a situation analysis by the IUCN Oil Palm Task Force.

Molotoks, A., Kuhnert, M., Dawson, T. P., and Smith, P. (2017). Global hotspots of conflict risk between food security and biodiversity conservation. *Land*, *6*, 67.

Mora, C., Tittensor, D. P., Adl, S., Simpson, A. G. B., and Worm, B. (2011). How many species are there on earth and in the ocean? *PLoS Biol*, *9*, e1001127.

Mora, C., Tittensor, D. P., and Myers, R. A. (2008). The completeness of taxonomic inventories for describing the global diversity and distribution of marine fishes. *Proceedings of the Royal Society B: Biological Sciences*, *275*, 149–155.

Myers, N. (1988). Threatened biotas: "hot spots" in tropical forests. *Environmentalist*, *8*, 187–208.

Myers, N. (1989). Extinction rates past and present. *Bioscience*, 39–41.

Nee, S. (2006). Birth-death models in macroevolution. *Annual Review of Ecology, Evolution, and Systematics*, 1–17.

Novotny, V., Miller, S. E., Hulcr, J., Drew, R. A. I., Basset, Y., Janda, M., Setliff, G. P., Darrow, K., Stewart, A. J. A., Auga, J., Isua, B., Molem, K., Manumbor, M., Tamtiai, E., Mogia, M., and Weiblen, G. D. (2007). Low beta diversity of herbivorous insects in tropical forests. *Nature*, *448*, 692–695.

Ocampo-Peñuela, N., Jenkins, C. N., Vijay, V., Li, B. V., and Pimm, S. L. (2016). Incorporating explicit geospatial data shows more species at risk of extinction than the current Red List. *Science Advances*, *2*, e1601367.

Ødegaard, F. (2000). How many species of arthropods? Erwin's estimate revised. *Biological Journal of the Linnean Society*, *71*, 583–597.

Pimm, S., Raven, P., Peterson, A., Şekercioğlu, Ç. H., and Ehrlich, P. R. (2006). Human impacts on the rates of recent, present, and future bird extinctions. *Proceedings of the National Academy of Sciences*, *103*, 10941–10946.

Pimm, S., Russell, G. J., Gittleman, J., and Brooks, T. M. (1995). The future of biodiversity. *Science*, *269*, 347–350.

Pimm, S. L. (2001). *The world according to Pimm: a scientist audits the earth.* McGraw-Hill.

Pimm, S. L., and Askins, R. A. (1995). Forest losses predict bird extinctions in eastern North America. *Proceedings of the National Academy of Sciences*, *92*, 9343–9347.

Pimm, S. L., and Brooks, T. (2013). Conservation: Forest fragments, facts, and fallacies. *Current Biology*, *23*, R1098–R1101.

Pimm, S. L., Jenkins, C. N., Abell, R., Brooks, T. M., Gittleman, J. L., Joppa, L. N., Raven, P. H., Roberts, C. M., and Sexton, J. O. (2014). The biodiversity of species and their rates of extinction, distribution, and protection. *Science*, *344*, 1246752.

Pimm, S. L., Jenkins, C. N., Joppa, L. N., Roberts, D. L., and Russell, G. J. (2010). How many endangered species remain to be discovered in Brazil. *Natureza & Conservação*, *8*, 71–77.

Pimm, S. L., and Joppa, L. N. (2015). How many plant species are there, where are they, and at what rate are they going extinct? *Annals of the Missouri Botanical Garden*, *100*, 170–176.

Pimm, S. L., and Raven, P. (2000). Biodiversity: extinction by numbers. *Nature*, *403*, 843–845.

Pimm, S. L., and Raven, P. H. (2017). The fate of the world's plants. *Trends in Ecology & Evolution*, *32*, 317–320.

Pouzols, F. M., Toivonen, T., Di Minin, E., Kukkala, A. S., Kullberg, P., Kuusterä, J., Lehtomäki, J., Tenkanen, H., Verburg, P. H., and Moilanen, A. (2014). Global protected area expansion is compromised by projected land-use and parochialism. *Nature*, *516*, 383.

Purvis, A. (2008). Phylogenetic approaches to the study of extinction. *Annual Review of Ecology, Evolution, and Systematics*, *39*, 301–319.

Pyron, R. A., and Burbrink, F. T. (2013). Phylogenetic estimates of speciation and extinction rates for testing ecological and evolutionary hypotheses. *Trends in Ecology & Evolution, 28*, 729–736.
Ray, D. K., and Foley, J. A. (2013). Increasing global crop harvest frequency: recent trends and future directions. *Environmental Research Letters, 8*, 044041.
Ricotta, C., Ferrari, M., and Avena, G. (2002). Using the scaling behaviour of higher taxa for the assessment of species richness. *Biological Conservation, 107*, 131–133.
Riggio, J., Jacobson, A., Dollar, L., Bauer, H., Becker, M., Dickman, A., Funston, P., Groom, R., Henschel, P., and de Iongh, H. (2013). The size of savannah Africa: a lion's (Panthera leo) view. *Biodiversity and Conservation, 22*, 17–35.
Robson, A. S., Trimble, M. J., Purdon, A., Young-Overton, K. D., Pimm, S. L., and Van Aarde, R. J. (2017). Savanna elephant numbers are only a quarter of their expected values. *PLoS ONE, 12*, e0175942.
Scheffers, B. R., Joppa, L. N., Pimm, S. L., and Laurance, W. F. (2012). What we know and don't know about Earth's missing biodiversity. *Trends in Ecology & Evolution, 27*, 501–510.
Solow, A. R., and Smith, W. K. (2005). On estimating the number of species from the discovery record. *Proceedings: Biological Sciences, 272*, 285–287.
Stork, N. E. (1993). How many species are there? *Biodiversity and Conservation, 2*, 215–232.
Stork, N. E. (2010). Re-assessing current extinction rates. *Biodiversity and Conservation, 19*, 357–371.
Taylor, D. L., Hollingsworth, T. N., McFarland, J. W., Lennon, N. J., Nusbaum, C., and Ruess, R. W. (2014). A first comprehensive census of fungi in soil reveals both hyperdiversity and fine-scale niche partitioning. *Ecological Monographs, 84*, 3–20.
Tilman, D., Balzer, C., Hill, J., and Befort, B. L. (2011). Global food demand and the sustainable intensification of agriculture. *Proceedings of the National Academy of Sciences, 108*, 20260–20264.
Valente, L. M., Savolainen, V., and Vargas, P. (2010). Unparalleled rates of species diversification in Europe. *Proceedings of the Royal Society B: Biological Sciences, 277*, 1489–1496.
Vijay, V., Pimm, S. L., Jenkins, C. N., and Smith, S. J. (2016). The impacts of oil palm on recent deforestation and biodiversity loss. *PLoS ONE, 11*, e0159668.
Vijay, V., Reid, C. D., Finer, M., Jenkins, C. N., and Pimm, S. L. (2018). Deforestation risks posed by oil palm expansion in the Peruvian Amazon. *Environmental Research Letters, 13*, 114010.
Wallace, A. R. (1855). XVIII.—On the law which has regulated the introduction of new species. *Journal of Natural History, 16*, 184–196.
Weir, J. T., and Schluter, D. (2007). The latitudinal gradient in recent speciation and extinction rates of birds and mammals. *Science, 315*, 1574–1576.
Westwood, J. (1833). On the probable number of species of insects in the creation; together with descriptions of several minute Hymenoptera. *The Magazine of Natural History and Journal of Zoology, Botany, Mineralogy, Geology and Meterology*, 6, 116–123.
Zabel, F., Putzenlechner, B., and Mauser, W. (2014). Global agricultural land resources–a high resolution suitability evaluation and its perspectives until 2100 under climate change conditions. *PLoS ONE, 9*, e107522.
Zhou, Z., and Zheng, S. (2003). Palaeobiology: The missing link in ginkgo evolution. *Nature, 423*, 821–822.

INDEX